Global Climate Change and Human Life

Global Climate Change and Human Life

M. A. K. Khalil
Portland State University, Portland, Oregon, USA

This edition first published 2022

Registered Offices
John Wiley & Sons, Inc., 111 River Street, Hoboken, NJ 07030, USA
John Wiley & Sons Ltd, The Atrium, Southern Gate, Chichester, West Sussex, PO19 8SQ, UK

Editorial Office
9600 Garsington Road, Oxford, OX4 2DQ, UK

For details of our global editorial offices, customer services, and more information about Wiley products visit us at www.wiley.com.

Wiley also publishes its books in a variety of electronic formats and by print-on-demand. Some content that appears in standard print versions of this book may not be available in other formats.

A catalogue record for this book is available from the Library of Congress

Paperback ISBN: 9780470665787; ePub ISBN: 9781118526149; ePDF ISBN: 9781118526156

Cover image: © pio3/Shutterstock
Cover design by Wiley

Set in 9.5/12.5pt STIXTwoText by Integra Software Services Pvt. Ltd, Pondicherry, India
Printed and bound by CPI Group (UK) Ltd, Croydon, CR0 4YY

C9780470665787_230522

Contents

Preface

At the start, I intended to write this book for all college students and practically everyone else with an interest in the global environment. My goal became increasingly difficult to attain as I progressed. I still think it will serve this community, but not everything will be understood by everyone. In studying climate change, and environmental science in general, there are many simultaneous and credible influences. Which are important and which can be ignored require the ability to associate numbers with the causes of each phenomenon. To provide that ability without resorting to complex mathematics was a daunting task, but I have managed to do so with the use of only basic algebra. Results from more advanced mathematics have been relegated to simple formulas without proofs, because these are not the debatable aspects of climate science, but whether they apply under the circumstances we are studying may be. So, my readers should be comfortable in proceeding to learn from this book, regardless of their majors, or academic backgrounds. I believe it will benefit the middle years' college students the most, and the exercises at the ends of the chapters are designed for them to attain a deeper understanding of global change science within a formal course. Resources are available from the publisher for university-level instructors that will considerably reduce the work needed to get a course up and running in any department.

The book itself takes a holistic view of global change science in which the earth's climate is a focal point. It takes established ideas from the basic sciences such as physics, chemistry, biology, and several social sciences and fuses them into a coherent framework using many new ideas and concepts that are needed to make the connections. These connections and ideas, and their consequences, are expected to evolve over time, as the science develops to serve the societal needs for managing the global environment and especially the climate.

Numerous colleagues, friends, students, and family members have contributed to the development of this book over many years. Major contributions came from my hundreds of students who were asked to write an essay at the end of the course titled "The single most amazing thing I learned from this course." It made me understand better what mattered to those who wanted to learn this subject. Colleagues who read the text and provided suggestions are: Drs. R.M. Mackay, P. Loikith, A. Rice, and C. Butenhoff. I had conversations about the contents of the book over many years with: Rei Rasmussen, Kathayoon Khalil, and Ed Immergut (at J.Wiley & Sons). Finally, my parents, wife, and children contributed in intangible, but crucial, ways just by being there.

M.A.K. Khalil, Professor
Portland, Oregon, USA
September 2021.

About the Companion Website

This book is accompanied by a companion website for Instructors:

www.wiley.com/go/khalil/Globalclimatechange

This website includes:

- Solutions to the Exercises in the book
- A downloadable MCQ test bank
- Figures from the book in PowerPoint
- General Tools and Course Elements
- Term Paper Project: A Guide for Students from the Instructor

1

Introduction

1.1 What Is Global Change Science?

In our time, the population has become large enough to cause perceptible environmental changes all over the world. With it, a science of global change has emerged, mostly as a practical matter to understand and manage the earth's habitability and create a sustainable environment for some time to come. This goal amounts to balancing the benefits of technological and societal advances with the undesirable potential side effects. Concerns started with the depletion of the stratospheric ozone layer and its possible adverse consequences on human health. It has, in recent decades, shifted to climate change driven by an observed ongoing global warming.

Global change, including climate, is a derivative, interdisciplinary science. It deals with how the earth's environment and particularly its climate interact with human life over periods of decades. The system it represents and describes is the earth's habitable surface environment, which is determined by the characteristics of the atmosphere, oceans, and the land, and the interactions between them. It uses long known and verified principles of the sciences. Still, of all the issues surrounding the scientific inquiry into our world, global change science is the most obscure, mostly because it is not just about physics, or chemistry, or biology, or sociology nor any other "ology"; it respects no such boundaries because it is about the real world, which functions according to all of them acting simultaneously.

There are two complementary aspects to consider. The first is traditional climatology that has developed over the last century. It deals with how climate is manifested in different parts of the world and its repeating annual cycles; it includes the science and models developed for weather forecasting. An enormous amount of observational data have been taken and combined with theories to achieve a solid understanding of the climate we experience, although climate science, like all others, will continue to evolve forever. The second aspect deals with climate change that can arise if one or another of the fundamental variables that control climate is altered. This takes us beyond the realm of climatology into new domains of the earth's environment for which there is less observational data and little human experience. Phenomena that are dormant in a stable climate suddenly come into play in a changing one, and these are even more complex to unravel than the natural stable climate. It is this second aspect that will be our interest. We will see how global change science goes beyond the comfortable understanding of climate and into the realms of a new science.

We will spend a considerable amount of time to understand the current climate and environment because it is the baseline from which future change will evolve, altered by

Global Climate Change and Human Life, First Edition. M. A. K. Khalil.

Companion Website: www.wiley.com/go/khalil/Globalclimatechange

natural and human forces. We will focus on expected changes over time scales of decades to perhaps a century or so. These are the periods that overlap with human life spans, and it is the times over which societies can cause global climate change and over which they may be able to manage it.

1.2 Current Global Change

We start by looking at the global observations which show that the earth is getting warmer (Figure 1.1a.). As this warming progresses, it will have consequential impacts on parts of the environment that affect human life in various ways, some of which will be undesirable. In this figure we see the readily observed direct environmental consequences such as increasing sea levels and decreasing high-latitude snow cover.

1.3 Raising Fundamental Questions

From the discussion so far, the following fundamental questions may be posed: Why are these global changes occurring? How will they affect our lives? If we find the effects undesirable, what should we do? We will create a path to answering these questions. For the first question we will need an understanding of how the composition of the atmosphere determines the climate; in turn, what determines atmospheric composition and then, how it can be altered. This is the best-known part of the science. To answer the second question, we will study how climate affects the environmental factors that are important to human life, such as crops, rain, and extreme weather events. The impact of the consequences is not uniformly global; it is often detrimental for some regions and negligible or beneficial for others, thus adding further complexity to finding unequivocal answers. The last question has no clear answers, and perhaps never will, but what we learn here will show how to accomplish the goal of managing climate, what it will take, and when it needs to be done. Such a management process has to be dynamic, making it more complex and less didactic, requiring changes in strategy to achieve a longer-term goal as our knowledge advances.

It is apparent that global change is expected to have concerning consequences for human life, but that will depend entirely on how big the stated impacts are going to be. It is not enough to merely say what climate change might do. It may occur, but may be so small that there is no cause for concern. Without the ability to understand the numbers associated with these changes, people, including you and me, have no clear way to accept or reject a plan for climate management or even a need for it. Building the numbers from the underlying science that is readily understandable is a major goal of this book. This inevitably requires models which take the theoretical or conceptual understanding and translate them into mathematical forms with the primary goal to calculate the values of variables that interest or concern us. The models we will use in this book are the simplest we can construct to serve our goals and represent global change science. They will lay out a holistic view of the science that develops and teaches the main principles, concepts, and conclusions. In the end, readers will be empowered to use science and the scientific method to decide how important and timely climate change is as a social issue and which solutions can succeed. But behind this practicality, a greater goal is to satisfy our curiosity about how the earth's climate works and our role in it – indeed, it is to understand the conjoined character of life on earth and its environment (Endnote 1.2).

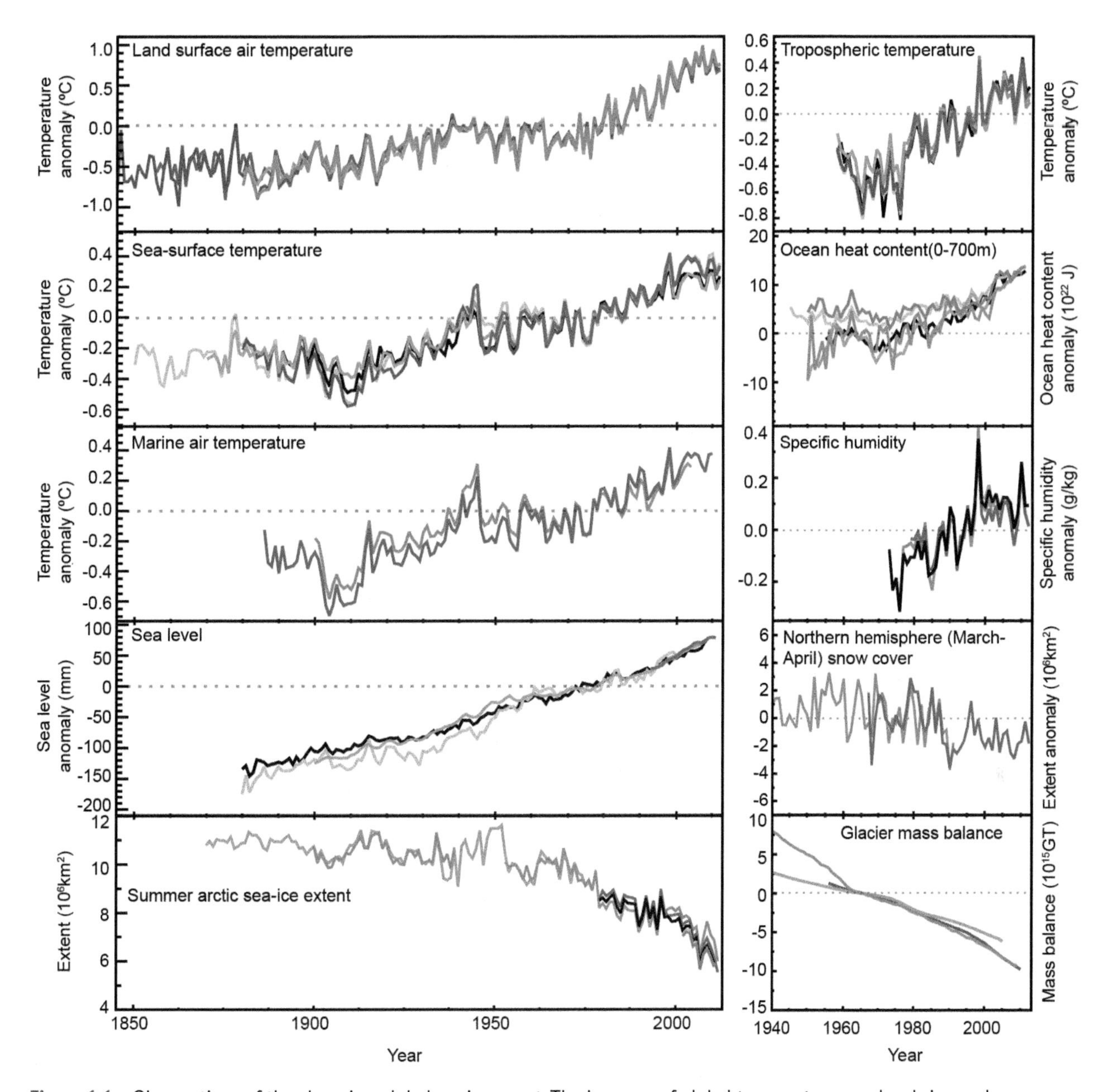

Figure 1.1 Observations of the changing global environment. The increase of global temperature, sea-level rise, and ice decline are shown (Endnote 1.1).

Endnotes

Endnote 1.1 Figure 1.1 is taken from IPCC's AR4. The IPCC is "The Intergovernmental Panel on Climate Change"; it is a United Nations organization charged with evaluating ongoing global climate change. They issue periodic consensus reports that review, synthesize, and document the published scientific literature on global change science and in recent years have stimulated research to facilitate such an analysis. The synthesis of research results from these reports is used in this book to illustrate aspects of global change, as needed. The reports are: FAR (First Assessment Report), 1990; SAR (Second), 1995; TAR (Third), 2001; AR4 (Forth), 2001, AR5 (Fifth), 2013 and AR6 (Sixth), 2021; all published by the Cambridge University Press, Cambridge, UK.

Endnote 1.2 It is useful to state the meaning of some terms that are commonly used in this book, and in everyday public discourse. *Global warming* is the increase of the average temperature of the earth that persists over time scales of a decade or more. It is narrower than *climate change*, which can include changes of rainfall, winds, humidity, and perhaps other attributes of climate. They may change because of global warming, or from some other cause. *Global change* is wider still. In our context, it includes climate change, environmental changes that can come from it, such as sea-level rise, and its effects on human life as well as the global societal responses to manage its adverse effects. The term "global change" may be used in other contexts that are not related to climate. "Greenhouse gases" is another commonly used term. It expresses an attribute of atmospheric constituents that have the special property, that they trap the earth's heat and warm the earth's surface which is stated as *the greenhouse effect*. These are the only gases that can cause global warming if increased by human activities or natural causes. Many other terms will be used in this book that relate to climate and global change; these will be defined and elaborated as we go along.

2

The Framework

2.1 The System

To study the global environment, it is useful to define the system by its major components and specify the scales of space and time that we will consider. The earth's environmental system may be divided minimally into three main components – the atmosphere, land, and oceans (Figure 2.1). These components are very different from each other in how they affect the human environment, habitability, and the climate.

The atmosphere is central to the workings of the climate system. Because of its composition, it causes the greenhouse effect that warms the earth's surface and is the source of global warming; it regulates the amount of ultraviolet radiation reaching us due to the ozone layer that supports life as it currently exists; it harbors clouds that reflect sunlight and cool the earth; and it regulates ocean acidity, which in turn determines the nature of ocean life. Atmospheric winds distribute heat and water between the land and the oceans, and drive surface ocean currents that move and mix the waters that sustain aquatic life.

The oceans buffer climate change, serve as reservoirs of water for rain and snow that falls on the earth; cause hurricanes; and they emit, absorb, make, or destroy atmospheric gases that affect the climate.

The environment on the land is most important for human life. It can be further subdivided into three aspects that have the most influence on climate – soils, plants, and the frozen regions, such as the poles (cryosphere). On land, soils and plants emit, absorb, and store greenhouse gases, and the frozen regions reflect sunlight, cooling the earth. The interactions of sunlight, heat, and atmospheric constituents in these three components determine the climate and global warming.

2.2 Scales of Action

Atmospheric phenomena occur over regions of various sizes and over various times. This distinguishing characteristic allows us to separate global climate change from other types of environmental conditions. Figure 2.2 shows three canonical environmental time scales. The first is cyclical behavior that ranges in time from day-night events, to seasonal changes, to climate cycles that repeat over thousands of years. Pollution in cities, for instance, follows a characteristic daily cycle in which high concentrations of particles and ozone, generated by the cycles of work, are observed during midday and generally cleaner air is seen at night. A

Global Climate Change and Human Life, First Edition. M. A. K. Khalil.

Companion Website: www.wiley.com/go/khalil/Globalclimatechange

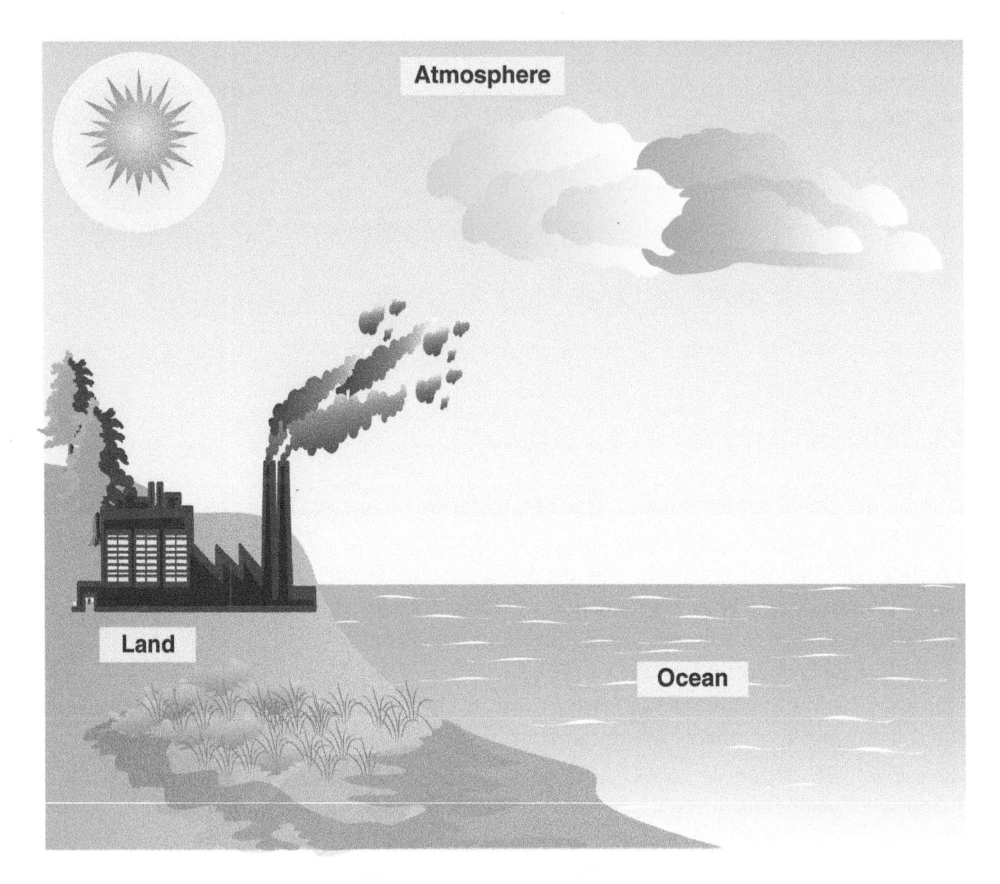

Figure 2.1 A minimalistic view of the global environment. The main divisions, or parts of the earth's environmental system, are shown here as three components: land, oceans, and the atmosphere. Each of these can be further subdivided as required by the questions being considered.

second scale is the time it takes for the system, including the climate, to come to equilibrium after one of its components is disturbed. If for instance, we reduce the emission of a gas by one kilogram per year, it will take some time before the amount in the atmosphere will adjust to a new equilibrium with the reduced source. This lifetime indicates the persistence of the gas in the atmosphere. If it persists for a long time, then the effect of our reduction will take a long time to complete. Finally, there is a transport time associated with the moving and mixing of pollution from its origins to more distant locations. In this situation the space and time scales of environmental phenomena are related as a type of proportionality in which the levels of pollution decrease as the spatial scale becomes larger. As an example, consider indoor air pollution. If smoke is generated, it can fill up the room rapidly, but it can be eliminated just as rapidly by opening a window. Urban air pollution, on the other hand, builds up and disperses over a day, and sometimes may last a week if the air gets stagnant. But in most cases, the amount of pollution in urban air is likely to be less than a smoke-filled room. Similarly, there are longer-lasting air pollution episodes that spread over thousands of square miles that can cause haze and obstruct scenic views for a week or two at a time. But the largest scale is the atmosphere of the whole earth. It is the extreme extension of our last time scale by which man-made and natural compounds are taken from their sources in the cities and agricultural regions of the world and from the various natural ecosystems, such as the wetlands and forests, to the far reaches of the atmosphere ranging from pole to pole and the surface to stratospheric heights. It takes years for emissions from the sources to spread over this scale, and this is the scale of global environmental issues such as climatc change, stratospheric ozone depletion, and ocean acidification. As we will see, all these time scales will play a role in the climate and especially climate change.

The time scales of Figure 2.2 can be used to define the study of climate change. Air is constantly moving, and for the small areas of the cities spanning few tens of kilometers, it can stay for at most a few days. Urban air pollution is caused by precursor gases, emitted from automobiles and other sources, that react within seconds to hours while they are still concentrated, causing harmful levels of ozone, nitrogen dioxide, sulfur dioxide, and small particles or aerosols. These products too are relatively short-lived, lasting between minutes to days. So, these pollutants cannot spread out over the whole world because that takes a much longer time, represented by the moving and mixing time scale, compared with their urban lifetimes, represented by the equilibration time. Similarly, gases such as carbon dioxide or the chlorofluorocarbons, which cause global warming or ozone depletion, have atmospheric lifetimes of decades and so they do spread out everywhere; therefore, they

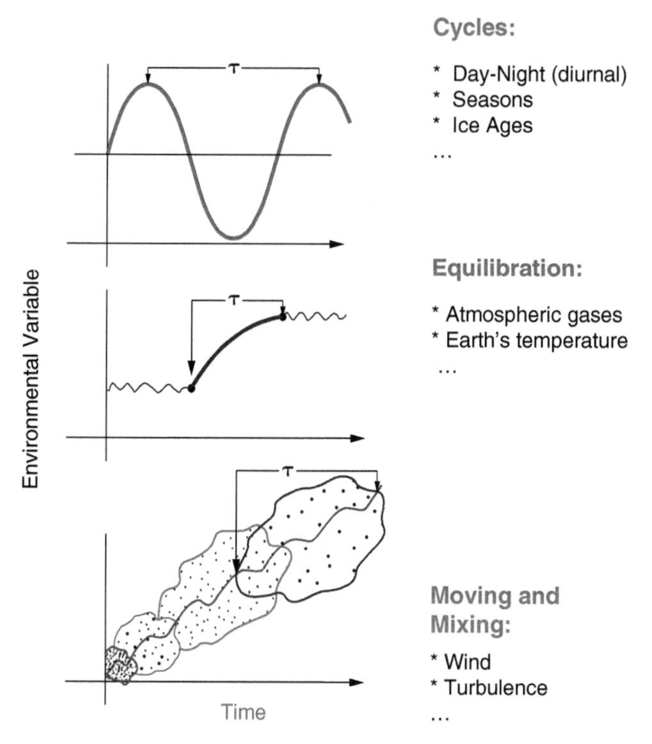

Cycles:

* Day-Night (diurnal)
* Seasons
* Ice Ages
...

Equilibration:

* Atmospheric gases
* Earth's temperature
...

Moving and Mixing:

* Wind
* Turbulence
...

Figure 2.2 Space and time scales in the environment. Three types of time scales are involved in environmental phenomena. Cycles repeat environmental conditions over one such scale in which it is the time between recurrences of the phenomenon. Repeating seasonal warm and cold temperatures are an example. Equilibration times are how long it takes for the system to adjust to a change that is made. It reflects the persistence of an atmospheric constituent or its lifetime. Transport time scales determine how long it will take to move and mix or dilute material within some region of the atmosphere.

cannot react over the time scales of air pollution of a few days. Urban areas may emit large amounts of these long-lived gases and may harbor moderately high concentrations, but they have little or no adverse health effects for humans or other life, and hence no effect on air pollution. Often the low overall chemical reactivity of long-lived gases also makes them less harmful for human health, even in the high levels that may occur in buildings or urban areas. Indeed, the known carcinogenic gases in urban environments such as benzene, ethylene oxide, formaldehyde, and trichloroethylene all have short lifetimes in the atmosphere and in other environmental and biological reservoirs.

The idea that *the lifetimes of the gases completely separate the causes of climate change and urban air pollution* is one of several guiding principles of global change science that we will articulate in this book. Man-made emissions, which cause air pollution, don't directly cause climate change and vice versa. This is disturbing because cleaning up city air doesn't solve the global warming issue and reducing greenhouse gases doesn't lead to healthier city air. Indeed, sometimes, fixing the air pollution problems may exacerbate climate change. Figure 2.3 further illustrates the principle.

2.3 What Determines Climate?

It is time to discuss what determines the earth's climate, or more specifically the earth's temperature (Figure 2.4). The first step in the process is incoming sunlight. The sun is the

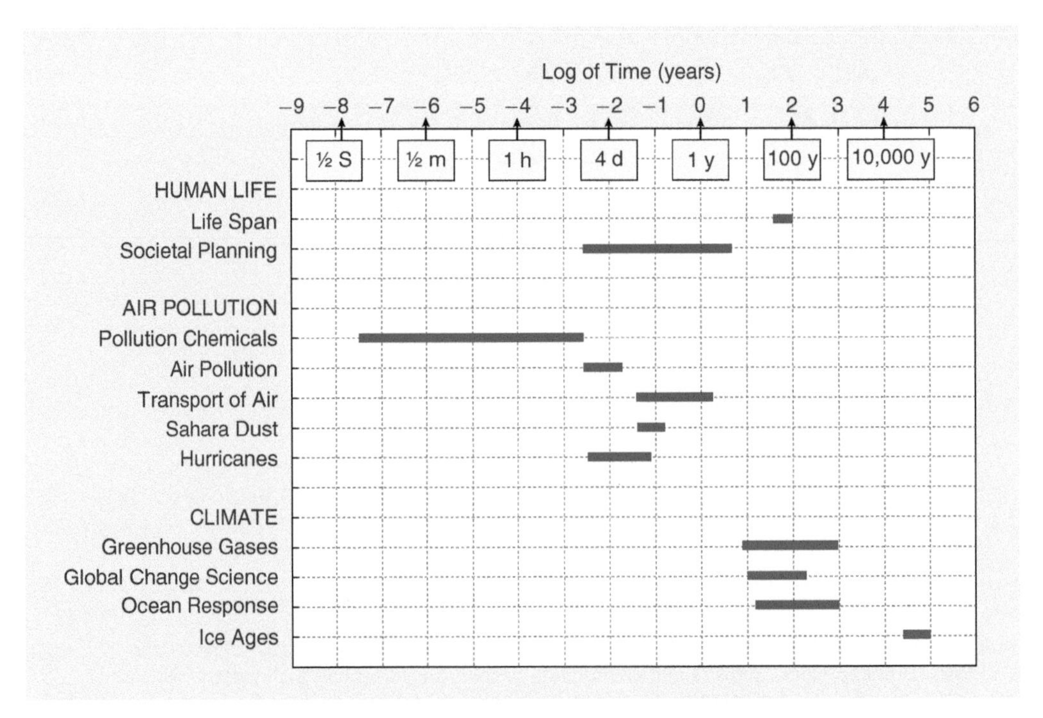

Figure 2.3 Time scales of air pollution, climate, and human life. The idea that air pollution and climate change are caused by different agents separated by their atmospheric lifetimes is illustrated in this chart. Included for comparison is the time scale for societal planning. It doesn't match well with the time scales of climate change – an important matter that we will return to later.

(b) Earth Radiation – Long Wave

Figure 2.4 Factors controlling the earth's temperature and climate. The process is described in three steps. (a) It starts with the action of incoming sunlight on the earth. Some is absorbed by the atmosphere and the earth, and some is reflected (albedo). (b) The earth's surface loses energy mostly by radiating heat but also by thermals and updrafts that carry warm surface air and evaporated water upward. Almost all of the radiated heat and all the heat of thermals is absorbed by the atmosphere. (c) Finally, the atmosphere too radiates heat toward the surface and out to space. The heat from the atmosphere coming back to the surface causes the greenhouse effect, and if it is increased, as by human activities, it will cause a global warming.

Figure 2.4 (Continued)

only external driver of climate and the only source of energy to power the machinery of climate. It is the only significant source of energy that reverberates throughout the environment and it is, in one form or another, all the energy that we utilize in our lives, whether it is electricity in the homes or the energy to run a marathon. The sun's radiation is electromagnetic waves that pack a lot of energy. The amount of energy that arrives at the earth's location in space is the solar constant and depends on the radiation power of the sun and how far it is from the earth. About 30% of the incoming radiation is reflected back to space mostly by clouds and to a lesser extent by dust, aerosols, air molecules, and the surface. This is the earth's *albedo*. The reflected radiation has no further direct role in the climate. The atmosphere absorbs about 20% of the solar radiation, including 3% in the stratospheric ozone layer and the rest mostly by clouds and water vapor. The dry and cloud-free atmosphere absorbs very little sunlight. So with the average persistent cloud cover, only about 50% of the sunlight arriving at the top of the atmosphere makes it to the surface, where it is absorbed entirely by the water, soils, plants, or other features of the surface. Small amounts of energy go into driving many environmental processes, but do not directly affect the climate. In the second step, some of the absorbed energy is used to evaporate water and some is transferred to the air at the surface through conduction. These cause convection by which some energy is sent back upward as rising warm air. More energy is sent to the atmosphere by evaporating water from the oceans and lakes, and by transpiration of plants. But the largest amount of energy is lost from the surface by heat radiation which is also electromagnetic in nature. Trace gases in the atmosphere have the capacity to absorb the heat energy of the earth's radiation. These greenhouse gases are mainly water vapor and carbon dioxide with lesser contributions from methane, nitrous oxide, and ozone. The absorption by the greenhouse gases, along with heat carried by rising air, warms the atmosphere. Finally in the third step, the atmosphere also balances the energy that it receives by emitting heat radiation, some of which goes up and out of the atmosphere and the rest goes back to the earth's surface. As these processes play out, the earth's surface and the atmosphere achieve stable, fixed temperatures releasing as much energy as they get. Moreover, the rising of warm air at the tropics creates a persistent motion of the air, resulting in a general circulation of the atmosphere that carries heat, water, and all sorts of gases and particles from where they are emitted to the furthest reaches of the earth.

The several new concepts introduced here will be discussed and explained in more detail in later chapters. For now it is important to note two points. First, that the temperature of the earth is determined by just three fundamental variables – solar constant, which supplies the energy; albedo, which reflects some of it; and atmospheric composition, which returns some of the earth's heat back to the surface. Second, that the earth's surface receives energy from two sources – the sun and the atmosphere. If the atmosphere did not contain the greenhouse gases and clouds, the surface would receive energy only from the sun and would be cooler by perhaps 30°–40°C! This warming by the atmosphere represents the *greenhouse effect*. It is a natural phenomenon that is large enough to prevent the earth from being frozen. It stands to reason that if we increase the atmospheric levels of the natural greenhouse gases or add new ones, then the temperature of the surface will increase.

While the greenhouse effect and global warming are anchored to the earth's temperature, the climate is more than that. It is defined variously, but we will take it to consist of three environmental elements – temperature, precipitation, and winds. These are selected because we can feel their effects directly and they affect life on earth in profound ways. We will see that life affects these variables too, forming a bond that is part of the internal engine of the climate system.

The sun's energy creates a temperature which is an expression of the heat stored in the components of the earth's system (land, oceans, and the atmosphere). This heat is shuttled around these reservoirs and creates internal cycles, many of which are not understood even now. The temperature, which represents the heat content, therefore becomes the central variable that defines the earth's climate. It determines how much precipitation we get, how humid the environment will be and which way the winds will blow. As the temperature changes, so do these climate variables. The earth has an average temperature that has changed and cycled over the millennia, due to both internal and external causes with various time scales. But the average temperature of the earth hardly occurs anywhere or at any time. There are two forms of departures from the average. One is a permanent difference between latitudes, between coastal and inland environments, and between seasons caused most significantly by the uneven distribution of solar radiation. The other is the presumed random fluctuations of the climatic variables. The latter is seen as follows: the average temperature at any location on a given day is usually the average of the observed temperatures over several decades; often 30 years is adopted by convention. It is a fact that on any given day the temperature is only occasionally the same as the average for that day; most of the time it is either higher or lower, balancing out to the middle only as an average value. It seems that we need more than averages to understand climate; we need a measure of variability such as a range, or standard deviation of the values that are experienced at any location or region. Although we do feel the "climate" at any location, we feel the variability even more, especially when extreme events occur. In hot environments the extreme event may make a day or several days unbearably hot one year and the same period pleasant another year. When the climate changes, we can characterize it unequivocally by a change of average temperature, such as the current global warming. But this global warming by itself is imperceptible in many ways and will remain so. What we will feel most is the effect of the variability including an inevitable increase of extreme events. The warming temperature will drive changes in precipitation and the winds, and these too will be manifested in changes of extreme events such as flooding and storms. After that, we will be affected by shifts of climate zones that naturally occur from warm to cold as we go poleward from the equator. Changes in seasonality round out the way in which the average rise of temperature or global warming is manifested in climate change that affects human life (Figure 2.5).

This description is a point of view in which climate is seen as a dynamic part of the earth's environment which changes over periods of time driven by an internal machinery that includes physical and chemical processes as well as living things. The changes create cycles of climate or take it from one stable state to another. Changes in precipitation, wind patterns, extreme events, and feedbacks are not merely effects of climate change, but are its very manifestation. Climate changes are known to also occur by external drivers, primarily in solar radiation that arise from long-term cycles related to the earth's orbit around the sun or the aging of the sun, but, although fascinating, these are far beyond the time scales of our focus here.

2.4 The Benchmark Average Climate

For several thousand years the average climatic conditions have been similar to the present. The temperature has two main dimensions – vertical and horizontal. In horizontal

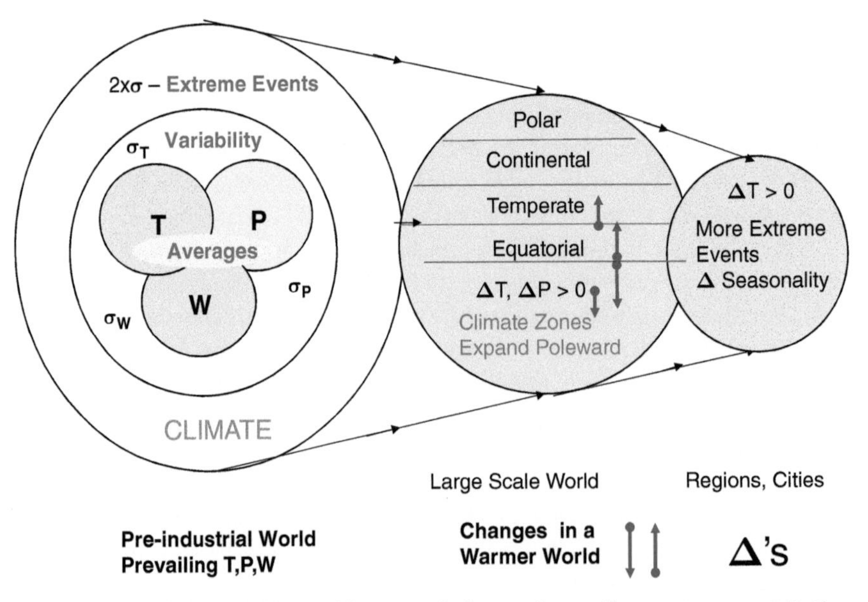

Figure 2.5 Defining variables and features of climate change. Temperature, precipitation, and winds describe climate. The variables are characterized by average values and a variability that measures departures from the average. In a warmer world, the large-scale climate zones can shift while at more localized levels the increase in natural extreme events driven by the variability would be felt. These are indicated by the arrows or the Δs.

bands the temperature is about 300°K at the equator and 250°K at the poles. This reflects mostly the effect of higher solar radiation and lower albedo at the equator, but it also includes the net transport of heat from the equatorial to the polar regions that reduces the discrepancy. In the vertical, the temperature undergoes several reversals in trends that define layers of the atmosphere. For up to about an average of 12 km from the surface the temperature decreases at a *lapse rate* of 6.5°C/km starting from the surface at about 288°K (Figure 2.6a). This is the *troposphere* and its upper boundary is the *tropopause.* The height of the tropopause varies with latitude and to a lesser extent, with seasons; it is highest at the equator (~16 km) and lowest at the poles (~8 km). In the average troposphere therefore, we can write the vertical temperature as: $T(z) = T_0 - l\,z\,(T_0 \approx 288°\text{K}, l = 6.5°\text{C/km})$.

As we move higher, the temperature is stable for a while but then it increases up to about 50 km. This region is the *stratosphere* and its upper boundary is the *stratopause.* The lapse rate is not quite simply linear, but if we look at the upper and lower temperatures it averages to about + 1.2°C/km. Higher than that, we enter the mesosphere and the temperature

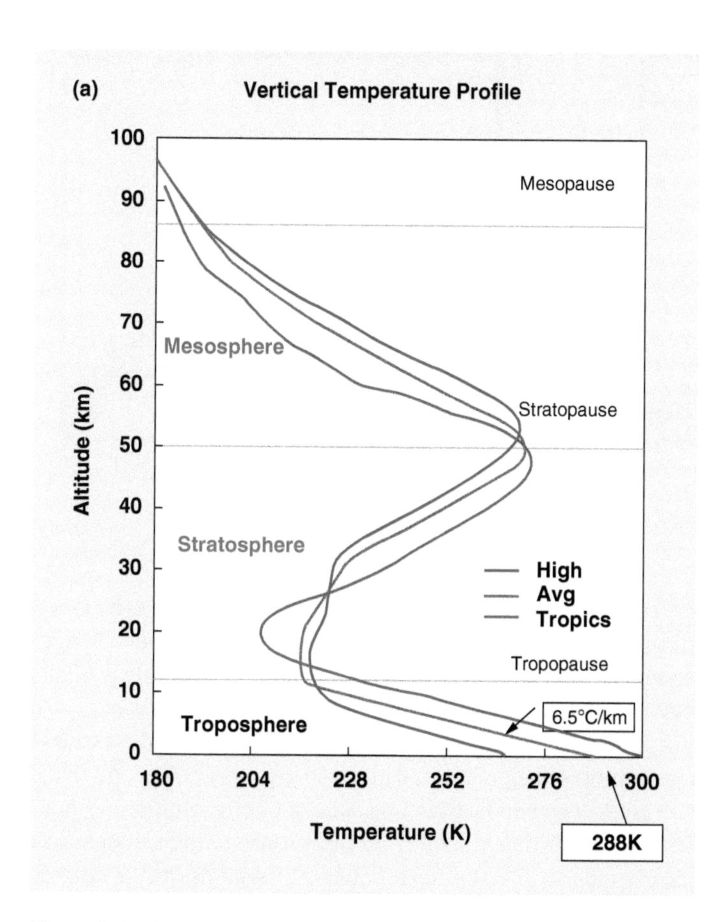

Figure 2.6 State of the Climate. The state of the climate as expressed in the temperature and precipitation is shown here. (a) Vertical temperature defines layers in the atmosphere. The troposphere is most important for climate and global warming, and the stratosphere for the ozone layer and its perturbations. The reasons will be discussed in later chapters. (b) Horizontal temperature defines climate zones driven mostly by differential amounts of solar radiation. (c) Global rainfall patterns with latitude. Figs 2.6 (b) and (c) are adapted from IPCC's AR4. The colored lines are model results described in AR4. See Endnote 1.1.

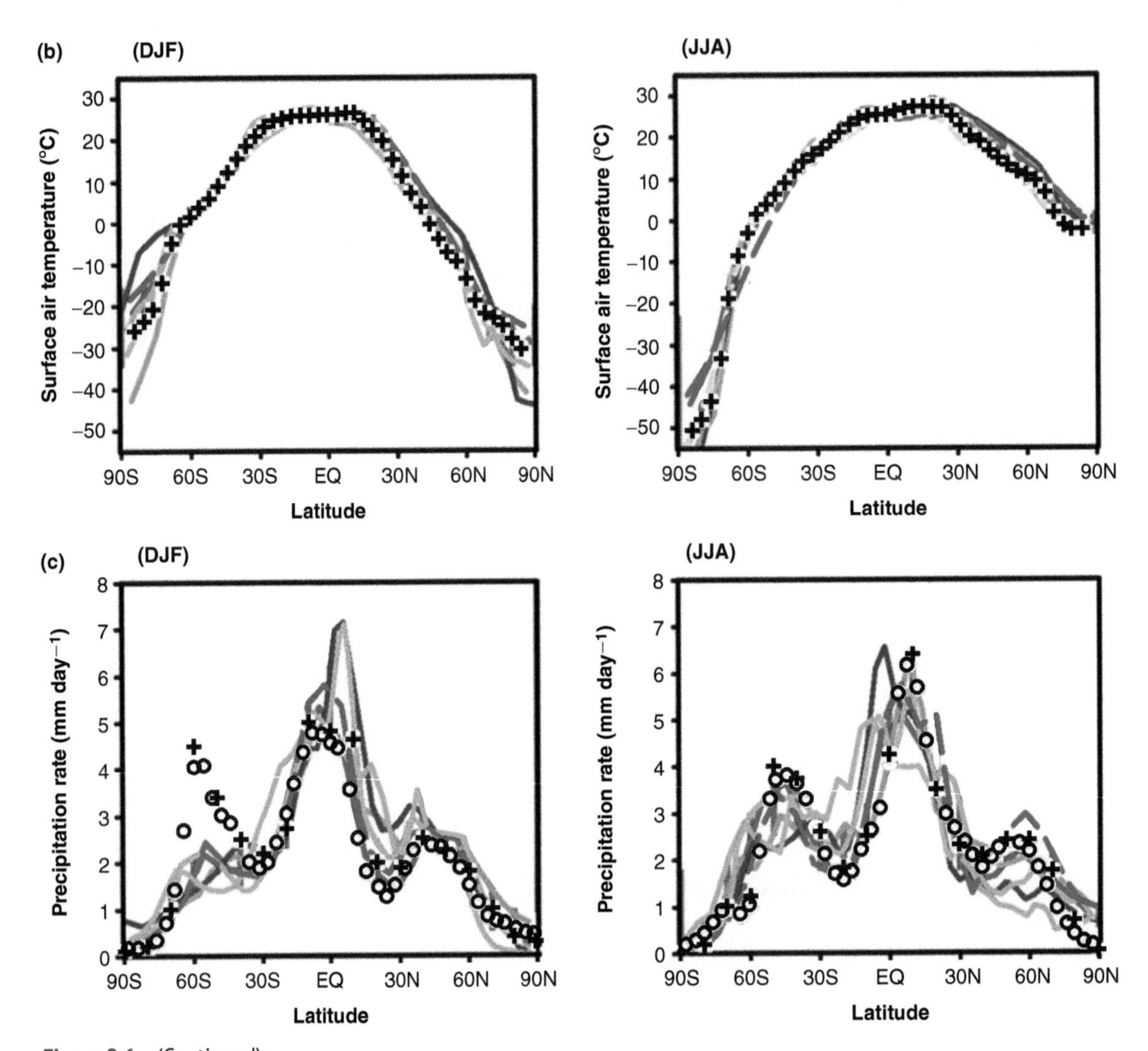

Figure 2.6 (Continued)

starts to decrease again to very cold conditions. For our interest, the troposphere is the most important region. Almost all the factors that affect the climate and cause climate change are vested here. The stratosphere contains the iconic ozone layer, but it has only a small influence on climate. The region above the stratosphere has no effect.

The precipitation is highest in the equatorial latitudes and decreases in mid-latitudes around 30 degrees. Then there is another belt of rainfall at 40–50 degrees and beyond that, it decreases to low rates at the poles. The winds drive this pattern of precipitation. Winds form permanent cells in the atmosphere that extend from the surface to the tropopause, causing north easterly surface winds in the northern equatorial regions and a mirror image in the southern hemisphere. Above 30 degrees latitude the winds form complex patterns that include a cell carrying air in southwesterly directions in the northern hemisphere (Figure 2.6b,c). These winds and precipitation are explained by the general circulation of the atmosphere and atmospheric thermodynamics as will be discussed in later chapters.

2.5 Irreducible Uncertainties

The matter of prevailing and potentially irreducible uncertainties requires discussion at this point so as to recognize from the start this important aspect of global change science. It can be illustrated by constructing a linear view of climate change (Figure 2.7). In this picture, the fundamental questions raised earlier are answered in a sequence starting with atmospheric composition, which is the best-known aspect. It can be measured with sufficient accuracy for all environmentally active gases, natural or emitted by human activities. Uncertainties start to enter as we connect the observed concentrations, and particularly the rates of increase to the sources of these gases and the ways in which they are removed from the atmosphere or "sinks." For instance, there are significant uncertainties as to where carbon dioxide released from burning fossil fuels is stored on land, a sink. For other greenhouse gases, disparate

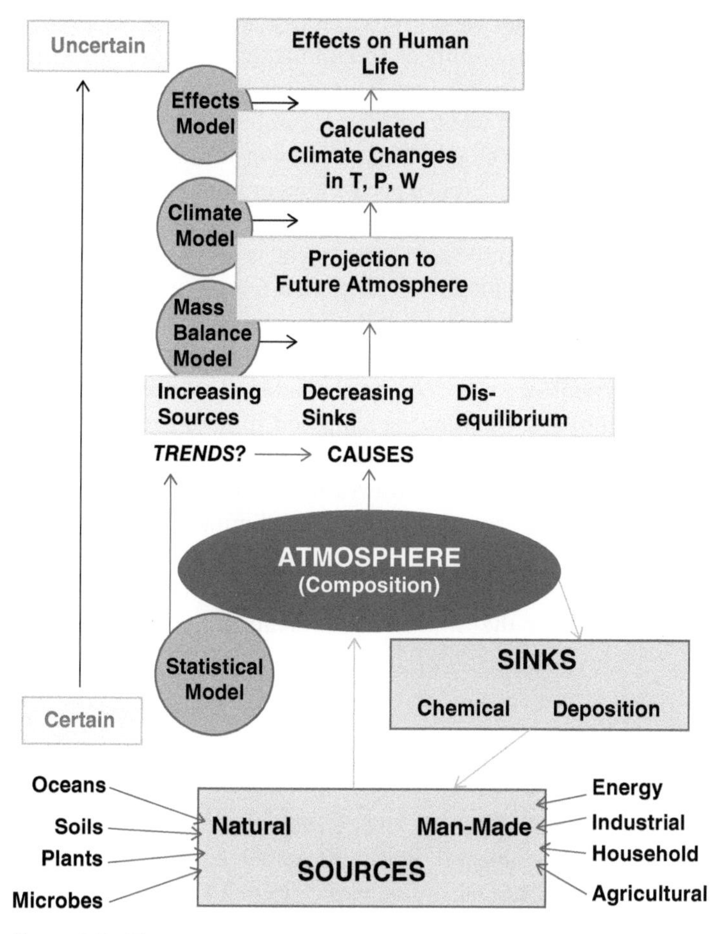

Figure 2.7 The process of global change science and inherent uncertainties. A view of how the science is built from atmospheric observations of changing composition to impacts on human life. Uncertainties arise at each step. The largest junctions of uncertainty are in determining future concentrations of greenhouse gases, climate feedbacks, and translating climate change into effects on human life such as crop yields, or health.

combinations of sources and sinks can explain the atmospheric concentration and its changes equally well, and estimates of future concentrations depend on which combination is correct.

Most of the concern about climate change, however, is for times several decades to a hundred years from now over which environmental control can be exercised and climate change managed. Moving upward in the figure, we encounter one of the most intractable uncertainties in determining a future climate – the concentrations of greenhouse gases to expect. The emissions of the greenhouse gases depend not only on our understanding of the sources now but even more on societal evolution and action over time. With current knowledge, it is impossible to make a credible case for any specific societal future, so scenarios are developed to determine what the range of future emissions and concentrations might be under a diverse range of possible conditions. But this creates a very large range, perhaps a factor of ~ 3 in the effective future greenhouse gas levels. As we progress further up the path, the greenhouse gas concentrations of the future have to be translated into climate changes, that is, global warming, rainfall, and wind patterns. Later we will see that climate change is subject to feedbacks, which are processes that can amplify or ameliorate the effect of the direct global warming by greenhouse gases. Currently models can explain the present climate very well, but are not able to agree on what the effect of some of these feedbacks will be, especially from possible changes in clouds. This matter can add another uncertainty factor of ~ 2 to ~ 3 to the assessment of the expected temperature change from a given increase in the greenhouse gases. But the ultimate questions we want to answer are the effects of climate change on human life such as crop yields in a warmer world, or the effect of hot weather on human health. These require a further translation by models that take the calculated climate change, from the previous step, and estimate the consequences on human life, leading to another similar range of uncertainty of at least a factor of ~ 2. These uncertainties cannot be easily combined to come up with how well we can establish the impacts of climate change on human life, but it is apparent that they are disconcertingly large. Even armed with a good quantitative theory of the present climate that matches with observations, some of these uncertainties are likely to remain irreducible based on the inherent limitations of science and scientific inquiry. It is ironic that in this chain of ideas, we can answer questions about how atmospheric composition is changing very precisely, but it doesn't matter to human life; what matters is whether the future environment can sustain us, and that we cannot answer.

2.6 The Plan

As a means of organizing the topic of global change, let's consider three areas consistent with the goal of addressing the fundamental questions posed earlier; these are greenhouse gases, climate science, and human influences (Figure 2.8). We start with how atmospheric constituents, including greenhouse gases, are generated and maintained. This is a fundamental part of environmental science that connects the unique processes that occur on the earth to its atmospheric composition. It leads us into the

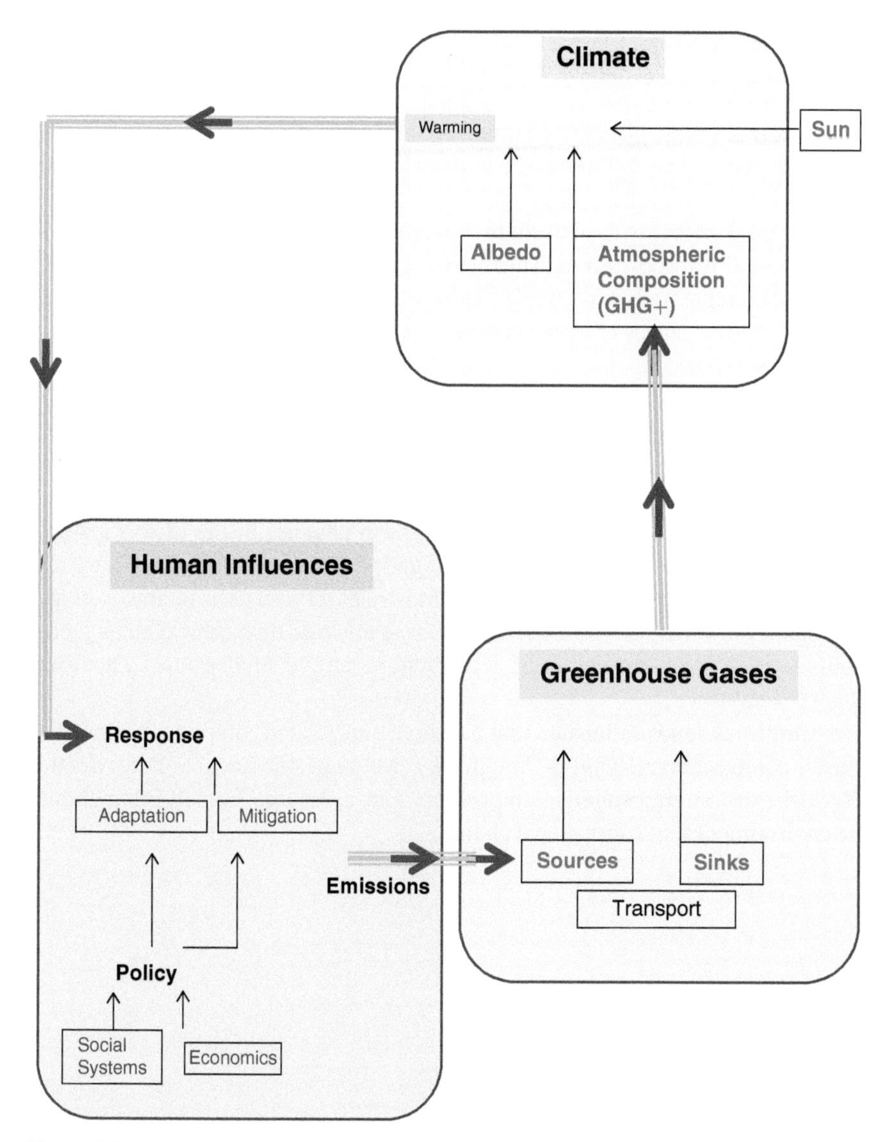

Figure 2.8 The plan and logic. Three main components of our interest are shown to build a logical framework for the discussion of climate change and human life. Atmospheric composition determines the earth's ability to absorb radiation and reflect sunlight; it then becomes the foundation for understanding how that creates the climate combined with the effect of the sun, which is the only external force in our diagram. The human influences both cause (emissions) and ameliorate climate change (mitigation).

second area that explains the climate in terms of atmospheric composition, albedo and sunlight. Solar radiation is regarded as almost constant and outside the direct influence of living things. After we have seen how the climate works, we look at how it will change and affect human life, that is, our impact on climate and the response to manage global climate change.

Review of the Main Points

1. The environmental system is separated into three minimal elements – the land, oceans, and the atmosphere. The environment is determined by their characteristics and interactions.
2. Three types of time scales are used to study the environment and climate: cycles, equilibration times, and transport times. These time scales separate the causes of various environmental changes and, in particular, show that the causes of urban air pollution are different from the causes of global climate change. Solving one of these problems will not solve the other.
3. Climate is defined. Temperature is a central variable. It is determined by the interactions between sunlight, reflectivity of the earth, and absorption of the earth's heat by the atmosphere. A benchmark present climate is articulated. It is the base from which climate change can occur.
4. The future state of the climate, affected by long-term changes in the earth's system caused by human activities, has several layers of uncertainties, some of which are irreducible. These uncertainties arise because of fundamental limitations to how well we can understand the earth's environmental system at any one time, how climate feedbacks will play out, and because human social systems and technology are difficult to predict with credibility.
5. A plan for the discussion is delineated that has three interrelated components: how the atmospheric composition relevant to the climate can change, the science of the climate itself that depends on atmospheric composition, and lastly, the human interactions with the environment that cause global change.

3

Atmospheric Composition

3.1 Trace Gases and Their Roles in Climate and the Environment

Average atmospheric composition at the earth's surface is shown in Table 3.1. Many of the longer-lived gases are present at the same percentages in much of the troposphere. The table is a little different from what appears almost everywhere else because water vapor has been included as an integral part of atmospheric composition at 0.5%. It should be noted that water vapor consists of separated molecules that are present in air which also contains condensed water as clouds, mists, and fog; and some is in solid form as ice crystals. Water vapor is invisible, although we feel it as humidity. Traditionally, dry air composition is tabulated and water is not reported because it is highly variable in both space and time; at some locations or times it may be a few parts per million and at other places it could be up to 5%, whereas dry air is thought to have the same composition everywhere. This is so for the major gases such as nitrogen, oxygen, and the noble gases, but trace gases generally vary by latitude and season. Water vapor, as we will see later, is the most important greenhouse gas and also contributes to global warming significantly. In a discussion about climate science, it is reasonable to include it directly in atmospheric composition to emphasize its importance from the very beginning, along with the less variable but also less concentrated greenhouse gases.

The picture we see is that 99% of the dry air is nitrogen (78%) and oxygen (21%) and yet they have almost no direct influence on the greenhouse effect or global climate change! Their role in climate is to be the winds, hold heat, carry water from the oceans to the land, and to transport greenhouse gases and pollutants away from the sources and spread them over the whole world. Nitrogen and oxygen are the only atmospheric gases whose existence we feel regularly, including our need to breathe. That leaves only 1% of the atmosphere, and most of it is made up of argon and the other noble gases that have virtually no role in the environment as it affects living things. We are left with less than 0.05% of the dry atmosphere that includes greenhouse gases such as carbon dioxide (CO_2), methane (CH_4), and nitrous oxide (N_2O); and the ozone-depleting compounds such as the chlorofluorocarbons (CCl_3F, CCl_2F_2). This small fraction, made up of myriad trace gases, have influences on the environment that are disproportional to their meager concentrations. Man-made gases implicated in the destruction of the ozone layer, even when taken together, never reached one molecule in a million of air. This leads to *the principle that living things gain the capacity to perceptibly alter the environment and the climate or even to control it, because*

Global Climate Change and Human Life, First Edition. M. A. K. Khalil.

Companion Website: www.wiley.com/go/khalil/Globalclimatechange

even small emissions of the trace gases can have hugely amplified impacts. This may be so on other planets as well, making it more probable that life can take hold in a wide spectrum of physical environments because of rare gases in the atmosphere.

There are a few more noteworthy characteristics that appear in Table 3.1. One is that most of the earth's atmosphere consists of very light gases of two or three atoms and the single-atom noble gases. Methane is the largest-sized gas of natural origins affecting climate. Many of the five to seven atom gases are man-made halocarbons with strongly bonded atoms. More complex organic compounds are produced in abundance by biological processes, but either do not evaporate into the atmosphere, or are so chemically reactive that they do not last long once they get there. Another matter is that once we go below about 0.000001% concentration there are many gases that are known to exist in the global atmosphere, perhaps thousands of them. Despite the amplified effects of some of them, they are of marginal environmental significance because of their extremely low concentrations. Table 3.1 lists some of these extremely rare gases because they are targets of major international agreements such as the Montreal Protocol, to prevent the depletion of the ozone layer, and the Kyoto Protocol to manage the climate.

Table 3.1 Atmospheric composition in recent millennia.

Atmospheric composition: Major, greenhouse and noble gases.[1]	
Gas	Volume(%)[2]
N_2	78.08
O_2	20.95
Ar	0.93
H_2O	0.5
CO_2	0.040
Ne	0.0018
He	0.00052
CH_4	0.00018
Kr	0.00011
H_2	0.000055
N_2O	0.000033
Xe	0.0000090
O_3	0.0000040
HCFCs	0.00000003
CCl_3F	0.00000002
CCl_4	0.000000013
PFCs	0.000000008
SF_6	0.0000000010

Notes 1) There are many more gases in the earth's atmosphere, but their roles in climate and global change are very small. 2) The total adds up to about 100.5% because of the inclusion of water vapour. The remaining concentrations are for dry air as traditionally reported.

Readers may find an analogy about the rarity of trace greenhouse gases both amusing and illuminating. Let's say we represent the different gases in the atmosphere by the color of M&Ms® in a container. The brown ones are common so they can be nitrogen and red ones can be oxygen. Let's make carbon dioxide blue, methane yellow, nitrous oxide purple, and carbon tetrafluoride the elusive pink one. We ask: how large a box would we need to find an M&M of the color that represents one of the trace gases. After much testing and consuming of M&Ms, it turned out that the volume of M&M is about 0.636 cm^3 and they fill up about 68% of the space, so each occupies 0.9 cm^3 of the box. To find one blue M&M representing carbon dioxide you will need a cubic box about 5 inches (13 cm) on the side. The rest will be brown or red assuming that the colors are evenly mixed. For methane, you will need a cubic box about 30 inches (80 cm) – that is already fairly large. Nitrous oxide's purple M&M will be one in a box that is 5 feet (143 cm) on the side; and to find a pink one, you will need a box the size of a very large warehouse 1/4 of a mile × 1/4 of a mile on the sides and 10 feet high! It is a marvel of nature that gases present in such small quantities, and many generated by living things, can have such a major effect on the earth's climate.

To get our bearings, it is useful to review which gases do what, so as to motivate a further study of these particular gases and how they can affect the climate, ozone layer, or other aspects of our environment (Figure 3.1). For climate, the gases that cause the greenhouse effect and global warming, in order of their importance are H_2O,

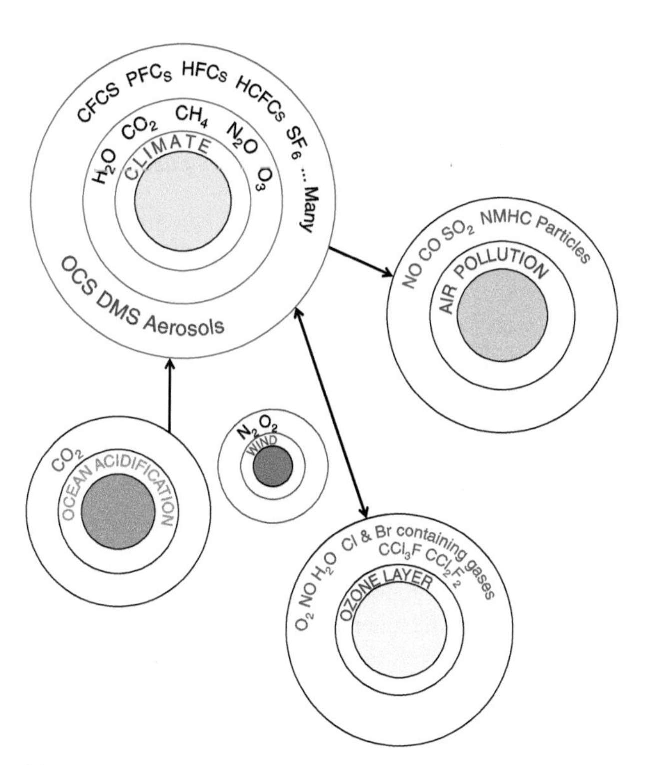

Figure 3.1 Which gases do what. In these pictures you see the role of each of the gases in the earth's atmosphere. The gases close to the core have the most effect and those further out have lesser impacts, primarily because of their lower concentrations. Gases that cause global warming and ozone depletion have long atmospheric lifetimes and are distinct from the gases that cause air pollution which have short lifetimes.

CO_2, CH_4, N_2O, and O_3. Of lesser importance, because of very low concentrations, are many technological gases including chlorofluorocarbons (CFCs), hydrofluorocarbons (HFCs), hydrochlorofluorocarbons (HCFCs) and sulfur hexafluoride (SF_6). There are sulfur gases such as sulfur dioxide (SO_2), carbonyl sulfide (OCS), and dimethylsulfide (DMS) that turn into light-scattering aerosols and, along with other sources of particles, can cool the earth's surface. Their effects from man-made emissions are overwhelmed in most places by the greenhouse gases mentioned earlier. For the depletion of the ozone layer the two most abundant man-made chlorofluorocarbons (CCl_3F and CCl_2F_2, also known as CFC-11 and CFC-12) are the most significant followed by many other rarer chlorine and bromine containing gases, most of which are entirely from human activities. The natural ozone layer chemistry is driven by sunlight, oxygen (O_2), and destructive molecular fragments from the breakup of nitrogen, hydroxyl, and halogen-containing gases (N_2O, H_2O, CH_3Cl, and others). Ocean acidification is almost entirely due to increased carbon dioxide. By contrast, urban air pollution is caused by NO, NO_2, CO, SO_2, and non-methane hydrocarbons.

We have discussed atmospheric composition as percentages, which is one form of a *mixing ratio*. For N_2 as an example, it says that for every one hundred molecules of air, there are about 78 molecules of nitrogen, or 78 parts per hundred (pph = percent = % = 10^{-2}). This idea is extended to a mixing ratio expressed as parts per thousand (per mil = $^{00}/_{0}$ = 10^{-3}), million (ppm = 10^{-6}), billion (ppb = 10^{-9}), trillion (ppt = 10^{-12}), and even quadrillion (ppq = 10^{-15}). For example, the mixing ratio of CO_2 as a percentage is 0.04%, but it can also be expressed as 400 ppm meaning that there are 400 molecules of it in a million molecules of air. We will use these units instead of percentage because we will be dealing with gases that are present in exceedingly small quantities in the atmosphere. The mixing ratio is the concentration of a gas in molecules per cubic centimeter to the total number of molecules of air in the same cubic centimeter.

Readers should note well that the action of gases on global change, whether it is the natural climate, global warming, stratospheric ozone depletion, or ocean acidification, is proportional to the actual number of molecules of the gas per unit volume and not to the mixing ratio. Likewise, the movement of gases from one place to another is determined by how many molecules are moved per second. The convenience of using a mixing ratio comes from the fact that for long-lived gases it is more or less constant in the troposphere, but the concentration in molecules per cubic centimeter is not. So if we say that methane is 1800 ppb, as we have done in Table 3.1, it is valid everywhere in the troposphere at least, but any expression as molecules per cubic centimeter requires associating a location with it, or more specifically an altitude, because that is where the concentration in molecules per cubic centimeter changes the most. The abundance expressed as a mixing ratio, although convenient, loses information about the action of a gas on the environment, which has to be recovered to evaluate most of the factors that determine climate, climate change, and chemical processes in the atmosphere.

3.2 Quantifying the Atmospheric Composition

From the preceding discussion it follows that we must know the density of air in order to convert the mixing ratio into the actual number of molecules of the gas in a unit volume of air, such as a cubic centimeter. Let's consider first the pressure of air in the atmosphere. Pressure is defined as the force per unit area. The permanent annual average pressure of air varies most with altitude and less so in the horizontal. Let's look at its vertical changes because that determines its role in environmental and climate sciences.

At the earth's surface the weight of the air above causes this pressure to be about 1 bar (1.013 bars), which is 14.7 pounds per square inch (P_0). This is taken as a given average state of the earth's atmosphere that has persisted for myriads of years. If the air is not moving up or down, then according to Newton's laws of motion, there must be no net force on it (Endnote 3.1). The gravitational force on an air mass caused by the earth, which pulls it down, and is always present, must be balanced by the force of the air below that pushes it up. This balance, the ideal gas law, and the approximation that the temperature is constant with altitude can be shown to establish an exponential decline of pressure and density of air with altitude as stated in the following equations (Endnotes 3.2 and 3.3):

$$N(z) = N_0 e^{-z/H} \tag{3.1}$$

$$P(z) = P_0 e^{-z/H} \tag{3.2}$$

$$H = \frac{RT}{M_A g} \tag{3.3}$$

In Eqs. 3.1 and 3.2, $N(z)$ is the density of air (molecules/cm^3) at a height z (km), e is the base of natural logarithms (2.71828..), H (km) is called the scale height, and N_o is the air density at the earth's surface. In Eq. 3.3, R is the universal gas constant, g is the acceleration due to gravity, M_A is the molecular weight of an average air molecule, and T is the temperature as in Figures 2.6 a,b. The fixed parameters are: $R = 8.314$ j/mole – °K, $g = 9.8$ m/s^2 and $M_A = 29$ g/mole (air). If we also take the temperature to be constant at, say, the surface temperature of 288°K, then $H \approx 8$ km. A constant H is required to derive Eqs. 3.1 and 3.2. The pressure is shown in Figure 3.2.

The scale height has a useful interpretation. It is the height above the earth's surface within which the entire atmosphere would be confined if it maintained the same density and temperature with altitude as at the surface. It is used to define an alternative scale for stating the concentrations of gases in the atmosphere as: $\delta = C\ H$ where C is the mixing ratio. If the entire atmospheric content of a gas was confined to a layer at the surface, its

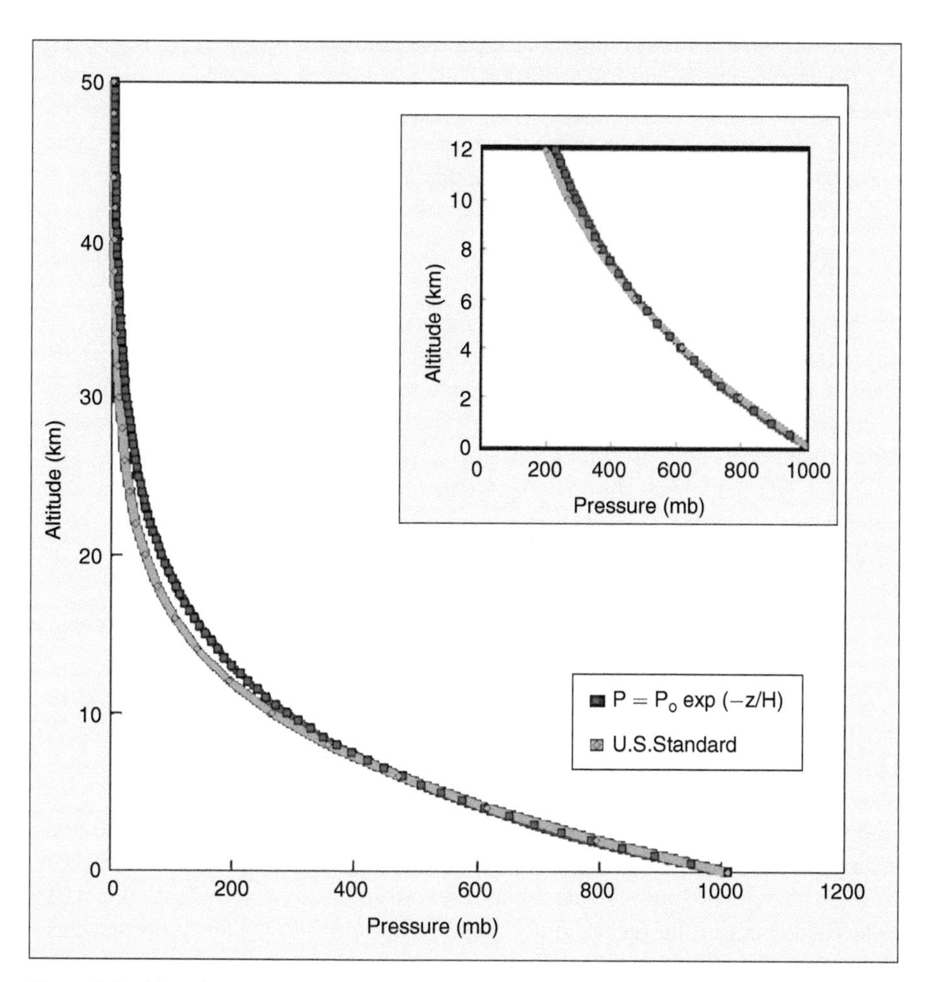

Figure 3.2 How the pressure of air varies with altitude. At any altitude, the weight of the air above is balanced by the push of the air from below to create a zero net force that keeps the air from moving up or down on average during a year. This balance requires the pressure of air to decrease with altitude exponentially. Applying the ideal gas law, the density of air follows the same curve with an appropriate shift of units.

depth would be δ. Such a measure is commonly used in ozone research where δ is measured in Dobson units: 1 DU = 0.01 mm.

Example What is the δ for CO_2 and CF_4?

Solution: From Table 3.1, CO_2 = 400 ppm = 400×10^{-6}. $\delta = (400 \times 10^{-6}) \times 8$ km $\times 10^5$ cm/km = **320 cm.** It says that if we took all the CO_2 in the atmosphere and put it at the surface, it would occupy a depth of 320 cm, or about 10 feet of pure CO_2 – not very healthy for us if that happened.

For CF_4: C = 75 ppt = 75×10^{-12}. $\delta = (75 \times 10^{-12}) \times 8$ km $\times 10^9$ μm/km = **0.6 μm**. This is an incredibly small depth as can be seen by comparing it to the diameter of human hair, which is about 100 μm.

The pressure decreases exponentially with altitude. Sometimes meteorological and climate variables are pictured and discussed in terms of the pressure rather than the altitude. Readers can readily convert these units to approximate altitudes by inverting Eq. 3.2 as $z = H \ln (P_o/P)$ where ln is the natural logarithm.

The ideal gas law needed to derive Eqs. 3.1 and 3.2 is of use both here and elsewhere in our discussion of climate. For our present purpose it connects the density of air and pressure. Suppose we put "*n*" moles of a gas in a box of volume *V*. The ideal gas law is a relationship between the pressure (*P*) and temperature (*T*) of this system expressed as:

$$PV = nRT \tag{3.4}$$

A mole of a gas, by definition, has $N_a = 6.02 \times 10^{23}$ molecules, which is Avogadro's number. If you divide both sides of Eq. 3.4 by the volume, you see that the pressure $P = \rho RT$ where $\rho = n/V$ is the density in moles/cm³, which can be converted to the more intuitive units of a number density in molecules/cm³, so that N (molecules/cm³) = ρ (moles/cm³) × N_a (molecules/mole). We can deduce that $P = N(RT/N_a)$. Or, for a constant temperature, the density and pressure are proportional. The fact that air satisfies the requirements of the ideal gas tells us that as the pressure increases the volume decreases according to Eq. 3.4, all else being the same. This causes the air near the surface to become more compressed, thus increasing in density.

Example What is the number density of air at the surface if the average pressure is known to be P_0 = 1013 mb and the average temperature = 288°K?

Solution: The relationship derived above says that $N_0 = P_0 (N_a/RT)$. Therefore, N_0 = 1013 mb ×100 (Nt/m²)/mb × (6.02 × 10^{23} molecules/mole)/(8.314 Nt-m/mole-°K × 288°K) = **2.5 × 10^{25} molecules/m³**.

The assumption of constant temperature required to derive Eqs. 3.1 and 3.2 may seem extreme or invalid for the atmosphere in view of its stated change with altitude (Figure 2.6a). If we take the temperature change with altitude into account a more accurate version of Eq. 3.1 can be derived. The differences in the predicted densities are not large, so for our purposes, this increased accuracy does not compensate for the complexity of the more accurate formula. The assumption works well because the average temperature of the atmosphere is around 260°K and the range in the troposphere is within about ± 10% of this. We will often follow my 10% rule – that we will accept any approximation that brings a calculation within 10% of the true value, or we will ignore influences that change our conclusions by less than 10%.

Eq. 3.1 contains a great deal of information that can be applied to understand trace gases in the environment and their trends. To start with, we see that to define a mixing ratio requires dividing by the continuously decreasing density of air as we go higher up in the atmosphere. Gases that live longer than the time it takes to move and mix them upward will end up with a nearly constant mixing ratio with altitude. That is, their density in molecules per cubic centimeter will fall at the same rate as that of air, so the ratio will be constant. Most of the longer-lived gases in Table 3.1 have nearly constant mixing ratios in the troposphere. Expressing the abundance of carbon dioxide as a density in terms of number of molecules or grams per cubic centimeter is valid for only some location and will vary significantly with altitude according to Eq. 3.1. For instance, the mixing ratio of oxygen, 21%, is nearly constant with altitude, so its density will decrease at the same rate as in Eq. 3.1. At the top of Mt. Everest therefore, the 21% oxygen means that there is only a third as many molecules in each liter of air as at sea level, making it difficult for people to breathe, even though the mixing ratio is the same as at the earth's surface.

We have the tools now to go further and calculate some important characteristics of the atmosphere that are both interesting and have practical consequences. Let's start with the total number of molecules of air in the various layers of the atmosphere, which will also tell us how many molecules there are in the entire atmosphere! It will turn out to be indispensable knowledge for understanding quantitative aspects of the global environment and climate change.

How many molecules of air are there in the atmosphere? To answer this question, consider Eq. 3.1, which tells us how many molecules there are per cubic centimeter of air at various altitudes. We can assume that the changes of the air density with latitude and longitude are relatively small and usually within our 10% rule if averaged over a year. But in the vertical we have a permanent decrease determined by the laws of physics as discussed, to arrive at Eq. 3.1. Imagine surrounding the earth with thin shells. The shells are shallow enough that the density of air does not change much within them and can be estimated by Eq. 3.1 for the altitude z_a in the middle of the shell. If we multiply this density by the volume of the shell, we will get the number of molecules of air in that shell. That is, $N(z_a)$ × volume from Eq. 3.1. The volume of the shell is $A\,\Delta z$, where A is the area of the earth and Δz is the thickness. Then we can just add up the results from each shell within the bounds of the altitudes we are interested in; and in all the shells if we want to know the total number of molecules in the entire atmosphere. It can be shown mathematically that the result of such a process, based on Eq. 3.1, would be expressed by this convenient equation (Endnote 3.4):

$$N(z_1, z_2) = N_0 HA\left(e^{-z_1/H} - e^{-z_2/H}\right) \tag{3.5}$$

where, $N(z_1,z_2)$ is the number of molecules of air between altitudes z_1 and z_2. This is a remarkable equation; let's put it to use. To calculate the number of molecules of air in the entire atmosphere we put $z_1 = 0$, representing the surface of the earth, and $z_2 = \infty$ to cover the entire atmosphere. You get the equation $N_A = N_o\,H\,A = (2.5 \times 10^{19}$ molecules/cm$^3 \times 8$ km $\times 10^5$ cm/km $\times 5.1 \times 10^{18}$ cm$^2) \approx 10^{44}$ molecules! Similarly if you put in the limits from the surface to the average tropopause, $z_1 = 0$ and $z_2 = 12$ km, you will get $\approx 0.8 \times 10^{44}$ and for the stratosphere $z_1 = 12$ km and $z_2 = 50$ km, you will get about $\approx 0.2 \times 10^{44}$ molecules. Although the troposphere is only 12 km thick it contains 80% of the atmosphere, while the stratosphere, which is more than 40 km thick, has only 20%. Half the atmosphere lies below 5.5 km, an altitude surpassed by many mountain peaks. This is the nature of the atmospheric environment by which most of the air is confined near the surface where it is used by living things and where the climate, the greenhouse effect, and global warming are generated.

Example What percentage of the atmosphere is above the stratosphere?

Solution: From Eq. 3.5, the percentage above the stratosphere = $[N(50\text{ km, infinity})/N_0 H A] \times 100\% = e^{-50\text{ km}/8\text{ km}} \approx$ **0.2%**, well below our 10% rule.

To make these ideas more useful, we can convert the number of molecules to masses. Each molecule has a weight. If you know how many molecules there are of a particular gas, you can figure out their collective mass in the atmosphere. The relationship between number of molecules in a volume of air and the mass of the air can be built like this: Let's say we want to calculate the mass of the whole atmosphere which has N_A molecules, that is, 10^{44} as found earlier. Convert these to moles first, which is: N_A/N_a = (molecules)/(molecules/mole), where N_a is Avogadro's number. For the whole atmosphere we get 1.7×10^{23} moles. Now convert the moles of air in the atmosphere to mass as: N_A/N_a (moles) $\times M_A$ (g/mole) = 5×10^{21} grams, where M_A is the molecular weight of air taking into account that ~ 79% is nitrogen and ~ 21% is oxygen (0.79×28 g/mole + 0.21×32 g/mole = 28.8 g/mole) . Similarly, the masses of the troposphere and stratosphere are about 4×10^{21} g and 1×10^{21} g. It follows that for any volume of the atmosphere the relationship between molecules of air and grams is: N(grams) = N (molecules) M_A (grams/mole)/N_a (molecules/mole) and similarly the density is: ρ (g/cm^3) = N (molecules/cm^3) M_A/N_a.

The conversions we have derived show an equivalence between tons or grams of a gas in the atmosphere and the mixing ratio or concentration (Endnote 3.5). This has a frequent practical use because in order to delineate a balance of any gas in the atmosphere we must use the same measures for all the components that determine how much is coming in and how much is going out – either tons in the atmosphere or a mixing ratio such as ppb. Usually, the annual emissions, whether carbon dioxide from fossil fuel burning or methane from agriculture, worldwide or from a specific country, are given as teragrams or petagrams per year but the observed concentrations in the atmosphere are given as ppb or ppm as in Table 3.1. From the discussion, the necessary conversions for concentrations and emissions are:

$$C(grams) = \frac{C(ppb) N_A M_G}{N_a} \times 10^{-9} \tag{3.6a}$$

$$S(ppb/y) = \frac{S(g/y) N_a}{M_G N_A} \times 10^{9} \tag{3.6b}$$

Here M_G is the molecular weight of the gas of interest. The 10^{-9} is needed to describe the mixing ratio in ppb and would change to 10^{-6} or 10^{-12} for gases that are expressed in ppm or ppt. These equations are written to express the quantity you want to calculate on the left-hand side and the information you are given, on the right-hand side.

Example We are told that currently the world's CO_2 emissions are about 10 PgC/y. How much does this add to the average CO_2 mixing ratio every year?

Solution: Using Eq. 3.6b, $S\,(\text{ppm/y}) = S\,(\text{g/y})\, N_a/M_G\, N_A \times 10^{-6} = 10 \times 10^{15}\,(\text{gC/y}) \times 6 \times 10^{23}$ (molecules C/mole)/[12 (g/mole) $\times 10^{44}$ molecules air] $\times 10^{6}$ ppm = **5 ppm/y**. The observed change is about half that, which illustrates that the actual increase is determined not just by how much we add each year, but also by other processes that remove it from the atmosphere. These concepts will be explored in the next chapter.

A practical complication arises when the outcome of these calculations is compared with observations. Measurements of gases in the environment are usually taken at the surface; however, the C(ppb) in Eq. 3.6a is the global average mixing ratio, and likewise for S: it is the global rate of emission that is being distributed over the whole atmosphere. Most gases have significantly lower mixing ratios in the stratosphere compared with the troposphere reflecting their losses or sinks as well as transport factors that slow down the transect to the stratosphere. To make the observed surface concentrations compatible, a generic factor can be applied as: C (ppb) = $0.95 \times C$(ppb at the surface) (see Endnote 3.6).

Example The average mixing ratio of methane from observations at various sites on the earth's surface is 1800 ppb. How many teragrams (10^{12} g) of methane are there in the atmosphere?

Solution: First we determine the average concentration C(ppb) to put into Eq. 3.6 as: 1800 ppb $\times$ 0.95 = **1710 ppb**. Then we use 3.6a to calculate the *global burden*, as it is called = (1710 $\times 10^{-9}$ molecules CH_4/molecule air) $\times$ (10^{-44} molecules air) $\times$ (16 g/mole)/(6.02 $\times 10^{23}$ molecules CH_4/mole) $\times 10^{-12}$ Tg/g $\approx$ **4560 Tg**. Note the usefulness of stating all the units and their cancellations to end with Tg.

Review of the Main Points

1. A little less than 99.5% of the dry air is nitrogen, oxygen, and noble gases that have no direct role in the greenhouse effect or global warming. The gases that do are water vapor at about 0.5% and carbon dioxide that is currently about 0.04%. Methane, nitrous oxide, and ozone have a small influence on the natural greenhouse effect due to their low concentrations.
2. The abundance of gases in the atmosphere is expressed by several units. The percentage is an expression of a mixing ratio that is defined as the number of molecules of the gas per cubic centimeter divided by the number of molecules of air in the same volume $\times$ 100%. This is a convenient unit, because it can reflect the abundance of the gas in most of the atmosphere, but it has to be converted back to molecules per cubic centimeter for most circumstances when we need to evaluate the impacts of the gases on climate, the stratospheric ozone layer, or their transport from one location to another.
3. Many trace gases that exist at very low concentrations have amplified influences on the natural greenhouse effect and even more so on climate change. The most important of these are CO_2, CH_4, N_2O, and O_3. The reason that human activities can cause perceptible global warming is because these gases are present at low enough concentrations that the relatively small emissions from human activities can increase their atmospheric levels significantly and sufficiently.

4 To quantify changes of atmospheric composition some key variables have to be understood. These are the density and pressure of air at various altitudes where their variation is the most, and it is also a permanent feature of the atmosphere. This determines how much of the greenhouse gas is at each altitude. It is important because global warming, ozone depletion, and all other global change phenomena are proportional to the density of gases expressed as molecules per cubic centimeter, or the grams present, and not as the mixing ratio. Since Because most of the air molecules are concentrated, in the lower part of the troposphere, near the surface, that is where the climate and the greenhouse effect are formed.
5 Based on the previous point, the conversion between atmospheric concentration (percent) and atmospheric burden (grams in the atmosphere) is necessary to understand the balance of gases in the atmosphere. Formulas were derived to represent these relationships that require the ideal gas law and the calculation of the number of molecules of air in the troposphere, stratosphere, and the whole atmosphere. It looks like there are 10^{44} molecules of air in the earth's atmosphere.

Exercises

1 Suppose that measurements of chlorofluorocarbon 11 (CCl_3F) show that is present in the atmosphere at 260 ppt. What is the concentration in micrograms/cubic centimeter at the surface of the earth? At the top of Mt. Hood?
2 How much oxygen is there in molecules per cm^3 at the top of Mt. Hood? How about at the top of Mt. Everest? What percentage of the atmosphere is oxygen at these two mountain tops? Will it be hard to breathe at either of these locations? Explain why?
3 An expandable helium balloon is to be sent into the upper atmosphere. At the surface it has a volume of 1 m^3. What will be its volume when it reaches 20 km? What will happen if the balloon is made up of rigid material? Assume that the balloon rises slowly so that the temperature inside is the same as outside. The temperature at 20 km is 216°K and the surface is 288°K. Use the ideal gas law $PV = nRT$.
4 The mixing ratios from surface measurements of the greenhouse gases methane (CH_4), carbon dioxide (CO_2), and nitrous oxide (N_2O) are 1800 ppb, 400 ppm, and 330 ppb, respectively. How many Tg of each of these gases are there in the atmosphere? Determine the conversion factors ppb/Tg for CH_4 and N_2O and ppm/Tg and ppm/PgC for CO_2.
5 An average person weighs 70 kg (let's say) and 18% of the body is carbon (this is true). The population of humans has increased by 5 billion over the last 100 years. How much extra carbon is being held in the human mass? Compare this to 10 PgC emitted each year by fossil fuel burning and other human activities. 1 PgC = 1 Petagram carbon = 10^{15} gC.
6 Calculate the average molecular weight of air molecules using only N_2 and O_2. Repeat the calculation by adding Ar and CO_2, the next two gases from Table 3.1, for dry air (not including water). Find the % errors introduced by ignoring gases beyond the N_2 and O_2. Can we ignore these gases based on the 10% rule? Take % Error = Difference/M (with more gases) × 100% where M is the molecular weight and Difference = [$M(N_2+O_2) - M$ (with more gases)].
7 Calculate the scale height at the surface and at the tropopause based on the difference of air temperatures at these locations. Compare with the scale height calculated using the average temperature of 260°K by calculating the % errors based on the previous problem. Comment using the 10% rule.

8 You are flying a small airplane. The altimeter uses pressure to figure out how high the airplane is. Suppose you took off at a pressure of 1000 bar and now find that the air pressure outside is 10 bar. How high are you flying? Should you be flying that high?

9 If the temperature of the earth increased by 6°C and the lapse rates remained the same, by how much will the pressure change at the top of Empire State Building?

10 Suppose the amount of nitrogen (N_2) in the atmosphere increases by 50%.
(a) Determine its new mixing ratio in the atmosphere in percentage.
(b) What will be the mixing ratios of O_2 in %, CO_2 in ppm, and CH_4 in ppb?

11 We usually use the observed concentrations in the non-urban atmosphere to state the concentration of a gas in the earth's atmosphere. Yet most gases come from urban areas where there is a high concentration of people and therefore a high concentration of these gases.
(a) Show that the global average concentration including the urban concentrations, should be: $C(\text{global}) = f(\text{non-urban})\ C(\text{non-urban}) - f(\text{urban})\ C(\text{urban})$ where the C's are the mixing ratios measured in the urban and non-urban environments and the f's are the ratios of the number of molecules of air in these places relative to the total number of molecules of air in the atmosphere.
(b) The mixing ratio of CO_2 in the non-urban atmosphere is 400 ppm in 2020 and in the urban areas it is variable, but let's say it is 500 ppm on average. Estimate the correction to the global average mixing ratio. (c) How high does the average non-urban mixing ratio have to be to make a difference of 2.5% in the mixing ratio? Use these data: The present urban area is estimated by the World Bank to be 3.6 million square kilometers. The average altitude in which urban pollutants mix vertically is $h = 1$ km, and the density or air in this region can be approximated as at the surface.

Endnotes

Endnote 3.1 Newton's laws of motion apply to all objects and all motion. Motion is caused by forces that are either being applied to the object or were applied at some earlier time. The second law connects the rate at which objects move, expressed as an acceleration, to the forces acting on the object. It is the iconic equation $F = m\,a$, where F is the total force, m is the mass, and a is the acceleration. Acceleration is, as your experience tells you, the increasing speed of an object. The third law says that if an object A exerts a force on an object B, object B will exert an equal and opposite force on A. We will mostly use the second law. If there are no forces acting on an object, the acceleration will be zero. This doesn't require that the object be stationary, only that it is not accelerating, which can also happen if it is moving at a constant velocity relative to you (this is expressed as the first law). A simplifying aspect is that it doesn't matter how you generate the force that acts on an object, by gravity, pressure gradients, or electricity, the resulting motion will be the same.These laws have an explicit dependence on direction and are often expressed in vector form to reflect that. Although we will not use vectors in this book, you nonetheless have to account for this nature in other ways. It means that merely knowing the force is not enough to tell you which way the wind blows, you also need to know the direction. Then the motion will be in the same direction as the net force. The vector nature causes an additional complication – that even if an object moves at a constant speed, it can be accelerating because the direction of the motion may be changing. This is what happens when the earth goes around the sun. The gravitational force of the sun's pull is balanced by the constantly changing direction of the velocity, even though the magnitude of the velocity at about 67,000 mph is more or less constant and has been for ever past.An important consequence of the law is that if objects, including air masses, are observed to be stationary, then by definition, net $a = 0$ and even velocity $v = 0$, which can only happen if $F = 0$. This again does not mean that there are no forces on the object, only that the totality of all forces add up to zero. The vector nature of the force is important here because a force in one direction can only be canceled by an equal force in the opposite direction, not merely a force that has the same magnitude or strength. This should not be confused with the third law in which the equal and opposite forces act on two different objects. Additional complications arise when the location from where you are looking at moving objects are themselves moving, and worse, if they are accelerating. The spinning of the earth is such an example. It generates motion that requires additional work to explain the observed movement of air masses and all objects in the earth's atmosphere.

Endnote 3.2 The exponential function $f(x) = e^x$ is used frequently in this book. As the name implies, its values increase as x increases and they get larger faster as x becomes larger. But when x is small, the function can be approximated as a straight line: $e^x \approx 1 + x$, which is easier to use if applicable. Similarly the value of the function $f(x)$ decreases if we have a negative x or e^{-x}. Then the function goes to zero rapidly as x gets large. We use the notation df/dx to mean the rate of change of a function f with respect to x. Simple equations with rates such as the density equation can be solved mathematically as in Eqs. 3.1–3.3. In this book the solutions will not be derived, but simply provided. These solutions are unequivocal expressions of the mathematics. The formulas you get are not debatable, but whether they apply to the environment in the way they are used may be.

Endnote 3.3 The following discussion will inform readers who want to know how the density and pressure (Eqs. 3.1 and 3.2) are derived from basic principles. Consider a mass of air in the box along with all the forces acting on it according to Newton's second and third laws, as shown in the associated figure. It has a horizontal area A and a vertical depth δz. As noted earlier, the pressure on the lower surface caused by the air below will be higher than the downward pressure on the upper surface caused by the air above (because there is less air there). This creates an overall upward pressure gradient force, that is: $F_{UP} = F(z) - F(z + \delta z) = P(z)A - P(z + \delta z)A = -A\,\delta P$, where δP is defined as $P(z + \delta z) - P(z)$, that is, the difference of pressures at $z + \delta z$ and z, and the force is always pressure times the area. Counteracting this net upward force is the downward force of gravity that is the weight of the air in the box or $F_{DOWN} = - M\,g$, where M is the mass of the air in the box; $M\,(\text{gm}) = \rho\,(\text{grams/m}^3)\,V\,(\text{m}^3)$. Using the fact that $V = A\,\delta z$, $F_{DOWN} = \rho g\,A\,\delta z$. For the air not to move up or down, the net force on the box must be zero, that is: $F_{UP} = F_{DOWN}$ or, $-A\delta P = \rho g\,A\,\delta z$; therefore, $\delta P/\delta z = -\rho g$. This tells us that the pressure decreases with altitude at a rate given by $-\rho g$. The ideal gas law implies $P = \rho\,RT/M_A$. If we substitute this into $\delta P/\delta z = -$

ρg we get: $\delta\rho/\delta z = -(M_A\, g/RT)\, \rho = -\rho/H$. Next if we make the thickness of the box δz exceedingly small ($\delta\rho/\delta z \approx d\rho/dz$), then the solution of this equation for the density is exactly Eq. 3.1 if H is constant. $d\rho/dz = -\rho/H$ and Eq. 3.1 have exactly the same content and meaning. This is a general mathematical result that says that if the rate of change of a dependent variable $y(x)$ with respect to the independent variable x is proportional to y itself ($dy/dx = \alpha\, y$, α = proportionality constant), then y must be an exponential function of x (that is, $y(x) = y(0)\, e^{\alpha x}$) and vice versa (if y is described as an exponential function of x then its rate of change $dy/dx = \alpha\, y$). The exponential function will arise in many contexts in this book.Figure: Forces on Air.

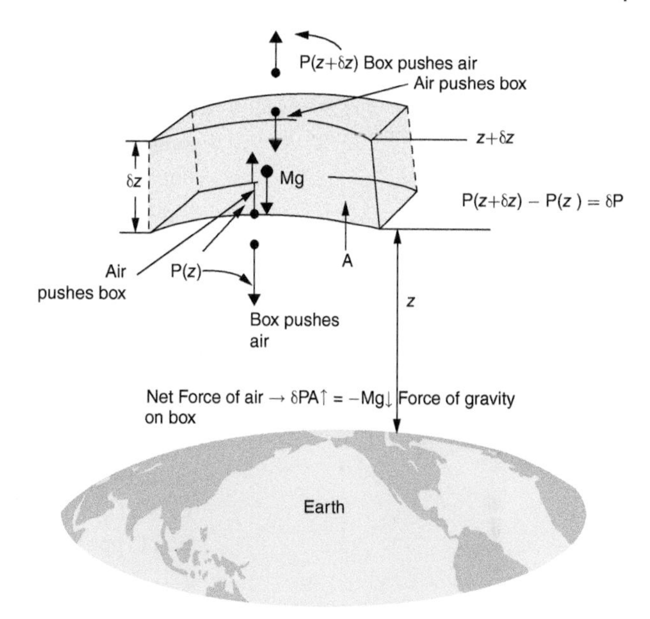

There are two forces on air masses at any altitude – the pushes from the surrounding air, and the pull of the earth's gravity. The only condition under which the air can be stationary in the vertical is if these two forces acting on the box are equal and oppositely directed. Since on an annual average the air does not move up or down, this balance exists. The explicit forces are shown.

Endnote 3.4 To arrive at Eq. 3.5 we have used the concept of integration from mathematics. We can plot any function of a single variable, such as the density of air that depends on the altitude z according to Eq. 3.1. In general, the area under the curve between any two points of the independent variables can be calculated by the process of summing the areas of the small segments over which the function does not change very much. Such an area calculation is widely applicable, where the "area" is needed. For many functions, the area under the curve is itself represented by another function that is the *integral* of the original. For our case the integral of Eq. 3.1 $\times A\, dz$ = integral of ($AN_0\, e^{-z/H}\, dz$). It is Eq. 3.5 and it represents the area under the curve of density $N(z)$ between the two points z_1 and z_2. It is a significant benefit to be able to derive Eq. 3.5 instead of having to calculate the number of molecules by creating many small shells and adding the number of molecules in each. We will use this concept of integration again in other contexts without explicitly deriving the final formulas but recognizing that these are exact connections.

Endnote 3.5 The mass of an object, including molecules, is an intrinsic property. It is the quantity of matter that is in this object. The weight of the same object is the force that the earth exerts on it. According to Newton's second law the two are related by $F = ma$ and $a = g$ where g is the acceleration due to gravity, which for near the earth's surface is a constant 9.8 m/s^2. So the weight and mass are related by a constant and we frequently use the concepts interchangeably (weight = mass × g). But note, that the same object transported to the moon will still have the same mass, but its weight will be considerably lower since the acceleration due to the gravity of the moon, when the object is on the moon's surface, is much less than that due to the earth, when it is here.

Endnote 3.6 Let's look at the relationship between the mixing ratio and the burden in Eq. 3.6a – it says that if the gas has an average mixing ratio of C(ppb) then the number of grams in the atmosphere can be calculated by the right-hand side of the equation. Measurements taken at one or even many locations may not be sufficient to determine the average for the whole atmosphere. Mixing ratios of trace gases can vary horizontally by latitude and longitude or vertically by altitude. For longer-lived gases the horizontal variations are not large, so we can set them aside for now, or assume that there are data at various latitudes to form an accurate surface average. In the vertical dimension, the gases may have approximately constant mixing ratio in the troposphere, but this is rarely the case in the stratosphere as mentioned in the text. We can always write the average stratospheric mixing ratio as a fraction α of the tropospheric mixing ratio, that is, $C_S = \alpha C_T$. It follows after some algebra that the average atmospheric concentration C(ppb) can be written in terms of the tropospheric mixing ratio as:

$$C(\text{ppb}) = \left[N_T C_T(\text{ppb}) + N_S C_S(\text{ppb})\right] / N_A = C_T(\text{ppb})\left(N_T + \alpha N_S\right) / N_A$$
$$= f \times C_T(\text{ppb})$$

The "*f*" is the factor that relates the tropospheric or surface mixing ratio to the global average that includes the stratosphere. The value of α, and hence of "*f*" too, is different for each gas and requires knowledge of stratospheric mixing ratios that are determined by how rapidly they are transported and destroyed there. It may also change in time, but we will take it to be constant. A generic α is difficult to estimate. Since it is between 0 (rapid sink) and 1(no sink), we can take it to be 0.5 giving an $f = 0.9$. But this factor is likely to underestimate the stratospheric concentrations for most gases, because the lower part of the stratosphere does not have a high destruction rate so gas concentrations here would not go to zero under any circumstance and is likely to retain a concentration closer to the tropospheric level. The upper limit of $\alpha = 1$ is possible, and is observed for gases such as carbon tetrafloride and approximately valid for CO_2. Based on observations we can estimate a more realistic generic factor by taking the mixing ratio as constant up to the base of the ozone layer at 20 km. After that it varies between 0 and the tropospheric concentration due to destruction by chemical and photochemical processes that are more active there. Then, based on the equation above and Eq. 3.5, $f = 0.9 + 0.5 \times 0.1 = 0.95$ giving a generic value of $\alpha = 0.75$ in the equation above. For more accurate calculations the value of α must be determined for the gas and the circumstances of interest. Using this factor ($f = 0.95$) will bring your calculation to within the 10% rule.

4

Mass Balance Theory and Small Models

4.1 The Components

So far, we have discussed which gases make up the atmosphere and established the state of the environmental system. We can now look at what determines and keeps this atmospheric composition in a stable balance and what causes change. Consider an area on the earth's surface and construct an imaginary box above it. The amount of any gas mixed with the air inside this box can be written as C, expressed in grams per cubic centimeter or more often as a mixing ratio, such as ppb. There are just three fundamental processes that can cause this concentration to change in time at a rate dC/dt; these are: the sources or emissions into the box by natural or human processes (S), losses caused by various chemical or physical reactions, the "sinks" (L), and transport processes (T) that can move material into or out of the box. The transport is driven by sustained winds and associated gusts that also cause mixing with the surrounding air (Figure 4.1). Transport, by its nature, involves processes that are occurring "outside the box." To start with we can write the relationship as follows and then examine the terms in more detail:

$$\frac{dC_J}{dt} = S_J - L_J + T_{NET_J} \tag{4.1a}$$

$$T_{NET_J} = T_{IN_J} - T_{OUT_J} \tag{4.1b}$$

It is evident that all the quantities in Eq. 4.1 must be expressed in the same units such as grams/yr or ppb/yr. To extend this idea to the global scale we can attach a multitude of small boxes to each other until the whole atmosphere is covered. The J is used to identify each box by assigning it a unique index value $J = 1,..,N$ when there are N boxes. When two boxes share an interface we can join them by realizing that the transport of material going across the interface, from one box is T_{OUT} for that box, but it becomes the T_{IN} for the adjoining box into which the air is flowing or mixing. The boxes can be extended deeper into the soils and oceans, but then must include complex cross-media transport processes to be discussed later. The system of N boxes tells us not only how the atmosphere changes globally but also how it changes in each region of the earth defined by the size of our boxes touching the surface. Models with many boxes or their equivalents are useful for research on details of global change. For our purposes we will be content with covering the earth with very few boxes – even just one! The complexities of sources, sinks, and transport processes will be taken up in the next two chapters (5 and 6) that will connect the general

Global Climate Change and Human Life, First Edition. M. A. K. Khalil.

Companion Website: www.wiley.com/go/khalil/Globalclimatechange

ideas presented here with the gases that can cause global change (Figure 3.1). For now, we will accept the idea, that in the specialized circumstances of our interest, the sinks and transport can be written as proportional to the mixing ratio in ppb, or burdens in grams of the gas in the box:

$$L_J = \frac{C_J}{\tau_J} \tag{4.2a}$$

$$T_{NET_J} = T_{NET_{J1}} + T_{NET_{J2}} + \ldots \tag{4.3b}$$

$$T_{NET_{JK}} = \frac{(C_K - C_J)}{\tau_{T_{JK}}} \tag{4.3c}$$

Here, τ_J is defined as the *lifetime* of the gas in the box J of Eq. 4.1 and the τ_T's represents *transport times* (Figure 4.1). In Eq. 4.2c the index K refers to boxes with which the box J connects, probably less than six; but for our purposes, it will be just one. The transport times can be defined in different ways, but how they are defined, must be

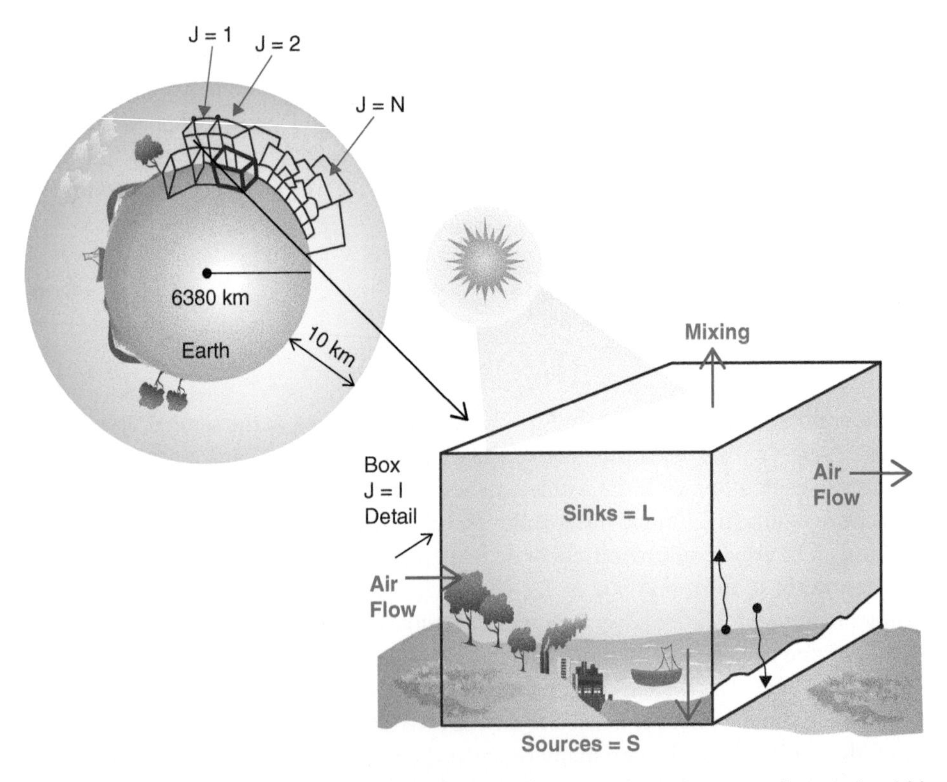

Figure 4.1 Components of the global mass balance. Concentrations of gases and aerosols within a region of our interest are determined by just three processes – emissions into the box (*S*), losses by chemical and physical processes (*L*), and the transport in and out of the box caused by the winds (*T*). The atmosphere, land, and oceans can be covered by such boxes to obtain a complete balance of where the gases come from and where they go.

specified to use the equations. For each pair of interacting boxes, we will define it as the time it takes to move the number of air molecules in box-1 to the adjacent box-2. Since our box is not gaining or losing air, it will get back the same number of air molecules from box-2 if it is an air exchange and mixing type of transport, or it will get the same number of air molecules from an upwind box if a steady wind is causing the transport (Endnote 4.1).

4.2 Global

Taking the atmosphere as one giant box is the simplest model we can construct from Eqs. 4.1 and 4.2. It is a canonical representation of the global atmospheric composition. In such a model, the transport by winds cancels out. Gases can still be transported or received from other components of the earth's system, such as oceans or land processes. We get the following minimalist mass balance equation for the atmosphere:

$$\frac{dC}{dt} = S - \frac{C}{\tau} \tag{4.3}$$

We can learn a lot about atmospheric composition from this equation by considering several prototype cases.

Case 1: Let's start with the simplest situation – that the sources (S) and sinks (C/τ) are in balance, or in steady state, which is the same as saying that the concentration is not changing, that is, $dC/dt = 0$, then we get:

$$C = S\tau \tag{4.4}$$

This important relationship tells us that the amount of a gas in the atmosphere in a steady state is proportional to the lifetime. For the same emissions in kg/y, long-lived gases will build up to proportionally higher concentrations than short-lived ones. Moreover, as we will see later, this equation is used frequently for short-lived chemicals that are mechanisms of sinks in the atmosphere (Endnote 4.2).

Case 2: What happens to the atmospheric concentration if we suddenly shut off all emissions? In that case $S = 0$ in Eq. 4.3 after the fact, and so, $dC/dt = - C/\tau$. This equation has a known mathematical solution assuming τ is constant:

$$C = C_0 e^{-t/\tau} \tag{4.5}$$

where C_0 is the mixing ratio in the atmosphere at the time we cut off the emissions and t is the time after that. We encountered a similar equation earlier in Section 3.2 and discussed the same solution (Endnote 3.2).

This case gives us an interpretation of the lifetime τ. We see that after we cut off emissions, the gas concentration will fall to $1/e$ or 36.7% of its initial concentration in one lifetime (Figure 4.2). Perhaps the concept of a half-life is more intuitive. It is the time when the concentration of the gas falls to half its initial value after we cut off all emissions. The lifetime τ is closely related to the half-life differing only by an absolute mathematical constant: $T_{1/2} = \ln(2)\,\tau = 0.693\,\tau$.

We may ask how long it takes for the gas to disappear from the atmosphere entirely. In Eq. 4.5 the concentration never quite goes to 0 except after an infinite amount of time. This is a shortcoming of the mathematics we have used. To be more practical we can apply the

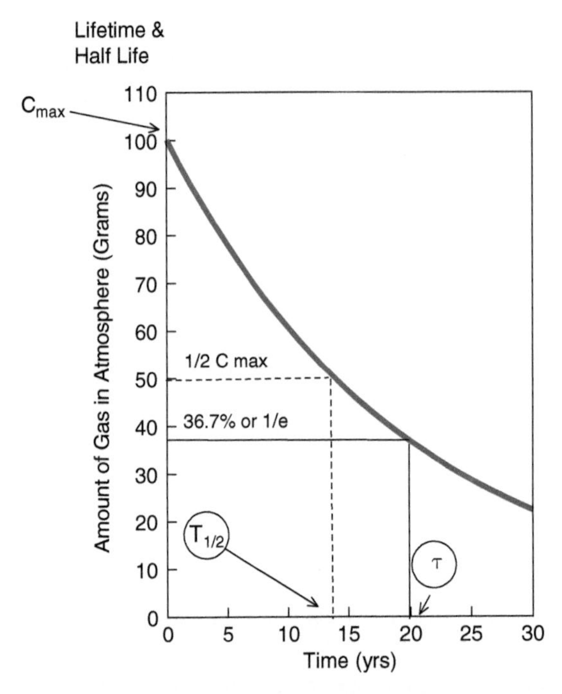

Figure 4.2 The meaning of lifetime. The figure illustrates how the concentration of a gas declines if you cut off emissions, assuming it follows Eq. 4.5. The behavior of the exponential function is apparent. The lifetime and half-life are shown to be closely related and represent the same concept in this situation.

10% rule, that is, we say that the gas has disappeared from the atmosphere for all practical purposes when there is only 10% of the initial concentration left. How long will that take? Inverting Eq. 4.5 tells us that this happens at a time, call it $T(10\%) = \tau \ln(10) = 2.3\,\tau$. So in round numbers it takes about two lifetimes for the gas to disappear from the atmosphere according to this criterion.

Example CFC-11 has a lifetime of 50 years. If it was 100 ppt at time zero, when will it reach 10 ppt? Suppose that is still too high for our liking, when will it reach 1 ppt?

Solution: Using algebra we can invert Eq. 4.5 to say that the time it takes to reach a concentration $C(T)$ is: $T = \tau \ln [C_0/C(T)]$. To reach 10 ppt, $T(10\%)$ = 50 years × ln (100 ppt/10 ppt) = 115 years. To reach 1 ppt, $T(1\%)$ = 50 years × ln (100) = 230 years.

The example illustrates two noteworthy points about the nature of sinks. First, it shows that to go from 100 ppt to just 10 takes the same amount of time as going from 10 ppt to 1, and second, that because gases involved in global change are long-lived, it takes a long time to reach targeted levels even if we cut off all emissions, but, most importantly, you now know how to calculate that time.

Case 3: What happens if we suddenly add a constant source? Now our Eq. 4.5 cannot be further simplified, except to note that for the constant source we can still solve it easily to get (Endnote 4.3):

$$C = S\tau(1 - e^{-t/\tau}) \tag{4.6}$$

This equation tells us how the gas will build up in the atmosphere. In the beginning, when t is near 0, the concentration is also near 0. After a while of continued constant emissions, the $e^{-t/\tau}$ approaches 0, as it did in the previous case, or 10% in 2.3 lifetimes, which creates a balance between sources and sinks resulting in the solution we found for Case 1. More to our interest, it tells us that if we decide to hold man-made emissions of a gas at current rates, the concentrations will continue to rise for a long time and still reach considerably higher levels than the present, as illustrated in the following examples:

Example Consider the case of nitrous oxide, a powerful greenhouse gas that is increasing due to human activities. This is what is known: $C_{Avg} = 315$ ppb (2300 Tg in the atmosphere), $\tau = 140$ y, and the total emissions are 24 Tg/y, of which 14 Tg/y are from natural sources and 10 Tg/y from human activities. Let's say we want to freeze man-made emissions at these levels to limit its effect on the future climate. What concentrations should we expect in the future?

Solution: (a) According to Eq. 4.6 which applies to this situation, the concentration will continue to rise until it reaches $S\,\tau = 24$ Tg/y $\times$ 140 y $= 3360$ Tg or 460 ppb, which is a lot more than at the start. On the plus side, it will take about 300 y before it reaches this ultimate value so whether this strategy is useful or not requires more analysis of how much reduction is desirable and at what time.

Example We can try to set a higher mark for controlling the effects of increasing nitrous oxide by demanding that we hold the concentration at today's levels. How much are we allowed to emit to accomplish this goal?

Solution: Now Eq. 4.4 applies. The total emissions must be held at $S = C/\tau = 2300$ Tg/140 y $= 16.4$ Tg/y. If we want to accomplish this by cutting man-made emissions, then we have to cut off about 80% of them.

The ideas we have discussed are represented visually in Figure 4.3 to solidify our understanding. A series of hypothetical gases are emitted into the atmosphere at a constant rate of 1 kg/yr, but with lifetimes ranging from about 2 months to 10,000 years. These emissions continue for a hundred years and are then cut off. The buildup and decline of the burden in the atmosphere is calculated over 200 years according to Eqs. 4.5 and 4.6. The salient points, clearly visible, are that there is very little buildup of gases with lifetimes of less than 10 years and that a constant concentration is reached soon after the start of emissions. Gases that cause air pollution may be in this category, but those involved in global change generally are not. Methane is at the upper boundary of this short lifetime range. Gases that live longer build up to much higher levels within the century of emissions. The very long-lived gases with lifetimes of a hundred years or more keep increasing throughout the emissions history and do not reach their potential even after a century, when the emissions are cut off. Once we stop emitting these gases they persist in large concentrations throughout the next century. If they are greenhouse gases, they will continue to cause global warming and climate change for this entire period and beyond. These long lifetime trajectories are expected for perflurorcarbons, sulfur hexafluoride, nitrous oxide, some of the chlorofluorocarbons, and a fraction of the emitted carbon dioxide that is not readily absorbed by the environment. Regardless of what we do now, these gases will continue to cause some global warming for the century or more to come.

These cases support the idea that the global lifetime of a gas has a major role in its environmental importance. With few exceptions, long lifetimes are required for the gas to cause

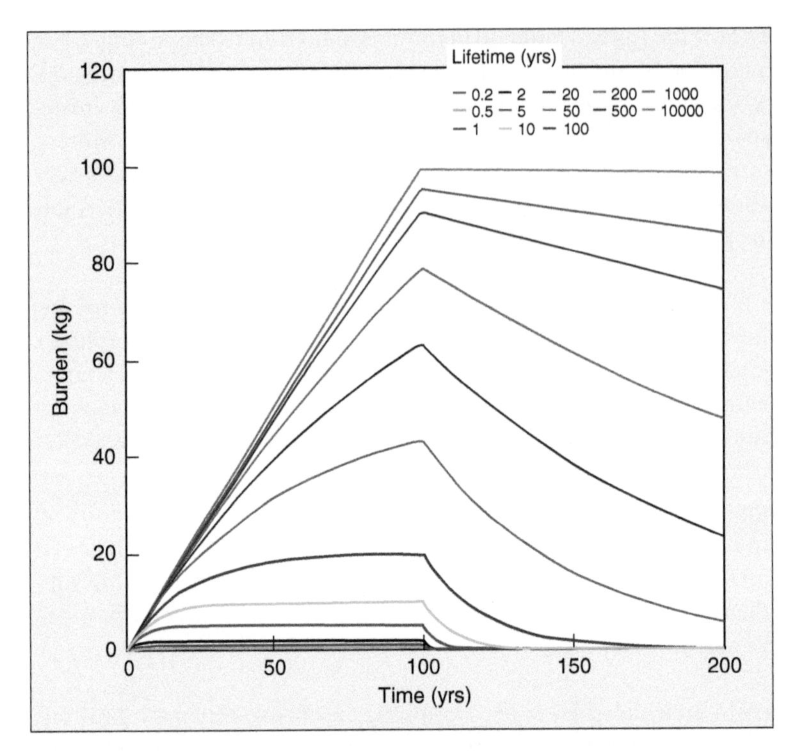

Figure 4.3 Buildup and decline of gases as determined by their lifetimes. Hypothetical gases are emitted at 1 kg/year with varying lifetimes in the atmosphere. Short-lived gases, characteristic of urban pollutants, do not accumulate. It takes lifetimes of several years, perhaps around ten, to see an appreciable buildup. On the other end of the spectrum, gases with lifetimes of 100 years or more, as characterized by some chlorofluorocarbons, nitrous oxide, perfluorocarbons, and sulfur-hexafluoride, do not reach their maximum potential and hardly decrease for decades to centuries after the emissions are cut off, defining their role as global greenhouse gases.

global climate change. Let's review some of the reasons for the environmental importance of the lifetime. First, it determines whether the gas will fill the whole atmosphere or exist mostly near its sources. Global change can occur only if the gas is spread over, well, the global scale, which is the whole atmosphere. Second, long lifetimes mean that even relatively small emissions will build up to high levels in the atmosphere. Global warming increases continually with higher concentrations. Third, once a long-lived gas has built up in the atmosphere to levels that we may consider undesirable, then even if we cut off further emissions, it will take a long time before the gas will disappear from the atmosphere – about two lifetimes long. Fourth, as we saw earlier, the lifetime creates a disconnect between gases that cause urban air pollution and those that cause global change, including warming temperatures. Fifth, gases that are long-lived in the atmosphere and therefore have low chemical reactivity are expected to be safer for human health upon exposure. Finally, the lifetime is a major ingredient in policies and international agreements to control global warming. It creates a means of trading between gases. Preventing a kilogram emission of a potent longer-lived gas is seen equivalent to preventing emissions of hundreds or thousands of kilograms of carbon dioxide. As we progress further, these aspects will be addressed in more depth.

A few more lessons can be learned from our global mass balance models. Concerns about global changes arise when the gases involved are found to have trends, most often seen as increases that can be linked to human activities. From Eq. 4.3 we see that atmospheric concentrations can increase only if the sources exceed the sinks. This leads to three basic possibilities: the sources are increasing; the lifetime is getting longer or the concentrations have not caught up with the present source. The last one is somewhat enigmatic because concentrations can rise, even though the sources and lifetimes are not changing! It has to be triggered by changes earlier that cause the present sink to be less than the source and so it represents a transient factor. The sink is C/τ and not just the lifetime. So, the concentration has to increase until C/τ becomes equal to the total source that was added at an earlier time, even though it may be constant since. This behavior is illustrated and represented by Eq. 4.6. Because there are several possibilities which may all predict the same result, when a gas is seen to be increasing, we cannot immediately decide what the cause is. For some gases and circumstances, such as the chlorofluorocarbons, it is apparent that the atmospheric buildup is caused by the emissions from human activities because experiments have shown that they are entirely man-made, but for many greenhouse gases such as methane and even carbon dioxide, natural and human processes may contribute to changes in both sources and sinks to varying degrees in their decade to century trends.

We have discussed the global lifetime as a single number that defines the character of the gas. In many real-world situations this lifetime comes about from several processes that remove the gas from the atmosphere. Each process, if it was the only one, would result in a lifetime characteristic of that particular process. How then do we represent the total lifetime τ in terms of the components τ_1, τ_2, τ_3,.., τ_N, each of which represents a different removal process, all occurring at the same time? The answer is:

$$1/\tau = 1/\tau_1 + 1/\tau_2 + \ldots + 1/\tau_N \tag{4.7}$$

The proof is that each year, process J removes C/τ_J of the gas where $J = 1,2,..,N$ for the N processes. All the processes together remove $C/\tau_1 + C/\tau_2 + .. + C/\tau_N$ and this has to equal the term C/τ in Eq. 4.3, which represents the effect of all processes. Equating the two, we get Eq. 4.7. A conclusion we can immediately draw is that the total lifetime is shorter than the shortest lifetime in the set of processes that remove the gas. The question of many important sinks becomes more relevant for gases with very long lifetimes. If we have accurate measured concentrations and if we can estimate the emissions, we can calculate the lifetime from Eq. 4.3. Then the challenge is to determine whether it can be explained by the known sink mechanisms using Eq. 4.7. If the lifetime is calculated by using a known sink mechanism, then the challenge is to figure out whether there are other sinks that can make the total lifetime much shorter.

Example It is calculated that the destruction of a chlorofluorocarbon, somewhat like CFC-11, occurs in the stratosphere leading to an overall atmospheric lifetime of 60 y and it is taken to be its actual lifetime. Sometime later it is found that it dissolves in the oceans at a rate that would give it a lifetime of 100 y if this was the only process. What is the correct actual atmospheric lifetime now?

Solution: $1/\tau = 1/60\,\text{y} + 1/100\,\text{y} = 0.027$. Inverting gives $\boldsymbol{\tau} = \mathbf{38\,y}$, leading to the conclusion that the oceans contribute significantly to the sinks.

As intuitive and useful as the one-box model is, it hides many aspects of atmospheric gases that we may want to know. The lifetime in Eq. 4.3 depends on two factors – the destruction rates of the gas at different locations on the earth, and the transport of gases from where they are emitted to places where they may be destroyed more or less effectively. To understand the connection, we must extend the model to more boxes. By doing this, we can see more clearly how the transport and sink processes work together to create the global average lifetime in Eq. 4.3. We will consider this next. The increase in mathematical complexity may be daunting to some readers; however, the useful equations and concepts are readily understandable and will add to your knowledge of the environment.

4.3 Hemispherical and Horizontal

The smallest models that can include the effects of transport have two boxes, either in the vertical or in the horizontal. The former is most useful when the lower box is the troposphere and the upper one is the stratosphere. It can then represent the transport of gases to the stratosphere where they are destroyed or deplete the ozone layer. The latter is best used to divide the atmosphere between northern and southern hemispheres which is applicable to man-made greenhouse gases that are emitted mostly in the northern hemisphere and have to transport to the south.

From the previous discussion using Eqs. 4.1 and 4.2 we can write the horizontal, hemisphere-sized latitudinal model as:

$$\frac{dC_n}{dt} = S_n - \frac{C_n}{\tau_n} + \frac{1}{\tau_T}(C_n - C_s) \tag{4.8a}$$

$$\frac{dC_s}{dt} = S_s - \frac{C_s}{\tau_s} + \frac{1}{\tau_T}(C_n - C_s) \tag{4.8b}$$

Here the "n" subscript is for the northern hemisphere and "s" for the southern and the Cs are mixing ratios. The net amount of material that is lost by transport in one box, $-1/\tau_T(C_n - C_s)$, is gained in the other. Models with a few boxes such as the ones in Eq. 4.8 work for gases with lifetimes that are much longer than the transport times between the boxes; otherwise, the representation of transport as the difference of the average concentrations in the adjacent boxes is ill-defined and inaccurate.

We can now explicitly include the idea that the sources and sinks can be different in the two hemispheres and see how this may affect the global lifetime. If we average these two equations, we will get our one-box model back, but with a formula in which the global lifetime will be explicitly separated into the sink part (τ_n, τ_s) and the transport part (τ_T). This is what we get when we compare the average of Eqs. 4.8a and 4.8b with the C and τ of Eq. 4.3:

$$C = \frac{1}{2}(C_n + C_s) \tag{4.9a}$$

$$S = \frac{1}{2}(S_n + S_s) \tag{4.9b}$$

$$\frac{1}{\tau}=\frac{1}{2}\frac{\alpha}{\tau_n}+\frac{(1-\alpha/2)}{\tau_s} \tag{4.9c}$$

where $\alpha = C_n/C$. Equation 4.9c shows that the global average lifetime of the one-box model depends on the rates of destruction in the two hemispheres determined by τ_n and τ_s and how the mass of the gas is distributed between the hemispheres represented by α, which depends on the transport time and the location of the sources. The sinks are often chemical in nature so this connection is then called the *chemistry-transport coupling*. Let's explore this concept further.

Consider the situation where the sources and lifetimes are *constant* in time but vary by location. Further, assume that sufficient time has passed since the source was turned on so that a steady state has been reached which means that the rates of change on the left-hand side are zero, greatly simplifying our task (Endnote 4.4). Here are some illuminating cases:

Case 1: The sources are only in the northern box; the lifetimes in the two boxes are the same:

$S_n > 0$; $S_s = 0$; $\tau_n = \tau_s$. From Eq. 4.9c, the effective lifetime will be the same $\tau = \tau_s$, and so transport will not contribute to it. After some algebra you will find that the ratio of the northern to southern concentrations C_n/C_s, call it R, is:

$$R=1+\frac{\tau_T}{\tau} \tag{4.10}$$

This case represents a common occurrence when the gases are emitted by human activities leading to emissions predominantly in the northern hemisphere. As we would expect, there is always more of the gas in the northern box compared with the southern because $R > 1$ even in steady state. Equation 4.10 tells us that the ratio will be larger for short-lived gases, because fewer molecules survive the transect and because the transport time τ_T is assumed to be the same for all gases.

Case 2: The sources are in the northern box and the sinks are in the southern box.

$S_n > 0$; $S_s = 0$; $\tau_n = \infty$; τ_s is finite. This time you will find the remarkable result that:

$$\tau = 2\tau_s + \tau_T \tag{4.11}$$

Here we see the most explicit representation of the global lifetime as it is determined by a combination of transport times (τ_T) and the effect of the sink (τ_s). The factor of two arises because the gas is destroyed in only one hemisphere – the southern, so it takes twice as long.

In the situation shown, if $\tau_s \gg \tau_T$, that is to say, the gas has a lifetime much longer than the transport time, then $\tau \approx 2\,\tau_s$. The transport time has a negligible effect on the global lifetime, which means that it is mostly "chemistry" which determines the global concentration making it a "chemistry controlled" case (assuming that the loss of the gas is due to chemical reactions in the atmosphere). If $\tau_s \ll \tau_T$ then $\tau \approx \tau_T$ and we have the "transport controlled" case; that is to say, the chemistry is so fast that it is the transport that limits how much gas can be supplied to the region where it is destroyed, so it controls the global lifetime; in fact, in this situation it is the lifetime.

Case 3: The sources are as before but the sinks are in the same box. The southern box has neither sources nor sinks. This case completes all disparate source-sink possibilities.

$S_n > 0$; $S_s = 0$; $\tau_s = \infty$; τ_n is finite. This time you will find that the concentrations will be the same in both boxes and the effective lifetime will be twice the lifetime in the northern box (Eq. 4.8):

$$\tau = 2\tau_n \tag{4.12}$$

This case corresponds to a situation when there is an inert reservoir where the gas can linger without being destroyed, causing the lifetime to be longer than what we would expect from its sink strength.

A further lesson can be learned from these cases by considering the somewhat more complex situation that exists when the sources are just turned on, long before the steady state is reached (Figure 4.4). For Case 1, since $\tau_n = \tau_s$, the effective lifetime is $\tau = \tau_n = \tau_s$ (Eq. 4.9c) and it is the same a long time later when steady state is reached, and therefore the same as the results of the one-box model with constant lifetime. As long as the sinks are uniformly distributed, transport does not affect the lifetime even in the transient states as

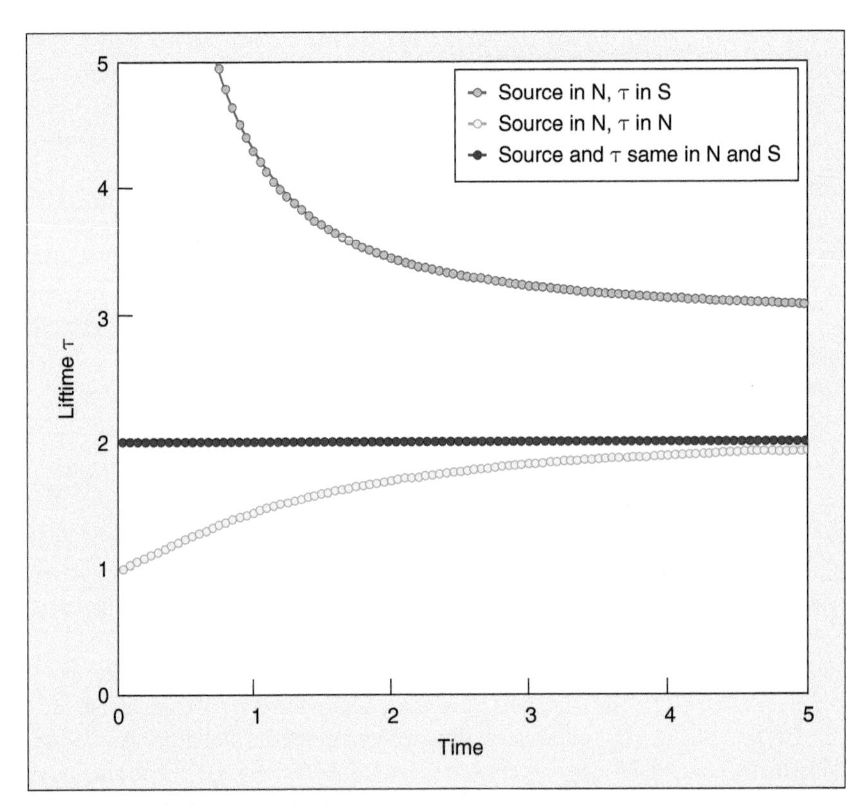

Figure 4.4 Change of global effective lifetime with constant emissions and destruction rates. The effective lifetime of a gas is shown under three cases discussed in the text. In all cases a constant source is turned on at time = 0. In the first case (*blue*), the source is in the north and the sink is the same in the north and the south; in the second case (*red*) the source is in the north and the sink is in the south and in the last case (*light blue*) the source and sink are both in the north. The graphs show the progression of the lifetime as steady state concentrations are approached in these three disparate scenarios. These play out to varying degrees for real gases and illustrate that the global lifetime may change when gases are not in steady state.

the concentration is changing. In Case 2 when we just start emitting the gas in the northern hemisphere, there is no destruction because the sinks are in the southern hemisphere and it takes time to get the gas there. So the global lifetime will be temporarily infinite! Alternatively, note that at the start $\alpha \approx 2$ for this situation, which also makes the global lifetime $\tau \approx \infty$ by Eq. 4.9c. But we see that when steady state is reached, the lifetime comes down to the value in Eq. 4.11. It is still higher than the other two cases, but finite. This is perhaps the best illustration of how the disparity of the location of sources and sinks can cause the global lifetime to change making the assumption of a constant lifetime unreliable during the transition. Finally in Case 3, α is once again ≈ 2 at the beginning, and since τ_n is finite, Eq. 4.10c says that $\tau \approx \tau_n$, but we see that in steady state it is $2\ \tau_n$, so this time, the lifetime gets longer as time progresses. This is because, in time, some of the gas has migrated to places where it is not destroyed, and lingers there, so the decline in its global abundance is slowed. From what we have seen, even if the sinks are constant but not uniformly distributed, the lifetime may change unless a steady state has been reached.

In the real atmosphere the sources and sinks of greenhouse or other gases are rarely at the same locations, uniformly distributed or in steady state, so the effect plays out to varying degrees.

Carbon dioxide has a complex mix of source and sink locations so that man-made emissions from the northern populated latitudes are dominant, but the ocean sink is further south toward the equator and southern oceans. The greenhouse gas, nitrous oxide, and ozone-depleting compounds such as the chlorofluorocarbons have the greatest disparity between source and sink locations: the sources are at the surface, but the sinks are in the stratosphere. These gases have to be transported to the stratosphere before they can reach the sink locations, thus greatly enhancing the source-sink disparity effect we have discussed here. To understand this situation, we need to re-formulate our two-box model so that the boxes are now the troposphere and the stratosphere.

4.4 Vertical

Writing the mass balance for the vertical model consisting of the troposphere and the stratosphere requires taking into account the different numbers of molecules in these two parts of the atmosphere (Endnote 4.1). The resulting equations are:

$$\frac{dC_T}{dt} = S_T - \frac{1}{\tau_T}C_T - \frac{1}{\tau_{ST}}\frac{N_S}{N_T}(C_T - C_S) \tag{4.13a}$$

$$\frac{dC_S}{dt} = S_S - \frac{1}{\tau_S}C_S + \frac{1}{\tau_{ST}}(C_T - C_S) \tag{4.13b}$$

Here the subscript T is for troposphere and S for the stratosphere and the C's represent mixing ratios; τ_{ST} is the stratospheric-tropospheric transport time. It is the time to mix the number of air molecules that are in the stratosphere with a like number of molecules in the troposphere. The time it will take to go the other way, that is, transfer the number of air molecules in the troposphere to the stratosphere, τ_{TS}, is much longer because there are more molecules to transfer; it is: $\tau_{TS} = \tau_{ST}N_T/N_S$. The τ_{ST} will turn out to be about 1.8 years as we will see later.

The most applicable circumstances of this model are similar to Case 2 discussed earlier for the latitudinal box model. The sources are only at the surface and the sinks are in the stratosphere. For the case when the source is constant and at the surface $S_T = S =$ constant; $S_S = 0$; $\tau_{Trop} = \infty$. A steady state will eventually prevail so that the left-hand dC/dt's can be set to zero (Endnote 4.4). After some algebra the ratio of the concentrations, $R = C_T/C_S$, similar to Eq. 4.11, is:

$$R = 1 + \frac{\tau_{ST}}{\tau_S} \quad (4.14)$$

In Section 3.2 we discussed the idea that the global average concentration of a gas is less than the concentration at the surface (C_T) because there is usually a sink in the stratosphere. α in Section 3.2 is $1/R$ in this specialized case (constant source and no tropospheric sink).

Of greater interest is the global lifetime that we would derive when the sinks are only in the stratosphere. The counterpart of Eq. 4.11 is:

$$\tau = \tau_S(1 + N_T / N_S) + (N_T / N_S)\tau_{ST} \quad (4.15)$$

This equation shows that the global lifetime is affected by three aspects of the situation: the lifetime due to the actual destruction of the gas τ_s, in this case due to photo-chemistry, the fact that the stratosphere, where the destruction takes place, has much fewer molecules of air compared with the troposphere expressed as τ_s (N_T/N_S), and there is the transport time effect (N_T/N_S) τ_{ST}. No matter how fast the gas is destroyed in the stratosphere, its global effective lifetime will be longer than 8 years obtained by setting $\tau_s = 0$ in Eq. 4.15. It is the global effective lifetime that is put into the one-box model. Accordingly, most gases that are destroyed only in the stratosphere will have long global lifetimes and will therefore build up to high levels in the troposphere. Such is the case for the chlorofluorocarbons, nitrous oxide, and the technological gases that are included as targets for managing global climate and the ozone layer. The lifetimes of such gases are transport controlled.

Example The global lifetime of CFC-11, τ, observed at the earth's surface is 50 y. Assume $\tau_{ST} = 1.8$ years. (a) How long does CFC-11 survive in the stratosphere? (b) How much do the transport and box size factors affect the observed global lifetime. (c) Estimate the α and f that connect the surface concentration to the global average.

Solution: (a) Inverting Eq. 4.15, and noting that $N_T/N_S \approx 4$ from before, $\tau_s = (50\text{ y} - 4 \times 1.8\text{ y})/(1 + 4) =$ **8.5 y ≈ 9 y**. (b) The contribution of the size factor in the first term of Eq. 4.15 = τ_s (N_T/N_S) = 8.5 y × 4 = 34 y and the transport factor = (N_T/N_S) τ_{ST} = 4 × 1.8 ≈ 7 y, which all add up to make the 50 year observed lifetime (9 + 34 + 7 = 50 years). (c) $R = 1 + 1.8/9 = 1.2$. $\alpha = 1/R = 0.83$. Therefore $f = 0.8 + 0.2\,(0.83) = 0.96$ and C(ppt) = 0.96 C_T(ppt), which is almost the same as our generic factor (Endnote 3.6).

We see that while CFC-11 can be destroyed quite rapidly in the stratosphere, its lifetime in the atmosphere as a whole is still long because it takes quite a lot of time to transport it to the stratosphere before it can be destroyed, and the stratosphere is not a "big box" compared to the troposphere, where it is emitted. We note further, for later reference, that the stratospheric lifetime τ_s determines how fast the ozone layer will be destroyed by CFC-11; for climate considerations, the composite lifetime of 50 years, which determines the accumulation in the troposphere, is most important. Gases such as CFC-11 therefore contribute efficiently to both ozone depletion and global warming.

When there is a major volcanic eruption some of the ash is injected into the stratosphere. Once the dust is there, the only way it can be removed is to first return to the troposphere, then settle to lower levels where it can be washed out by rain or moved to the surface. The smaller particles settle very slowly because the viscosity of air resists their downward motion as they are pulled to the surface by gravity. Continued effects of mixing in the troposphere can add more time to the settling of the dust. This air resistance and mixing add some time to the final removal of the volcanic particles, but the major impediment is the slow transport time across the tropopause (transport controlled). The exchange time of 1.8 y therefore becomes the approximate lifetime of the dust in the stratosphere causing it to stay up there for 2–5 years depending on how deeply the aerosol had penetrated the stratosphere during the eruption. The volcanic dust spreads out horizontally over large sections of the stratosphere, and because it is good at scattering light, red sunsets are observed for years. This phenomenon has an important connection with climate. The scattering by volcanic dust in the stratosphere increases the earth's albedo and thus causes a cooling effect. There are many records of unusually cold years over large regions following major volcanic eruptions. A decrease of global average temperatures is often recorded. It is this role of the stratosphere in the earth's climate that we will consider again when we look at possible ways to reduce global warming by using geo-engineering.

It is noteworthy that we can use the model in Eq. 4.1 to find the part of the mass balance we know the least. If we look at the simplified Eq. 4.3, we see that it is a relationship between the concentrations of a gas and its sources and sinks. Usually, to solve such an equation mathematically means that we find the concentration as a function of time in terms the sources (S) and lifetime (τ) as we did in the specialized cases considered earlier. However, for most long-lived gases the concentration is the best-known part of the mass balance because it can be obtained from global measurements at various locations or from satellites. We can use Eq. 4.3 instead to solve for either the sources or the sinks instead of concentrations. This situation is appropriately called "inverse modeling" because we know the answer to the mass balance equation (C) and we want to find out how it came to be. The method is applied in research using complex chemistry-transport models, but we will use it later to considerable success with the small models discussed here to find the total emissions, the man-made emissions, or the global lifetime of various gases of interest.

The foundations and theory of mass balance models was laid out in this chapter. In it we saw that the global effective lifetime is one of the most important attributes of a gas that determines what it will do in the environment. Although it represents the sink, it is made up of not only the destruction rates due to chemical or physical processes, but also the transport times in the atmosphere. Moreover, it is not exactly constant if the gas is in disequilibrium. Long-lived gases for which the mass balance is mostly determined by tropospheric processes tend to be "chemistry controlled," so the lifetime is dominated by the "actual sinks" and less by transport. Methane is an example. For gases, emitted at the surface, that are primarily destroyed in the stratosphere, and by extension in the oceans or deeper in soils, the effective atmospheric lifetime in Eq. 4.3 is often transport controlled. Nitrous oxide, carbon dioxide, and the chlorofluorocarbons are examples. The fact that the lifetime may not be constant is a manifestation of disequilibrium conditions, which exist for most gases of our interest since they are increasing. However, if the increase is slow compared with the other terms in the mass balance, then this effect would be small because the gas is always near equilibrium with the sources and sinks. In that case the lifetime can be taken to be constant.

We are now ready to discuss in more detail each of the terms in the mass balance equations in the next two chapters. We will start with transport processes and then consider sources and sinks. After that we will apply the results to gases of interest in climate and global environmental science.

Review of the Main Points

1. We moved from theoretical ideas about the structure and composition of the atmosphere to constructing a mass balance model of a gas in a box over a patch of the earth's surface. It consists of three elements: sources, sinks, and transport of the gas into and out of the box. Sinks are processes by which a gas is removed from the atmosphere. The sinks and transport processes can be taken to be proportional to the concentrations – concepts that will be explored in later chapters. The mass balance is a step toward a quantitative analysis of environmentally important gases in the atmosphere and to figure out which sources or sinks are important and which are not. This understanding is crucial in evaluating the possible futures of the earth's climate and the environment.
2. The most instructive models consist of one or two boxes. The one-box model internalizes the transport processes and gives us a straightforward picture of how the changing emissions and the sinks of a gas contribute to its concentration in the atmosphere.
3. The lifetime of a gas is its single most important attribute that determines whether it will cause air pollution, alter our climate, deplete the ozone layer, or affect human health.
4. The two-box models can divide the atmosphere into northern and southern hemispheres, or the troposphere and the stratosphere. These models add considerable information about the behavior of the gases that make up the atmosphere, and particularly about how the global effective lifetime works.
5. We find that the global lifetime is a combination of the local lifetimes due the destruction of the gas, the transport times, and the relative sizes of the boxes that make up our model. This effective lifetime is important because it determines the impact of man-made emissions on global warming, but it cannot be understood entirely in terms of the chemical or physical sink processes.

Exercises

1. Suppose that the emissions of CFC-12 are now zero, and the concentration is 500 ppt. If the lifetime is 70 years, plot the concentrations over the next 50 years. After how many years will the concentration reach half its initial value?
2. Show that the half-life is related to the lifetime as: $T_{1/2} = 0.693\ \tau$.
3. Suppose there are 4000 Tg of methane in the atmosphere at a given time, the lifetime is 8 years, and the emissions are 450 Tg/year. What is the rate of change of the mixing ratio in the atmosphere in ppb/yr ? Is your answer consistent with current observations? If not, what could be wrong?

4 Since 1965 an exotic gas has been emitted from a plastic manufacturing process at a constant rate. This industry is the only source. Atmospheric measurements show that the globally averaged concentration was 260 ppt in 1965 and has been rising since then; however, the rate of increase (change per year) has decreased over the years. When the yearly rate of change is plotted, it fits the following simple equation: (t in years): $dC/dt = 20$ (ppt/yr) exp (–0.75t).
   (a) What is the atmospheric lifetime of this trace gas?
   (b) Discuss the relative importance of this global pollution from the plastics industry.

5 From the two-box mass balance models derive Eqs. 4.10 ($R = C_n/C_s = 1 + \tau_T/\tau$) and 4.14 ($R = C_T/C_S = 1 + \tau_{ST}/\tau_S$).

6 A gas has been emitted from an industrial process at a constant rate for many years and mostly from the Northern Hemisphere middle latitudes. It has a molecular weight of 140 g/mole. When measurements were taken, it was found that the concentrations are 120 ppt in the northern hemisphere and 100 ppt in the southern. What are its annual emissions in Gg/y and what is its atmospheric lifetime?

7 Consider a situation of a greenhouse gas, like methane. It has an initial concentration of 1800 ppb. Use the one-box model.
   (a) Take the source to be S (ppb/y) = 230 + 5 $t^{1/2}$ and calculate the concentrations.
   (b) Take the lifetime to be τ (y) = 8.3 + 0.3 $t^{1/3}$, and calculate the concentrations.
   (c) Provide a brief comment on your results. You will need to use the numerical method described in Endnote 4.3. A spreadsheet will facilitate the calculation.

8 The lifetime of CF_4 is estimated to be 50,000 y due to high atmospheric processes above the stratosphere. No other sinks are known. However, the air that goes through high temperature combustion as in engines or power plants is hot enough that gases such as CF_4 may be destroyed.
   (a) If this is so, estimate the lifetime of CF_4 due to this sink. Assume that about 10 PgC/y of CO_2 is burned each year and CF_4 is at 80 ppt (although you may not need this).
   (b) What is the new composite lifetime? Is it within the 10% rule? Hint: Find how much air has to go through the high temperature combustion engines to generate the annual CO_2 emissions. The O_2 in the CO_2 must come from the air, so first figure out how much O_2 is processed and that will tell you how much air goes through. Assume all the CF_4 in this air is destroyed. Proceed.

9 Construct a two-box model of the atmosphere in which the boxes are the urban area, and the rest of the atmosphere, which represents "the background." An average city is represented by a box of dimensions L, W, h = 50 km × 50 km × 1 km. Follow the construction of the vertical and horizontal two-box models in the text and endnotes.
   (a) Show that the mass balance equations for emissions (S) from the urban areas of the world are:

$$dC_U/dt = S/(N_0 h) - 1/\tau_T(C_U - C_B) - (1/\tau_U)C_U$$
$$dC_B/dt = (N_U/N_B)(1/\tau_T)(C_U - C_B) - (1/\tau_B)C_B$$

   Subscripts B = Background, U = Urban. $\tau_T = U/L$ is the transport time by winds; U = average wind speed. N_0 = number of molecules of air in the atmosphere. N_U = number of molecules of air in the global urban boxes.
   (b) Assume that the lifetimes of the gases are the same in the two environments (τ). Show that the solutions in steady state are:

$$C_U = (1+f)S\tau / D; \quad C_B = S\tau / D$$
$$f \equiv (N_A / N_U)(\tau_T / \tau) \quad and \quad D \equiv (N_0 h)(\tau / \tau_T) f + (1+f)$$

(c) Calculate the ratios of concentrations of gases in the urban and background regions as a function of the lifetime (τ). Use the following information. Typical wind speeds are U = 1–3 m/s, total urban area = 3.6 million square kilometers (estimated). Plot as a function of lifetime, the fraction of the emissions that remain in the cities from where they are emitted, relative to the total amount in the global atmosphere. At what lifetime does the gas have negligible influence on the global environment? At what lifetime is it expected to have negligible influence on air pollution?

Endnotes

Endnote 4.1 We can write the mass balance in terms of the burdens (grams in a box), concentrations (grams/m^3) or mixing ratios (ppb). In the mass balance Eq. 4.1, the units of transport processes have to match the units of the other terms. The transport or exchange of a gas from any region into another can be described unequivocally as the number of molecules entering (or leaving) the box per unit time. The transport time is how long it will take to transfer all the molecules of air that are in box-1 to box-2. In the models we are considering in this chapter, it applies to situations where mixing processes are prevalent and air is constantly being exchanged between adjacent boxes. Let's say we have two boxes: box-1 and box-2 with mixing ratios of a gas C_1 (ppb) and C_2 (ppb) and containing N_1 and N_2 molecules of air. We can always take the exchanges or movement of molecules from our box of interest, to and from the boxes that interact with it, one at a time, or pair-wise. Let's make $\tau_{T\,1\rightarrow 2}$ the time it takes to transfer the air molecules in box-1 to box-2, that is, N_1 molecules. During the same time, transport will have to bring back the same number of molecules of air from box-2 (N_1, not N_2) to box-1 to preserve the state of the atmosphere if the exchange is by mixing, which is our focus here. Then, box-1 will receive C_2 N_1 molecules of the gas and lose C_1 N_1 molecules during the time $\tau_{T\,1\rightarrow 2}$ and therefore, $T_{\mathrm{NET}\,1}$ (molecules/y) = $(C_2 - C_1)\,N_1/\tau_{T\,1\rightarrow 2} = -T_{\mathrm{NET}\,2}$ (molecules/y) where $T_{\mathrm{NET}\,1}$ is the net transport for box-1 and $T_{\mathrm{NET}\,2}$ for box-2. But to have consistent units in the mass balance equation (4.1) we need to convert this transport to the effect on Cs and so we want T_{NET}'s in ppb/y (or ppm or ppt as needed). For box-1 we must divide the transport $T_{\mathrm{NET}\,1}$ by the number of molecules of air in box-1 or $T_{\mathrm{NET}\,1}$ (ppb/y)/$N_1 = (C_2 - C_1)/\tau_{T\,1\rightarrow 2}$ and for box-2 we must divide $T_{\mathrm{NET}\,2}$ by the number of molecules of air in that box, or N_2. So $T_{\mathrm{NET}\,2}$ (ppb/y) = $(N_1/N_2)(C_1 - C_2)/\tau_{T\,1\rightarrow 2}$ and $T_{\mathrm{NET}\,2}$ (ppb/y) is no longer minus $T_{\mathrm{NET}1}$(ppb/y). Indeed, the transport time from box-2 to box-1 is not the same as from one to two. We see that the transport time $\tau_{T\,2\rightarrow 1} = \tau_{T\,1\rightarrow 2}\,(N_2/N_1)$ which will then represent the time it takes to transfer the number of molecules of air in box-2 to box-1.

For other situations such as using concentration units of g/cm^3, or burdens, appropriate adjustments and definitions of transport times may be needed. For our purposes, knowing the impact of different sizes of the boxes, and different number of molecules of air in the boxes will be enough.

We will try to avoid even that by making our boxes the same size whenever possible as by taking them to represent the Northern and Southern hemispheres. In the vertical model of the troposphere and stratosphere, the discrepancy is unavoidable, but the adjustment follows as a direct application of the derivation above.

Endnote 4.2 The concept of a steady state can make our equations much simpler when it is applicable. There are several forms. For persistent gases, it means that the sources and sinks are constant over a long period of time, perhaps decades, when averaged on an annual basis. Then, Eq. 4.4 will hold. More often, it can be used as a valid approximation depending on the questions being addressed. The budgets of greenhouse gases such as CH_4 and N_2O have significant uncertainties in the sources and sinks and the terms S and C/τ are much larger than the trends dC/dt. In such a situation $C \approx S\,\tau$ is still approximately valid and is the usual way to balance the budgets.

At the other extreme when the lifetime is very short, $C = S\,\tau$ is also applicable with additional assumptions, even though the sources or the lifetimes are changing. This is called a pseudo-steady state that we will use later, especially in Chapter 6. We saw that if you suddenly change the source, it takes about two lifetimes before the sinks catch up and balance the new source. During the transition, the gas concentration continues to increase and C is not equal to $S\,\tau$. Similarly, when the sources and sinks are located apart, it takes time for an equilibrium to be established since the gas has to be transported from where it is produced to the location of the sinks causing the concentration to change for some time. For short-lived gases, say, with a lifetime of less than a half hour, the concentrations would re-balance within the hour. Equivalently, it can be said that for changes of concentration over times much longer than the lifetime, dC/dt is much less than either the source or the sink and so can be set equal to zero in the mass balance Eq. 4.1. The transport processes can be neglected because the gas does not get very far before it is destroyed. Concentrations at one location do not influence the concentrations at another. So, for any times of interest longer than the lifetime the concentrations can be represented by $C = S\,\tau$ at each location.

Endnote 4.3 The cases of long-lived gases we have studied in Section 4.2 are informative, but do not cover the circumstances when the sources or lifetimes are changing in time. Year by year changes in the emissions from human activities are a common circumstance. Formulas such as Eqs. 4.4–4.6 can be derived for some cases with changing sources and sinks, but in general, even for the one-box model, this is not possible or the formulas are very complex and therefore not useful for an intuitive understanding of what is happening. In such cases, numerical solutions are invoked, which are always possible for any mass balance equation you can write and with any number of boxes. For Eq. 4.3, we can write the $dC/dt \approx [C(t+\delta t) - C(t)]/\delta t$, where δt is a suitably short time compared with the changes in the sources and lifetimes. Readers should note that this is the definition of dC/dt when $\delta t \rightarrow 0$ and is approximately valid when it is not 0 as may happen when it is estimated from actual observations that are taken some finite time (δt) apart, which is often weeks, months, or even a year. Then knowing the concentration at time $= 0$ as C_0 we can write Eq. 4.3 for C at a time δt as: C

$(\delta t) = C_0 + S(0)\,\delta t - C_0/\tau\,(0)$, where $S(0)$ and $\tau\,(0)$ are the emissions and lifetime at time 0. Now, we can bootstrap this process to get the concentration at a time $t = 2\,\delta t$ as: $C\,(2\delta t) = C\,(\delta t) + S(\text{at }\delta t) \times \delta t - C(\delta t)/\,\tau\,(\text{at }\delta t)$ and so on, until you have the solution for any length of time you need at intervals of time $= \delta t$. At each step, the emissions or the lifetime can be different than the previous step. Each increment in time requires the calculation from the previous time step.

Endnote 4.4 The two-box model equations can be solved for useful and instructive cases similar to those discussed for one-box. In the text we used mostly the ratios of the concentrations or burdens in the two boxes. Here are the equations for the mixing ratios in each box and the assumptions needed to derive them.For the horizontal model, with sources only in one hemisphere, say, S_n, and equal lifetimes in the two hemispheres ($\tau = \tau_n = \tau_s$), the solutions are:

$$C_n = S_n\tau\left[\frac{1+\tau/\tau_T}{1+2\tau/\tau_T}\right]$$

$$C_n = S_n\tau\left[\frac{\tau/\tau_T}{1+2\tau/\tau_T}\right]$$

For the vertical model, for sources (S_T) at the surface, into the troposphere, and sinks only in the stratosphere (lifetime there $= \tau_S$), the solutions are:

$$C_S = S_T\tau_S\frac{N_S}{N_T}$$

$$C_T = S_T\tau_S\frac{N_S}{N_T}\left(1+\frac{\tau_{ST}}{\tau_S}\right).$$

5

Transport Processes

Atmospheric transport processes play an important role in the earth's environmental system. They are the winds over large areas that spread and mix gases and particles far beyond the places where they are emitted from natural events and human activities. Vertical transport by convection cools the earth's surface and constitutes an important component of the climate. Horizontal winds move heat from the tropics to the poles both directly and by generating ocean currents, making a more uniform and livable climate. The winds also cause storms, move water to land by rain, and remove urban air pollution.

Transport processes can be divided into mean motion – the winds, and turbulence that forms swirls, gusts, and eddies. Gases are emitted mostly from patches of land and oceans. The human sources are particularly confined, coming from the cities and larger expanses of agricultural fields, but also from narrowly focused tall chimneys of power plants and factories. These emissions are moved away by the winds, but the mixing of gases and particles with the surrounding air is caused entirely by turbulent processes.

As a general principle, winds are generated only when the air pressure at one location is different from that at nearby places which represents *pressure gradients* in the atmosphere. These winds span a wide range of space and time scales from local winds that change from day to day to permanent winds that blow over entire hemispheres driving the general circulation of the atmosphere. In all these cases, pressure gradients cause inceptive forces on air parcels in the direction from high to low pressure. On the earth, inertial forces caused by its rotation act simultaneously on air parcels, so the air may not always move from high to low pressure, but rather it moves in more complex ways determined by the combination of pressure gradients and the rotation of the earth, as for example, in the swirls of cyclones and hurricanes. We will explore these phenomena further.

5.1 Vertical Transport and Convection

Consider vertical motion first. One way to understand vertical buoyancy forces that push warm parcels upward is as follows. Imagine a hypothetical volume of air within the large stationary atmosphere. This air mass is also stationary because, as we saw earlier, the force of the pressure gradient pushing it upward, generated by the air surrounding our imaginary volume, is balanced by the force of gravity pulling this volume of air down (Endnote 3.1). If we now replace the air inside our volume with helium at the same pressure and temperature, then the downward force of gravity will decrease because the same volume of helium

Global Climate Change and Human Life, First Edition. M. A. K. Khalil.

Companion Website: www.wiley.com/go/khalil/Globalclimatechange

is lighter than air, but the upward force is unaffected because nothing has changed outside the volume. This will cause a net force directed upward and so the parcel with helium will rise as illustrated in Figure 5.1.

Now let's apply this idea to natural atmospheric processes. A dark patch of ground may get heated by the sun more than the surrounding surface. The warmed land transfers some of this heat to the air in contact by conduction. The heated air will then expand according to the ideal gas law into a larger volume until the pressure inside the parcel is the same as the air outside. At that point we see that the warm air now occupies a volume V that previously contained more air when it was cooler, or we can say that the same number of moles that were in some initial volume now occupy a larger volume. Either way, the density of this warmed parcel will therefore be less than the surrounding air which has not warmed. Lower density means that it weighs less than the same volume of colder air, so the downward gravitational force, which is synonymous with its weight, can be written as: $F_{DOWN} = -mg = \rho_W V g$ (ρ_W is the density of the warm air in the parcel). But the upward force on it, supplied by the surrounding air, is the same as it was before our parcel warmed; it is: $F_{UP} = \rho g V$ (ρ is the density of the cool surrounding air). This produces a net upward force $F_{NET} = (\rho - \rho_W) g V$ causing the parcel to rise. It can generate thermals that may be dry air or contain moisture that can produce clouds and

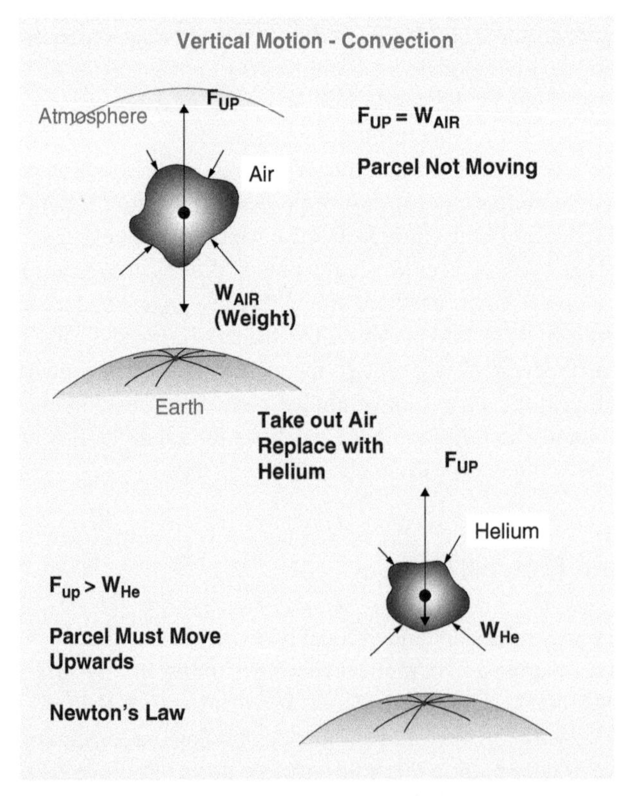

Figure 5.1 Convection and the vertical movement of air. Air is constantly rising and mixing, more in some places and less in others. This is caused by air parcels near the ground getting warmer by extracting heat from the surface. Warm air is less dense and lighter than the surrounding air, so it rises in the same way as balloons filled with hot air or helium rise.

rain. Either way, heat is transferred from the surface to the atmosphere, and gases near the surface are mixed into the larger atmosphere. The rising air is a common phenomenon everywhere and can be detected by finding soaring birds such as vultures and eagles who use it to stay up expending very little energy as they search for food below. On a global scale, this upward movement of air by convective processes has an important role in the climate as you would imagine. It is a constantly ongoing and variable process that causes cells of air flow and mixing along the way.

As a parcel of air rises, its pressure adjusts to that of the surrounding air. Since the atmospheric pressure falls as we go upward, the ideal gas law tells us that the parcel must expand. In doing so, it has to push the surrounding air out of the way which requires energy or work. This energy comes from the temperature of the air molecules causing the parcel to cool as it rises. The conservation of energy in this situation is expressed by the first law of thermodynamics. Under the assumption that the rising parcel of air does not exchange heat with its surroundings, called adiabatic conditions, the law says that the change in internal energy is balanced by the work done from the expansion, written as: $C_V\ \delta T = p\ \delta V$, where C_V is the specific heat capacity of air at constant volume, and δT is the change of temperature of the parcel because it did work on the surrounding air. The left-hand side is the heat extracted from the parcel to do the work. The right-hand term ($p\ \delta V$) is the mechanical work done (Endnote 5.1).

Heat capacity is a new concept that needs elaboration because it will also be useful later. The specific heat capacity is a property of materials including gases that tells us how much heat energy we must add to a kilogram of the gas to increase its temperature by 1°C and can be expressed in j/kg-°C. One form of energy stored in a gas is thermal; it consists of the kinetic energy of the molecules as they speed around and bump into each other or bounce off obstacles. The temperature is a measure of this average kinetic energy. When the expanding parcel does work on the surrounding air, under adiabatic conditions, the energy has to come from this internal resource, thus reducing the temperature.

It is a remarkable fact that we can calculate the rate at which a parcel of air cools as it rises by using the relation for the change of pressure with height discussed earlier ($\delta p/\delta z = -\rho g$), the ideal gas law ($pV = nRT$) and the first law of thermodynamics ($Cv\ \delta\ T = p\ \delta V$) (Endnote 5.2). From this, it turns out that dry air parcels will cool as they rise at a rate of about 10°C/km. When the rising air contains water, the behavior of the parcels is complicated by condensation processes that release additional heat into the parcel (Figure 5.2). This has to do with the fact that when a parcel of wet air was formed, water was evaporated from the surface such as the oceans. The evaporation causes molecules of water to be separated from the liquid and this requires energy called the latent heat of vaporization expressed in joules/mole. As the air rises, it cools causing the water vapor to condense back to liquid, forming clouds and causing rain. When water condenses, it must release the same amount of heat per mole that was added to evaporate it – a latent heat of condensation. This is because we have found that energy has to be conserved in all processes that occur, at least in our part of the universe. The release of this heat reduces the lapse rate because the parcel gets warmer than a dry parcel would be at the same altitude. For warm, wet air the adiabatic lapse rate varies depending on the water content and can go down to 3°C/km. The global average lapse rate of 6.5 C/km that we discussed earlier is a result of many parcels of air rising, sometimes dry, and sometimes carrying various amounts of water, reflecting the present state of the atmosphere averaged over the world and a year. There are other processes that affect the lapse rates at various latitudes and seasons, but this rising and sinking due to expansion and compression is the major cause averaged over the year (Figure 5.2).

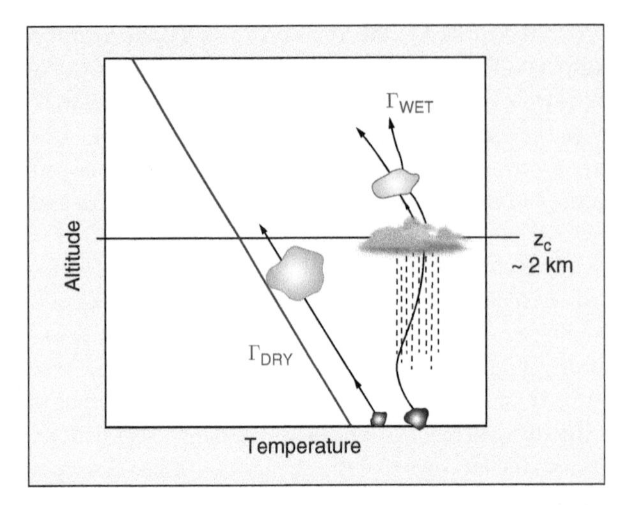

Figure 5.2 Rising air parcels. Dry air parcels cool more as they rise. Wet air parcels, as they cool cause the water to condense, which adds heat to the parcel making it warmer than the dry parcels would be at the same altitude. These processes transfer and mix heat and emitted gases into the upper atmosphere and move heat from the surface to the atmosphere contributing to the natural average temperature and climate of the earth.

We are now in a position to delineate how air moves in the vertical and what it has to do with the transport of gases or the structure of the global atmosphere. From our previous discussion, we can deduce, based on Newton's second law of motion, that the parcels of air will have an upward acceleration due to their buoyancy given by this formula (Endnote 5.3):

$$a_b = \frac{g[\Delta T_0 + (\Delta - \Gamma)z]}{T_{0_{Air}}} \tag{5.1}$$

$\Delta T_0 = T_{0\text{-Parcel}} - T_{0\,-\,\text{Air}}$, where the T_0's are the temperatures of the parcel and surrounding air at the surface, Γ is the adiabatic lapse rate, either dry or wet, that the parcel will follow as it rises and is due to the parcel's properties, and not the surrounding air. Λ is the prevailing lapse rate in the surrounding air at the location where the rising motion is taking place and is not affected by the rising parcel (adiabatic conditions). It is only 6.5°C/km as the global annual average, but at any time and place it can vary considerably depending on the movements of the winds and local atmospheric conditions. There are three possible cases: $\Lambda = \Gamma$, $\Lambda < \Gamma$ or $\Lambda > \Gamma$, as illustrated in Figure 5.3. It should be noted that the actual motion of the parcels is affected by more forces than shown in Eq. 5.1, including resistance to airflow, that are not of our interest at this point in our discussion.

In the first case ($\Lambda = \Gamma$) when a parcel of air is heated at the surface it will be warmer than the surrounding air by the same amount at all altitudes according to Eq. 5.1 because the second term containing (Λ- Γ) will be zero (Figure 5.1a). In the second case, ($\Lambda - \Gamma) > 0$, the parcel will be increasingly lighter than the surrounding air as it rises. This will make it rise even more vigorously (Figure 5.1b). In the last case (Figure 5.1c), when the atmosphere

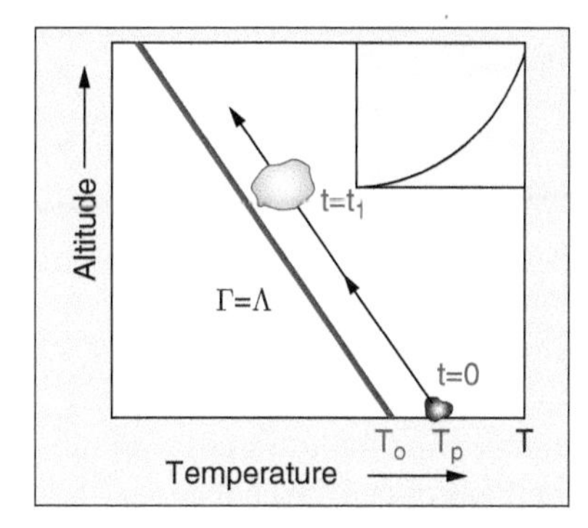

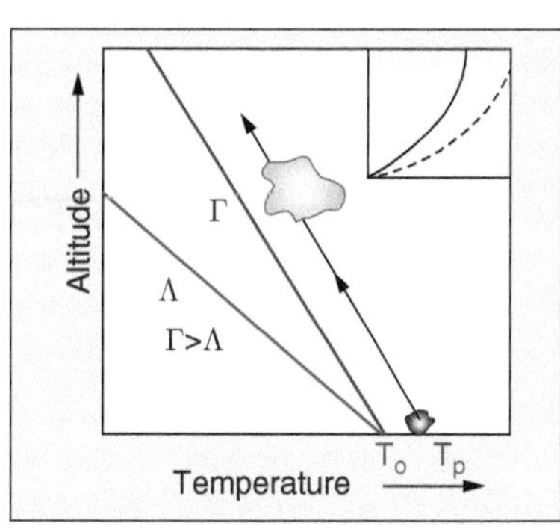

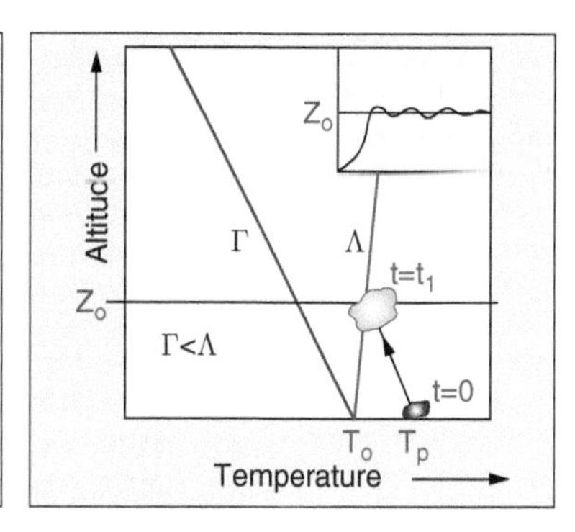

Figure 5.3 Possible vertical temperature conditions and resulting air motions. The rate at which a rising parcel cools is the adiabatic lapse rate, wet or dry, shown as Γ. The lapse rate in the atmosphere at any given time may be different from Γ or not; it is shown as Λ. When it is the same or lower than Γ air parcels move up and when it is larger they stagnate, reaching at most an altitude z_0.

cools more slowly than the parcel, we see that the (Λ- Γ) term in Eq. 5.1 will counteract the first term and will reduce the net force as the parcel moves up. The parcel will rise at first but it is inevitable that it will stop rising at an altitude $z_0 = \Delta T_{0/}(\Lambda - \Gamma)$. At this point $a = 0$ and the net force will be 0 as the temperature of the parcel matches that of the surrounding air. If it was moving upward, it may overshoot this mark and venture into the lighter air above and oscillate about this altitude, but it won't go any higher. A special case of this condition is when the actual atmospheric lapse rate is temporarily negative; that is, the temperature increases with altitude instead of decreasing, or a layer of warm air exists over the cooler surface air that has the same effect. In urban air pollution such conditions are called "inversions" and cause high levels of contaminants to build up in the cities, especially when the horizontal winds are also light and can't flush out the stale air.

These basic results have remarkable implications for the transport of gases and particles in the global atmosphere. In looking back at Figure 2.6a, we see that this lapse rate of decreasing temperature is replaced by a transition to a positive lapse rate in the stratosphere. We will see later that this reversal is due to the ozone layer that adds heat to the stratosphere, which is not available in the troposphere. For now, we can conclude that the troposphere on average behaves like the cases of adiabatic or sub-adiabatic lapse rates shown in Figure 5.3a and 5.3b, resulting in rapid transport of warm air upward and subsequent mixing along the way. Once the air reaches the tropopause, it encounters the temperature inversion similar to the conditions of Figure 5.3c. It is no longer lighter than the surrounding air and therefore cannot move higher. This is a *vertical thermodynamic barrier* to transport of air from the troposphere to the stratosphere, and it has a major influence on global atmospheric transport of air, and, by extension, on greenhouse gases and particularly the ozone-depleting compounds. Furthermore, since the lapse rate is constant or positive in the stratosphere, vertical mixing there is much slower than in the troposphere. The slow exchange of trace gases, slow vertical mixing, and the fast destruction higher in the stratosphere cause the mixing ratios of most gases to decrease with altitude in the stratosphere as discussed earlier.

5.2 Horizontal Motion and the General Circulation

The horizontal winds that blow across the surface of the earth are familiar in everyday and historical human experiences. Whenever air is moved upward by convection described earlier, it leaves a patch of lower pressure at the surface creating a horizontal pressure gradient. These smaller scale winds feed gases emitted from surface patches that may be cities, agricultural fields, lakes, or forests, into the large general circulation of the earth's atmosphere. How does that come about? We will see that it is a permanent planetary scale feature of the earth and has been playing out in nearly the same ways for hundreds of thousands of years. It is an important part of the climate and perhaps of climate change.

The origin of the general circulation is at the equator where the solar energy is most intense. If we measure the amount of energy (joules) arriving per second (watts = j/s) over each square meter of the surface (j/s-m^2 = w/m^2), it would be a maximum at the equator and decline systematically as we head toward the poles reaching very low levels there. The intensity of the sun's radiation, or irradiance, can be represented as rays that are nearly parallel when they arrive at the earth's surface. We see in Figure 5.4 that a set number of these rays, because of the spherical nature of the earth, will fall on a smaller area at the equator, while the same number will fall on a larger area at higher latitudes, where the land is tilted with respect to the incoming rays. The solar intensity, therefore, is highest at the equator and lowest at the poles. The absorption of the intense energy by the surface components such as oceans, land, and forests at the equator causes the air in contact with the surface to be warmed and to rise as discussed before. In this case, it is a very large-scale phenomenon and is more or less permanent because the sun constantly delivers energy to drive it. The upward trajectory is interrupted by the temperature inversion at the tropopause, but it does push it up and make it higher at the equator than at other latitudes where the intensity is less. Since the rising air cannot go higher, it spreads out in both the northerly and southerly directions under the tropopause heading toward the poles. The low pressure created at the surface around the equator causes a horizontal pressure gradient that, according to the discussion earlier, generates a net force: $F = - \delta P\, A = M\, a$, which causes an acceleration "a." Using $M = \rho_0 A\, \delta x$ – the mass of a parcel of air, $a = -1/\rho_0\, \delta P/\delta x$ where x represents the transect along longitudes for air moving from high latitudes toward the equator. As you see, it is proportional to the pressure gradient defined as: $\delta P/\delta x$. The poleward moving air at the upper levels must eventually sink by becoming drier, cooler and denser. The motion of the surface air toward the equator must draw from the poleward moving air that started at the equator. This would generate a closed cell of air movement from the equator to the poles in each hemisphere. At the surface, however, there are frictional forces that impede the flow causing slower and more constant winds than predicted by the acceleration $a = -\, 1/\rho_0\, \delta P/\delta x$. Furthermore, this picture of atmospheric motion is significantly altered by the rotation of the earth, which is quite rapid as we know from everyday experience – the entire earth going around once every 24 hours. The rotation generates an inertial force on the air parcels coming to the equator; this is the Coriolis force. Here is how it comes about:

Imagine a situation where a parcel of air leaves from a point at a higher latitude and heads toward the equator along a straight line represented by a longitude. If the earth was not rotating, it would arrive at a target that is on this longitude as shown in Figure 5.5. If the earth rotates, the point of origin of our air parcel goes around in a circle represented by its latitude of radius $R = R_e \cos \theta$, where θ is the latitude and R_e is the radius of the earth. The target point goes around the larger circle of the equator (R_e). This means the target

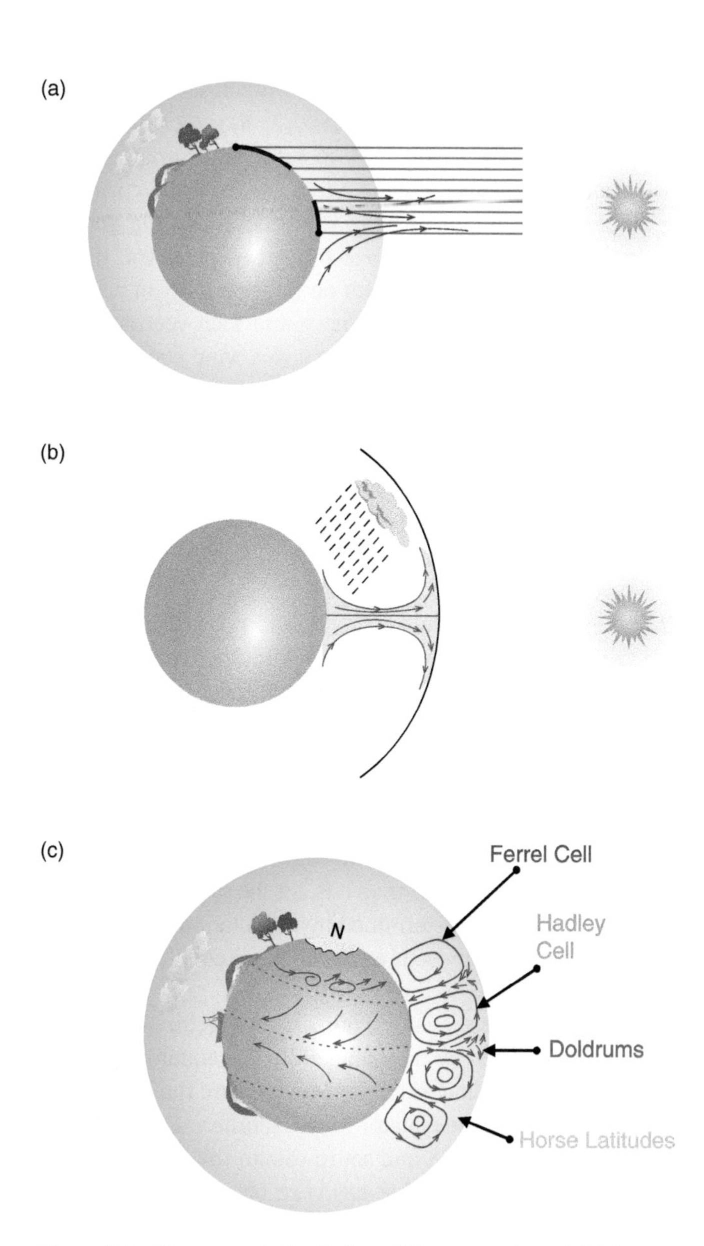

Figure 5.4 The general circulation of the atmosphere. (a) Solar energy is more intense at the equator than higher latitudes. This causes the convective motion carrying heat, water, and atmospheric gases upward at the equator. (b) The water condenses when the air rises and cools causing rain and drying the air. The air then spreads poleward. (c) The poleward moving air masses return to the surface at mid-latitudes due to the rotation of the earth and cause smaller cells of motion at middle and higher latitudes.

point is moving faster west-east, in the direction of the earth's rotation, than the point of origin since points on both, and indeed points on all latitude circles, go around once every 24 hours. Imagine now that this parcel does not interact with the surface or the rest of the air, but just moves toward the equator. Then, since there is no force acting on it in the east-west direction, it will retain the speed it had in that direction when it started its transect. So, when it arrives at the equator it will have gone $(2\pi R_e \cos \theta/24\text{ h}) \times \Delta T$ (km) $= V \times \Delta T$ in the west to east direction where ΔT is the time it took to get from its origin to the equator. The target point would have gone $(2\pi R_e/24\text{ h}) \times \Delta T$ (km). Since the cosine is always less than 1, clearly the parcel would lag behind the target point, having gone a shorter distance from west to east. It will therefore end up to the left of the target. If you were looking at this phenomenon standing at the target point on the equator, you will see the parcel was heading straight for you when it started to move and then veering away toward a westerly direction and land to your left. It would appear as a wind coming from the east (because it lands to your west when it gets to the equator). This explains the nature of the Coriolis force. A real air mass will interact with the surface and the rest of the atmosphere in which it is embedded. The surface will be moving faster as the parcel heads south, and so will the rest of the air. This interaction will impede the rest of the air and give the parcel some more speed in the west to east direction, but energy considerations show that the parcel will still be slower than the target point on the surface, and will still appear as a wind from the east and land to your left.

With many parcels, constantly moving toward the equator due to the lower pressure there, an east-west wind is generated relative to the surface. This motion can be represented by a force that pushes the air parcel to the left. It is clear from the illustration that such a force is generated by the rotation of the earth that causes a point on the surface to be constantly accelerating because the direction of its motion is changing. In general, such inertial forces, as they are called, arise every time the observer is anchored to a place that is itself accelerating. The Coriolis force is one manifestation in a rotating reference frame. It is noteworthy from Figure 5.5 and our description that the Coriolis force depends on how fast the earth is rotating. If the earth was spinning faster than it is, for the same initial speed the air parcel would deflect more. What we see is determined by the rotational speed and the pressure gradients that are generated under present environmental conditions of the earth and solar radiation.

There are two ways in which the Coriolis force changes our original picture of a large hemispherical-sized cell of circulating air driven by the rising air at the equator. First, it shows that the surface winds in the cell will not flow from the north toward the equator in the Northern Hemisphere or from the south to the equator in the Southern Hemisphere. Instead, they will veer and come from the north-east and south-east in the Northern and Southern Hemispheres, respectively. This is indeed what has been observed and experienced by us for centuries as the trade winds, so named because sailing ships used them to carry goods across the world. The second effect of the Coriolis force is even more dramatic; it causes the equatorial cell to collapse at middle latitudes around 30 degrees in each hemisphere instead of making it to the poles. We can focus the description of what happens to one hemisphere since similar effects occur in the other. In the Northern Hemisphere, the upper air moving poleward gets so deflected by the Coriolis force that by the time it reaches the middle latitudes, it moves mostly west to east. Since it cannot go further north, or into the stratosphere, is cold and dry, bunching up at middle latitudes, it moves to the surface causing a high pressure in the same way that it caused a low pressure at the equator by moving upward. This higher pressure drives surface winds both north and south of these latitudes. The southerly part are the trade winds, and the northerly part is the Westerlies in the Northern Hemisphere representing surface component of another cell-like motion. The large equator to mid-latitude cell is called the Hadley circulation, and the second one

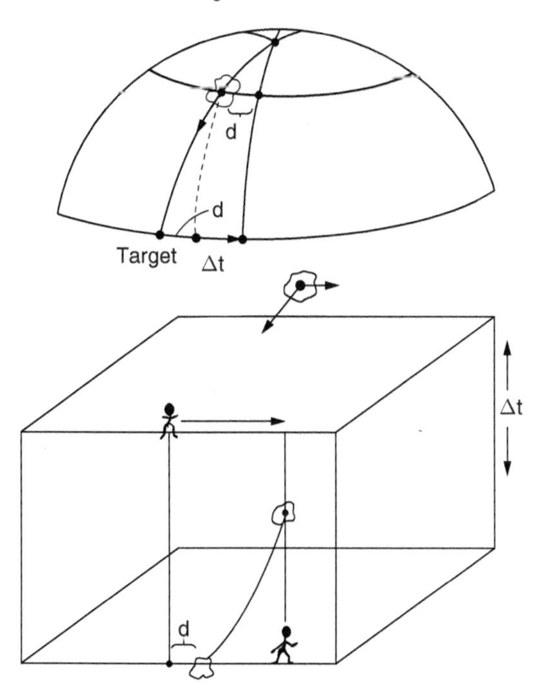

Figure 5.5 The Coriolis force. The rotation of the earth causes air masses coming toward the equator to be deflected westward.

at middle latitudes is the Ferrel cell after the discoverers of this theory of the general circulation. As the air in the Hadley cell descends it drags the more northerly air in the vicinity creating the downward arm of the Ferrel cell. This cell completes its rounds near the polar regions and affects a third region of connected large-scale motion at the poles. It stands to reason that the Coriolis force will also affect the surface winds in the Ferrel cell but in the opposite direction compared with the trade winds because the air is moving toward the poles instead of the equator. If we reverse the directions of our air parcel in Figure 5.5, we see that when the air leaves the southerly latitudes it is traveling faster in the east-west direction than a target on the same longitude further north, so the target will lag behind when the air arrives at this latitude causing it to appear from the west, or Westerlies. This cellular representation of the general circulation is summarized in Figure 5.4c.

From the preceding discussion it is apparent that the general circulation of the atmosphere can be seen as bands of air motion that affect the climate and weather as a function of latitude. The bands at the surface are bounded by low pressure at the equator, high at about 30 degrees and low again at about 60 degrees. The banding structure is a universal feature of atmospheres on spinning planets. The reason we have two cells, or possibly three in each hemisphere, is because of the specific features of our environment embodied in the size of the earth, the amount of solar radiation reaching here and the rate at which the earth spins. If it were spinning faster, there would be more bands because the Coriolis force would be stronger. If it were closer to the sun, the heating would be greater; the Hadley cell would be stronger and larger. It may be noted that for air moving horizontal to the earth's surface, the Coriolis force is zero at the equator and increases poleward. Hurricanes spin around due to the Coriolis force and hence cannot form at the equator or even at close by latitudes of a few degrees north or south. They form a little higher and hence affect tropical and subtropical zones of human habitation.

The Hadley cell in particular moves heat from the equator toward higher latitudes supplemented by similar movement of energy by the motion of the oceans, causing a cooler equatorial region and warmer upper latitudes. This has a significant effect on global habitability for many species including us. It is part of the climate characteristics of the earth affected by the "winds." We will return to this matter in Chapter 9 after some more ideas have been developed.

The discussion so far is the foundation for several phenomena that represent the state of the climate. As everyone knows, the tropics are warmer than higher latitudes because they get more sunlight per square meter than other places (Figures 2.4a, 5.4a). The rainfall patterns are more subtle. When the air rises in the equatorial arm of the Hadley cell, it is heavily laden with water that is evaporated from the oceans which cover much of the surface at these latitudes. The warmth of the air at the equator contributes to a high water content – a concept we will return to later. By the mechanisms we have discussed, rising air cools and water condenses to form clouds and eventually rain. In Figure 2.4 we see that the heaviest annual average rain is centered around the equator as a consequence of this mechanism. By the time the Hadley cell sinks at middle latitudes it is dried out. Moreover, as it sinks, it is compressed by the higher pressure near the ground causing it to get warmer as required by the ideal gas law. This constant pouring of dry air creates desert zones at these latitudes. Since part of this air moves northward in the surface arm of the Ferrel cell, it readily picks up water from the surface since it is both dry and warm. The rising arm of the Ferrel cell again takes the air higher up, causing it to expand and cool and dump the water as rain representing precipitation peaks at about 50 degrees in both hemispheres (Figure 2.4).

It seems that the cellular nature of the general circulation creates barriers to horizontal north-south transport at the equator and the middle latitudes of both hemispheres. These barriers are well known as "equatorial doldrums" and the "horse latitudes." The names come from the difficulty sailing ships had, and still do, in crossing these latitudes because of lack of forward winds. Legend has it that ships coming to the Americas carrying horses would get stuck around 30 degrees latitude for days or weeks, so sailors would eat the horses as rations ran low; hence the horse latitudes. While sailing ships have to wait for the winds to shift and meander their way, air and greenhouse gases are constantly, but slowly, getting across the barrier. An ever-present mechanism is turbulent transport which we will come to shortly.

We have looked at the large-scale movement of air and hence of the greenhouse gases, particles, and other materials that are emitted into the atmosphere and move with it. We should connect it to the mass balance in Eq. 4.1. The concept of flux (φ) is often used to describe the movement of material in the environment and especially by the transport processes. Here it represents the number of molecules that cross into or out of the box per square meter per second, taking the surface to be perpendicular to the flow. For the mean wind flowing into our box (Figure 4.1), the flux at the left boundary, where the air enters from the outside, will be φ_{IN} (molecules/cm^2-s) $= C_{UW}$ (molecules/cm^3) $\times$ U (cm/s) where U is the wind speed and C_{UW} is the concentration of a gas of our interest on the "upwind side" of our box. Similarly, if the wind is constant across the box, the flux leaving the box on the right-hand side in Figure 4.1, will be: $\varphi_{OUT} = C\,U$, where C is the concentration inside the box, so the net flux $\varphi_{NET} = (C_{UW} - C)U = \Delta C\,U$. The effect this flow has on the concentrations in the box (dC/dt)(due to wind, in molecules/s) $= \varphi_{NET}\,A = \Delta C\,U\,A$, where A is the cross-sectional area of our box perpendicular to the wind. We have to go a step further if we want to convert this result to the change of mixing ratio in the box. Divide both sides of the equation by the number of molecules of air in the box = N(molecules/cm^3) V (cm^3) = $N\,A\,L$, where ρ is the density of air in molecules/cm^3, V is the volume, and L is the length of the box. Recall that: C(ppb) = C(molecules/cm^3)/ρ (molecules/cm^3) $\times 10^9$. Then, T_{NET} (by winds) = $(dC/dt)_{Wind}$ (ppb/s) = $[\Delta C\,(\text{ppb})/L]\;U$ (Endnotes 4.1, 5.4). This can be written equivalently as: $(dC/dt)_{Wind} = \Delta C(\text{ppb})/\tau_T(\text{seconds})$, where τ_T, the transport time, is defined as L/U. We see that it is just the time it takes for the wind to cross the box.

5.3 Turbulent Transport

Turbulence is a prevailing feature of the atmosphere; it not only helps to get material across the barriers, but it is the singular mechanism that mixes the gases emitted from focused sources into the larger atmosphere. It does not arise by itself, but is generated by the motion of the winds described earlier, in particular by wind shear or by buoyancy. It transports air and the embedded trace gases in a direction perpendicular to the motion of the mean winds and with an efficiency proportional to wind speed and shear. Shear means that the wind speed changes as we go in a direction perpendicular to the flow. We see in Figure 5.6a flow that has a core with shear. Since the air is fluid the interaction of the core with the rest of the flow breaks off chunks of air masses and moves them laterally. These chunks are swirls or eddies that, by the same processes, create smaller swirls around them until the air from the core is thoroughly mixed with the surrounding air and the trace gases it may be carrying

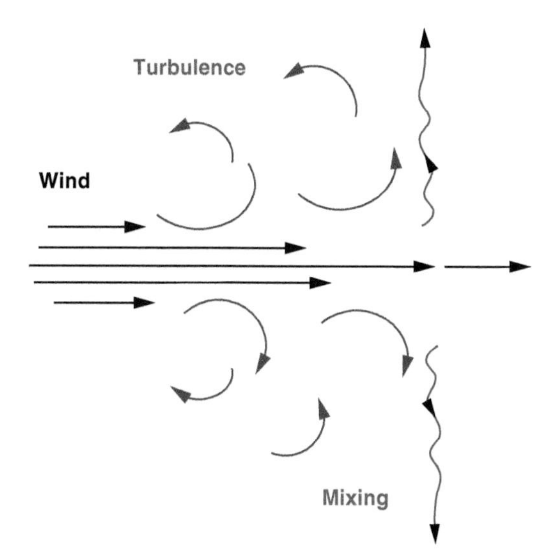

Figure 5.6 Turbulence and mixing of gases. Turbulence is generated by blowing winds and mixes the content of the air with the surrounding environment, particularly in the perpendicular directions to the flow.

are mixed likewise. Turbulence converts the kinetic energy of the wind eventually into heat, thus constituting a form of friction.

Thinking about it will convince readers that turbulence processes must always be present everywhere in the earth's atmosphere. At the equator the upward motion of air is confined to a band of latitudes with a strong flow at the center that stops when we reach the edge of the region of low pressure. This overall vertical flow is not very fast, so the turbulence it generates is weaker than at other places. At the descending arm of the Hadley cell, the wind speeds are high including the west-east component causing a greater efficiency of turbulent mixing. In the main part of the Hadley cell, at the surface, the wind is moving from north-east toward the equator but in the opposite direction near the tropopause to complete the circuit. There has to be shear as we go upward since the horizontal wind has to slow down and change directions to complete the cells. Near the surface the winds meet the unmovable surface and its obstructions, causing slowing. So throughout the Hadley and Ferrel cells, gases picked up from the surface sources are continuously mixed vertically into the troposphere by turbulence caused by the shear as well as mixing by the rising of hotter air into the colder atmosphere. Most trace gases of interest in climate change live longer than the time it takes to mix them vertically and horizontally, at least within each cell. But the impedances of the barriers cause the concentrations to be higher in the northern latitudes where most of the emissions are.

Turbulence works to mix materials in all sorts of circumstances, some of which we encounter in everyday life. If you put a tea bag in a cup of hot water, you see that most of the brew remains at the bottom near the tea bag, but some swirls develop reflecting movement of the water possibly because of small temperature variations. If you want the tea to mix in the cup before it gets cold, you have to stir it. This stirring action results in a circular flow with shear, which produces turbulence that causes the brew to mix evenly as you watch. The faster you stir, the faster the tea mixes.

5.3.1 Turbulence Illustration Model

Let's look more deeply at how turbulence works to mix gases into the atmosphere. If we take frequent measurements of the wind in either the forward or lateral directions the result will be an unsteady trace with fluctuations caused by the eddies breaking or arriving at our point of measurement. It is traditional to represent the wind speed (u) and variables affected by it, as a mean value (u avg) and fluctuations (u'), that is, $u = u$ avg $+ u'$. This decomposition is used to separate the effect of turbulent motion (u') on various phenomena of interest from the effect of the mean sustained winds we feel and see (u avg).

A simple model will illustrate the role of turbulent transport in our mass balance approach of Eq. 4.1. We look at a hypothetical band of atmosphere divided between north (N) and south (S) with no wind across the divide. However, there is an updraft that creates fluctuations of the wind in the N-S direction by turbulence as in Figure 5.7. We look at what happens in each increment of time δt. The gusts moving N → S or S → N, move dots back and forth. These may represent a trace gas of our interest. To start with all the dots are in the N box. The outcome of the turbulence is that even though there is no wind blowing steadily from N to S, the dots get to the S after a while when there were none to begin with. It is assumed that when a dot moves to one side, there is a good chance that it is carried away from this interface and mixed into the larger box, perhaps by a similar random fluctuation in the adjacent region, or a more sustained wind or one followed by the other. By a process

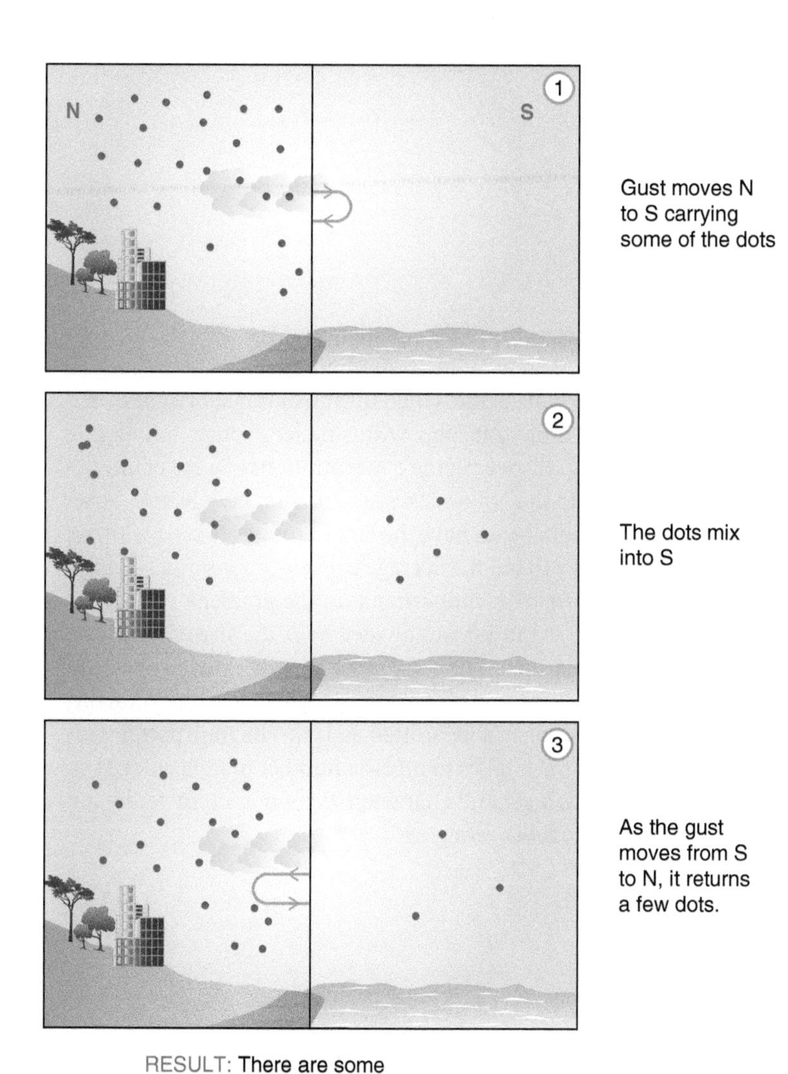

Figure 5.7 Illustration of turbulent transport. A model for a hypothetical atmosphere is created to show how turbulence can mix gases. Two regions are considered in which the air moves back and forth at the boundary by random gusts of varying magnitudes. This moves gases from where there is a higher concentration to where it is lower.

such as this, gases emitted from northern sources can get mixed all the way to the South Pole moving incrementally.

We can represent the situation mathematically as follows: If in time δt a swirl goes from N to S, it will transfer an amount of air = $u'\delta t A$, bringing in $u'\delta t A\, C_n$ (molecules/cm^3) molecules of the gas (dots) into the S box, and the same number are lost in the N box. The box has length L and cross-sectional area A over which the transport occurs. Dividing by the volume gives us the change of concentration this event generates as $(u'\delta t/L)\, C_n$. This will be added to the concentration that already existed in the S-box causing an increase in the concentration and likewise a decrease in the N-box concentration. This process repeats for each increment of time and can be described as:

$$\begin{aligned}
&u' < 0: \ \ N\,to\,S \\
&C_n(t+1) = C_n(t) - (u'\delta t / L)C_n(t) \\
&C_s(t+1) = C_s(t) + (u'\delta t / L)C_n(t) \\
&u' > 0: \ \ S\,to\,N \\
&C_n(t+1) = C_n(t) + (u'\delta t / L)C_n(t) \\
&C_s(t+1) = C_s(t) - (u'\delta t / L)C_n(t)
\end{aligned} \tag{5.2}$$

For illustration let's calculate 1000 increments of time, which produce the result shown in Figure 5.8. In it, the fluctuations u' is generated as random numbers that have an amplitude we can adjust, so two cases are considered – high turbulent transport, characterized as bigger average fluctuations, and slow transport with smaller average amplitudes (Endnote 5.5). When we put these into Eq. 5.2, we get the concentrations of a gas of interest in the N and S boxes starting with 100% in the N-box. We see that after a while both boxes have nearly the same concentrations. If both boxes have the 50% of the gas to start with, we see that there is some sloshing back and forth but the average concentrations don't change. These results suggest that turbulent transport is proportional to the gradient, that is, the difference of concentrations between the N and S boxes divided by L, the size of the boxes. When the gradient is not zero, embedded material, dots, move readily from high to low concentration areas, but when it is similar the net transport stops. The proportionality constant, which has the units of inverse time, can be written as $1/\tau_T$. The transport time τ_T can be interpreted as the characteristic time it takes to mix the number of molecules of air in one box with those of the other. This is a general characteristic of mass transfer by turbulence. We can therefore write the process *in average* as:

$$\frac{dC_n}{dt} = T_{NET}(n \to s) = -\frac{1}{\tau_T}(C_n - C_s) \tag{5.3a}$$

$$\frac{dC_s}{dt} = T_{NET}(s \to n) = +\frac{1}{\tau_T}(C_n - C_s) \tag{5.3b}$$

In Eq. 5.3, T_{NET} is the portion of the change of concentration in a box that can be attributed to turbulent transport in our original mass balance Eq. 4.1 and there are no sources or sinks during the times of interest. In the case shown in Figure 5.8, we can apply Eq. 5.3 with fast and slow transport times to calculate the concentrations in the two boxes. We see that gradient-driven transport with constant transport times τ_T can accurately represent the concept of transport driven by randomly fluctuating winds in Eq. 5.2.

Another prevalent description of turbulent transport can be constructed by assuming that the flux is proportional to the gradient of the *mixing ratios* and the proportionality "constant" is K (Endnote 5.6). Then, the flux is: φ (molecules/cm^2-s) = K(cm^2/s) N (molecules/cm^3) ΔC(ppb)/L (cm) where ΔC(ppb) = C - C(outside). T_{NET} (molecules/s) = φ A, where A is, as before, the cross-sectional area. This is the net number of molecules that are coming in due to turbulent transport, or leaving if the net flux is negative. Converting to the change of mixing ratio in the box, we divide by the number of molecules of air in the box ($N\,V = N\,A\,L$), and get:

$$T_{NET-TT}(ppb/s) = \frac{K}{L^2}(C - C_{Out}) \tag{5.4}$$

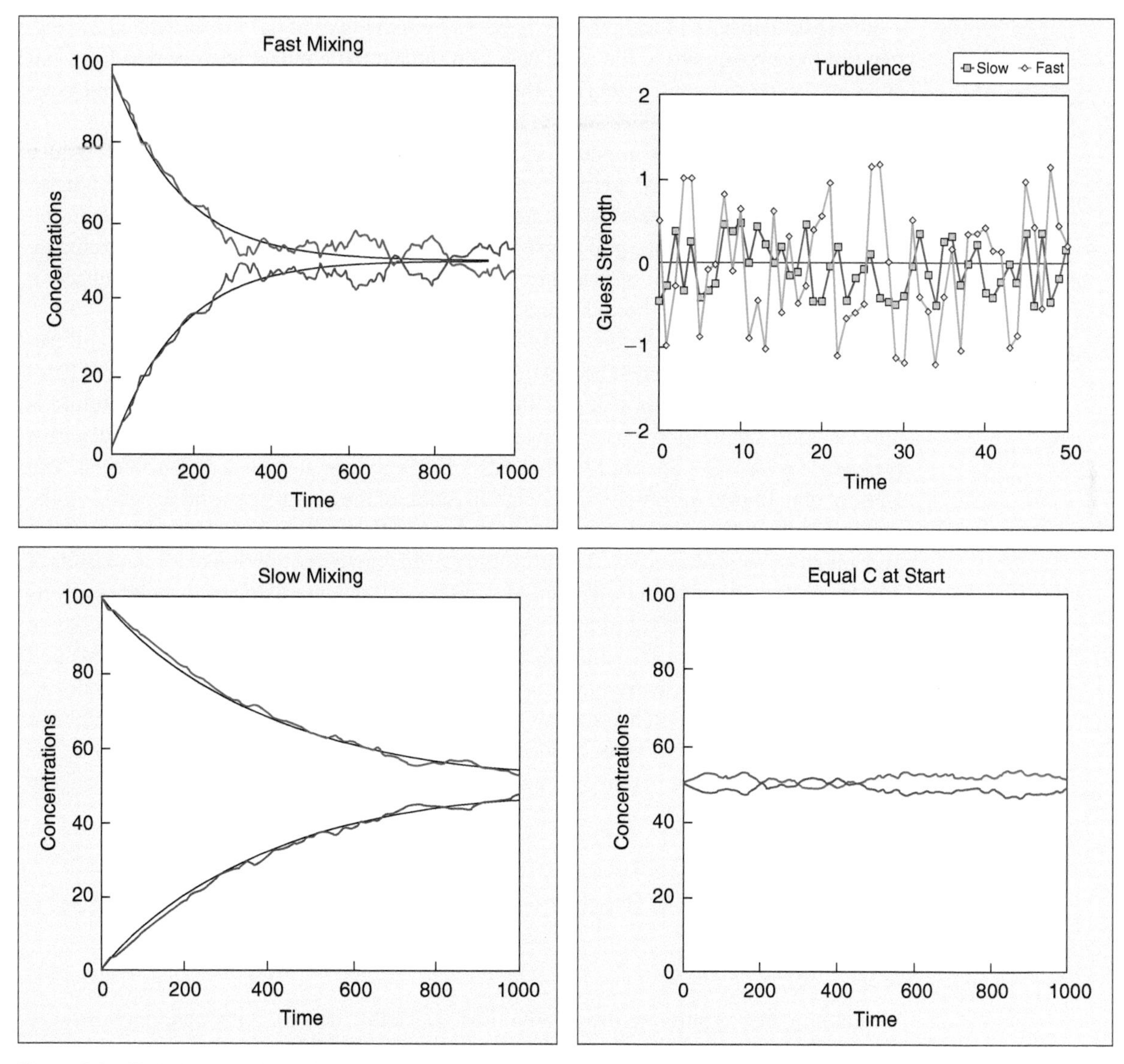

Figure 5.8 Turbulence and gradient driven transport. The fluctuations of the wind in the upper right panel representing u', describe turbulent mixing or materials in the atmosphere such as greenhouse gases. The results can be represented equivalently in average form as a transport that is proportional to the gradient of the flux in adjacent regions with the proportionality constant representing the inverse of a transport time. The lines show such calculations and their close match to the idea of transport by fluctuating wind components.

Comparing Eq. 5.4 with Eq. 5.3, we see a representation of the transport time in terms of the transport coefficient K and the size of our region L:

$$\tau_T = \frac{L^2}{K} \tag{5.5}$$

These connections extend to general conditions for turbulent transport. The transport time is inversely proportional to the strength of the turbulence, and K is proportional to it (Endnote 5.5).

Since turbulence is a local process, it works by moving material from one shallow segment of the atmosphere to the next. Based on the heuristic model just discussed, we can attribute a net turbulent transport for the box of Figure 4.1, as the exchange of air and gases of our interest across the sides and the top, perpendicular to the flowing wind.

It is useful to note some implications. We see that turbulent transport processes reduce concentrations where they are high and increase them where they are low. Keeping in mind that turbulent processes work in conjunction with the mean winds, for air pollution issues, they serve to dilute pollutants resulting in cleaner air in the cities. For climate change, they mix greenhouse gases into the entire atmosphere, thus creating an impact on the global temperature rather than just on local conditions.

The dispersal of smoke and contaminated air from chimneys of power plants and factories has been widely studied because of its air pollution impacts on nearby areas of habitation and crops. The main features are that the pollution experienced on the ground is highest along the line down wind, but as we get further it is more and more diluted, with a symmetric cross-wind distribution that falls off like a normal statistical distribution. The amount of pollution in this perpendicular direction of the prevailing wind depends on the rapidity of the turbulent processes. A similar experience comes from forest fires that are over much larger spatial scales. The prevailing winds can carry the smoke for thousands of kilometers, but although the lateral cross-wind spreading and mixing by turbulence affects regions along the way, it is to a much lesser extent, than places directly in the path. These common experiences of tall chimneys and large fires, with many available photos and videos, illustrate the ideas we have discussed here: that winds carry the pollutants downstream rapidly, but turbulence spreads them in cross-wind directions, both horizontal and vertical. It also spreads them in the down-wind direction, but this is overwhelmed by the action of the mean wind in that direction, so no accounting of turbulent effects is generally needed.

5.4 Quantifying Transport Processes

5.4.1 Transport Times

We are now in a position to discuss the time scales that control the global mass balance described in Eq. 5.4 and the discussions in the previous sections where we derived formulas for the effects of the mean and turbulent processes on the mixing ratios in the regions of our interest. Within these regions bounded by the barriers, the transport and mixing are quite fast. Typical values of the trade winds are ~ 6–8 m/s and they blow over transects from 0 to 30 degree latitudes, which is a distance of about 5000 km, giving transect times of ~ 10 days, and similarly for the higher latitudes (Time = distance/speed). Mixing times across barriers can be estimated by inverting Eq. 5.4. Average values of K across the equator is ~ 10^{6} m^2/s giving $\tau_T \approx 0.8$ yrs and across the middle latitude barrier it is ~ 4–5 × 10^6 m^2/s giving $\tau_T \approx 0.2 - 0.3$ yrs for 5000 km space scales representing the distances between horizontal barriers. These values are supported by observations of trace gases and tracers that have precisely known sources and mixing ratios. At the tropopause vertical transport K values fall to ~ 0.2 m^2/s, and remain low in the lower stratosphere. This causes a slow mixing, even though the distances involved are also small at ~ 8 km scale height (Endnote 5.7). Observations and calculations establish an exchange time of about 1.8 y between the stratosphere and troposphere. This is how long it takes to mix the number of molecules of air that are in the stratosphere to the troposphere and replace them with an equal number of molecules from the troposphere. The barriers and results are summarized in Figure 5.9. From these results we see that the mean winds are much faster at moving material around than turbulent processes, but the importance of the latter is that they are the mechanism for mixing everything in the atmosphere even when the winds are not blowing in that direction, making them a key factor in determining atmospheric composition and the nature of the environment, both global and local.

5.4.2 Boxified Transport

We are now in a position to describe the transport processes in box-form. The goal is to create a more concrete picture in our minds about how air, gases, and aerosols are moved from one hypothetical atmospheric box to another as we expressed at the start in Eq. 4.1 and Figure 4.1. For long-lived greenhouse gases, the effects of taking into account the details of the transport processes will not be of concern for us.

Consider, first, the mean wind transport of the Northern Hemisphere Hadley circulation in box-form. The least number of boxes needed is four. There are two boxes over the surface that cover the horizontal extent of the circulation, that is, from the equator to the

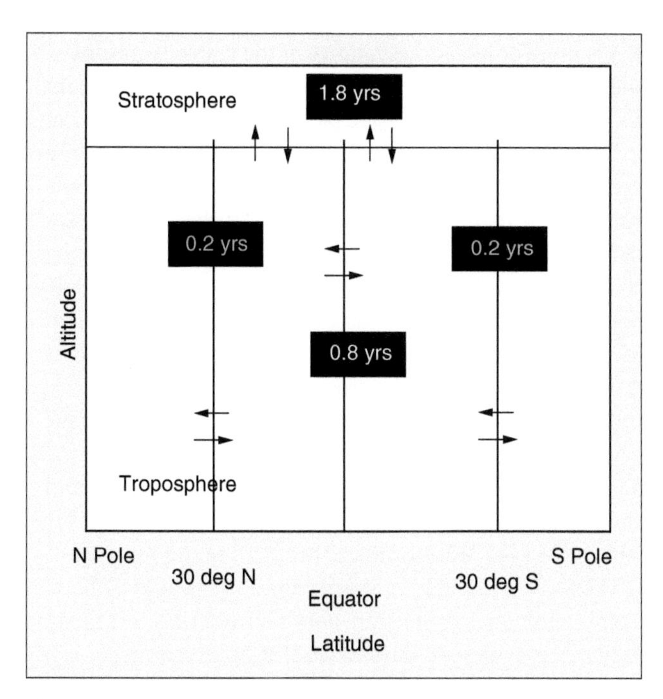

Figure 5.9 Transport barriers and average transport times. As discussed in the text, there is a vertical thermodynamic barrier at the tropopause that slows mixing from the troposphere to the stratosphere. In the horizontal the general circulation of the atmosphere creates barriers at the equator and at middle latitudes that slow down the transport of gases. The time scales associated with these barriers are shown. Each represents the time it takes to transport the number of molecules of air in the box to the adjacent box and receive a like number of molecules from the adjacent box. This mixing carries emitted greenhouse and other gases across the atmosphere.

sub-tropical boundary and from the surface to some point above where the winds reverse direction. And then two boxes above that cover the rest of the atmosphere up to the tropopause. The Hadley circulation is described by an upward motion of air from the lower equatorial box to the upper one, from the upper equatorial box to the upper tropical box, from the upper tropical box downward to the lower tropical box and from there to the lower equatorial box. If any gases or aerosols are emitted anywhere in these boxes, perhaps at the surface, they will get transported around this whole segment of the atmosphere and, as discussed earlier, thoroughly mixed due to turbulence within this circulation. This situation is analogous to the "tea bag in a cup" situation mentioned earlier. The tea gets moved around the cup by stirring and gets mixed in the hot water by this circulation, but is confined to the cup due to its rigid walls. The gases and aerosols in the Hadley cell would behave likewise and get mixed but stay within this region. The boundaries of the Hadley cell, however, are not rigid. The gases will mix perpendicular to the wind across the boundaries of the cell by turbulent transport. In the southerly direction, they will mix with the Southern Hemisphere Hadley cell; at the northern edge, they will mix with the Northern Hemisphere Ferrel cell; at the upper boundary, they will mix with the stratosphere and at the lower boundary, they will mix with the oceans and soils as cross-media transport, to be discussed later.

The discussion so far, gives us insight about the two-box models we have considered in Chapter 4. For the latitudinal model, our assumption is that the effect of the Hadley and Ferrel circulations is to just evenly mix any gas within the hemisphere into which it is emitted, resulting in an average concentration that is more or less representative of the whole hemisphere. The greatest impedance to horizontal transport is at the equator, where only slow turbulent processes work to transfer the gases across, and once there, are picked up by the symmetrical southern Hadley circulation. The transport time describes this turbulent process. For the vertical model, we have assumed that mixing of gases in the troposphere results in some approximately homogeneous distribution due to the combined mean and turbulent processes that are faster than transport across the vertical tropopause barrier, which then represents the main impedance to further vertical transport. It is noteworthy that these two-box models do not explicitly utilize mean-transport processes. They address the effects of only turbulent transport, and internalize the mean-transport. That is one step further than the assumptions underlying the one-box model that internalizes all transport processes.

The two-box models do not fully include the turbulent transport processes we have described. To do so would require at least five boxes: four in the troposphere to describe the impedance at the equator and the horse latitudes, and one between the troposphere and the stratosphere. To implement the circulation cells, we saw that you need four boxes for each cell. There are four strong cells in the earth's atmosphere, the Hadley and Ferrel cells in each hemisphere. That makes 17 boxes – 16 in the troposphere and 1 in the stratosphere. Then you will have a minimal representation of the transport representing the mean circulation, and cross-wind turbulence processes we have discussed here and in Eq. 4.1. That's a lot of boxes already and we have still only covered two dimensions, latitudinal and vertical. Simple formulas, such as the ones we have discussed, are untenable. If we want to go further and take into account land and ocean effects, we will need more boxes dividing the atmosphere along longitude circles. With at least 4 there, we are up to 65 boxes. This type of model would be a three-dimensional representation of the atmosphere, the 17-box model is two-dimensional, our two-box models are one-dimensional, and the globally averaged model is zero-dimensional.

Review of the Main Points

1 Transport processes take gases from the sources and move them over the entire atmosphere mixing them along the way, causing a global impact from localized emissions. Indirectly they affect the residence time of a gas by taking them to places where the destruction may be faster or slower than average.
2 Transport is divided into vertical and horizontal, and in each of those dimensions it is divided into the winds and turbulence. The wind can move material over very large distances, but turbulence is needed to mix gases in the atmosphere; it operates on more local scales. Convection causes vertical transport by warming of surface air. This causes thermals, clouds, and rain. Horizontal transport is driven by large-scale cells in which air rises at the equator, and while aloft, spreads toward the poles coming down at mid-latitudes completing the cell by joining surface winds. This is the Hadley circulation. It moves heat, moisture, and atmospheric constituents within tropical latitudes. Indirect Ferrel cells do the same in middle latitudes.
3 In the vertical there is a barrier at the tropopause that is explained by the warmer air in the stratosphere that prevents the cooler and thus heavier air from passing the tropopause. In the horizontal there are barriers at the equator and middle latitudes at the boundaries of the large Hadley cell. Because of the barriers, greenhouse gases and other emissions from the surface take about two years to get from pole to pole and surface to the stratosphere. Transport of gases is driven by the difference of concentrations and fluxes in adjacent regions. Once gases get across the barriers, the faster-moving winds can pick them up and take them further and mix them along the way. The efficiency of turbulent transport can be represented by the time it takes to mix gases τ_T, or by a transport factor K that describes the flux.

Exercises

1 Suppose we can represent the air quality in a city by using Eq. 4.1 making a box with dimensions $L, W, H = 20\text{ km} \times 20\text{ km} \times 1\text{ km}$. A steady wind is blowing from one side at $u = 3$ m/s. There are many sources of pollution that together generate $S = 1000$ kg/s emissions of CO_2, a greenhouse gas. There are no sinks of CO_2. After a while the emissions and the transport balance to create a constant amount of CO_2 in the city.
(a) Show that the residence time of CO_2 will be $\tau = L/u$.
(b) Find the amount of CO_2 that is contained in the city's box.
(c) How much CO_2 from this city enters the global atmosphere each year?
2 Derive the dry and wet adiabatic lapse rates from the three laws – hydrostatic balance, ideal gas law, and the first law of thermodynamics.

3 A parcel of air with a volume of 1000 liters and at a temperature of 300°K at the earth's surface, absorbs heat from an industrial process. At the end, it has expanded to 1500 liters.
   (a) What is its expected temperature?
   (b) How much heat has it stored in the process.
4 The magnitude of the Coriolis force can be calculated as $F = 2\, m\, \Omega\, v\, \text{Sin}\, \varphi$. Here m is the mass of the object moving in the atmosphere, Ω is the rotational speed of the earth, that is, $2\,\pi$ radians in 24 hours, v is the speed of the object, and φ is the angle between earth's axis of rotation and the direction of motion. The direction of the Coriolis force is perpendicular to both the direction of motion of the object and the direction of the earth's axis of rotation, which is taken to point up from the North Pole.
   (a) Determine the Coriolis force on an air mass that starts moving at 45 degrees N latitude toward the equator,
   (b) an air mass that starts moving in a southerly direction at the equator, and
   (c) an air mass that rises upward at the equator. What are the values for this force per kg mass of the air mass for velocities of 10 m/s in each of these cases?

In everyday life the Coriolis effect may have a small influence on objects that fall to the earth, or are otherwise in motion near us for short distances and times. Nonetheless, it is there. One way to look at this is to consider a tall building at the equator (for convenience in visualizing the effect). As the earth spins, the top of the building is going around faster than the bottom! (Because it is further from the center of the earth.) It can be shown that if you drop a ball from the top of such a building, then the deflection of the ball, due to the Coriolis force, will be given by this formula:

$$\Delta x = \left[\left(2^{3/2} / 3\right)\left(\sin\theta / g^{1/2}\right)\Omega\right] h^{3/2}$$

where h is the height of the building, g is the acceleration due to gravity (9.8 m/s^2), and θ the co-latitude of your location measured as 0 at the pole (the geographical latitude = o-latitude – 90°). Determine the magnitude and direction of the deflection of a lead ball dropped from the Taipei 101 Tower (508 m) at 65° co-latitude.

Endnotes

Endnote 5.1 Forces and energy are closely related. If you exert a force F on an object and make it move a certain distance D, the energy it takes is called work and it is force times the distance moved ($W = F \times D$). It is complicated a little by the fact that only the motion in the direction of the force counts, so if the object moves in some other direction at the same time, then work = force × distance × cosine of the angle between the direction of motion and the force.In our case of an expanding air parcel, we can write the work as the pressure × the change of the volume which is the same concept ($W = p\,\delta V$). Consider a spherical parcel of air as an illustration. The pressure inside is p and the surface area is $4\,\pi\,R^2$ where R is the radius. After it expands the radius becomes larger by δR, so the change of volume $\delta\, V \approx 4\,\pi\,R^2\,\delta R = A\,\delta R$. Therefore $W = p\,A \times \delta R$; since pA is the force we get the result that $W = F \times \delta R$ or force times displacement. The F and δR are in the same outward direction.

Endnote 5.2 Interested readers can find the derivation of the dry and wet adiabatic lapse rates in many textbooks, for example, "Atmospheric Chemistry and Physics of

Air Pollution" by J. Seinfeld (J. Wiley and Sons, N.Y., 1986). It may be useful to look at the resulting formulae here and calculate some values. The dry lapse rate is: $\Gamma_{DRY} = g/C_P$, where g is the acceleration due to gravity and C_P is the specific heat capacity of air at constant pressure. The value is (9.8 m/s)/(1006 j/kg °C) = 9.7°C/km. For wet air parcels, as discussed in the text, you have to add the latent heat of condensation that raises the temperature of the parcel at altitude. $\Gamma_{WET} = \Gamma_{DRY} - (H_C/C_P)(d\,q/dz)$. Here H_C is the latent heat of condensation per gram of water and q is the mass mixing ratio of water vapor as g H_2O/g air. The ratio $H_C/C_P = 2250$°C is fixed because it represents the properties of water vapor. The dq/dz represents the level of water in the air at a time and place and can vary from 0 to 0.003/km in the tropics, implying that the lapse rate can be as low as ~ 2–3°C/km.

Endnote 5.3 The formula in Eq. 5.1 can be derived by writing the temperature as a function of altitude using $T = T_0 - \Gamma z$ for adiabatic conditions and $T = T_0 - \Lambda z$ for actual atmospheric conditions that may occur for some period of time. The net force on a parcel of air is $F = (\rho_{Parcel} - \rho_{Air})\,Vg = M_{Air}\,a = \rho_{Air}\,V a$. Therefore $a = (\rho_{Parcel} - \rho_{Air})/\rho_{Air}$. From the ideal gas law, we can write $\rho = MP/RT$ where M is the average molecular weight of air molecules. Substituting this in the previous equation using the temperature functions above gives Eq. 5.1. Velocity-dependent impeding forces are also present and will cause the parcels to slow down to some terminal velocity. The point being addressed here does not depend on them, so these are not discussed in the text.

Endnote 5.4 We use the idea that if a wind U, or a fluctuation of the wind, moves into our box it brings in a flux, in molecules/cm^2-s = C(outside) (molecules/cm^3) U (cm/s). This is understood by recognizing that during a time δt, a volume of air moves into our box from the outside. This volume is $\delta V = A\,\delta x$ where x is the direction from which the wind is coming and is taken to be perpendicular to the side of the box with area A. This volume contains C(outside) (molecules/cm^3) × A (cm^2) δx (cm) = molecules of the gas. So the number of molecules that are brought in per unit time is C(outside) × $A\,\delta x/\delta t$. The $\delta x/\delta t$ is the velocity of the wind in the x-direction by definition. The number of molecules brought in per unit time is therefore C(outside) A U, and the flux is just this divided by the cross-sectional area or $\varphi_{In} = C$(outside) U as we wanted to prove. The same reasoning is applied to the molecules that move out of the box in time δt so, the flux moving out is $\varphi_{Out} = CU$. In general the Us may not be the same at the two ends of the box if the wind is twisting around, but such complications can be taken into account if necessary.

Endnote 5.5 The turbulence model is solved in a spreadsheet and can be easily duplicated, as are all simple models used in this book. The wind fluctuations u' have mean 0 and standard deviations, σ's = 0.29 for slow and 0.72 for fast transport in length/time units. The L is 100 length units. τ_T = 320 time units for the larger fluctuations and = 760 time units for the smaller fluctuations. The units are otherwise arbitrary.

Endnote 5.6 The concept of the transport coefficient K is mathematically similar to diffusion of gases in various media, including air. It states that the transport is proportional to the gradient of the concentration. Turbulent transport has been called "atmospheric diffusion" and "turbulent diffusion." It is important to note that turbulent transport does not share any similarities of mechanisms, or the physics, with gas diffusion. The speed of molecular diffusion is governed by a diffusion constant (D) for which there is a fundamental theory. It is different in predictable ways for each gas and the medium in which it is diffusing; it also depends on the temperature. Turbulent transport has a similar "constant" K but the theory for it is not similar to that for D and is incomplete in many ways. Observational data and semi-empirical models give us values for various prevalent atmospheric circumstances. In turbulent transport what is "diffusing" are not molecules but large chunks of air, which cascade to smaller and smaller chunks in a mechanistically unique process not found in ordinary diffusion. This leads to values of K that are more than ten orders of magnitude faster at mixing gases than the diffusion constant would be for the same gases! Moreover, the mixing by turbulence, and hence the K, is related to the motion of air itself rather than the characteristics of the gases it is moving, so it mixes all gases with about the same efficiency. The usual gas diffusion does not require bulk movement of material and is fundamentally related to the characteristics of the gas that is diffusing.

Endnote 5.7 For further reading about the topics discussed here and in later chapters dealing with atmospheric chemistry readers may consult Peter Warneck's, "Chemistry of the Natural Atmosphere" (Academic Press, San Diego, 2000). It is a nearly 1000-page compendium and synthesis of the state of established environmental science.

6

Mechanisms of Sources and Sinks

Concentrations of gases in the atmosphere are determined fundamentally by the sources and sinks as most clearly expressed in the 1-box model, while transport processes play an intermediary role in determining the concentrations, but are crucial in moving, mixing, and distributing the gases globally. Source and sink processes may be considered to occur in three groups of increasing scope and complexity: "pure gains and losses," "exchanges," and "cycling." Let's look at some examples to establish the context for further discussion.

A pure source or sink is distinguished from the cycling or exchange processes in that a gas is added to or removed from the atmosphere; if removed, it does not come back and if added, it stays in the atmosphere until removed by a process independent of the source. It is the simplest of the mechanisms. Examples of pure sources and sinks are the chlorofluorocarbons which were manufactured compounds released from their many consumer and industrial uses. Once in the atmosphere they are removed by stratospheric chemical processes. There is no chance that a chlorofluorocarbon molecule, once broken apart by ultraviolet sunlight, will return again to the atmosphere, or anywhere else. All technological gases are in this group and even methane and nitrous oxide may be included.

Exchange processes occur when gases move across media. A case in point is the gas exchange between the atmosphere and the oceans. For any given concentration in the atmosphere, gases can only dissolve in the oceans up to a certain capacity. If there is less in the ocean than this capacity, the gas will move from the air to the oceans and the oceans will act as a sink, and if it is the other way around, the oceans will release the gas to the atmosphere making it a source. Methyl halides (CH_3Cl, CH_3Br, and CH_3I) and some sulfur gases of interest in climate and environmental sciences (DMS, OCS) are examples of this ocean "source" group. An important example of exchange processes acting as a sink is for carbon dioxide, which is a focal point of global warming. Emissions of carbon dioxide into the atmosphere, as from burning fossil fuels, generate an imbalance allowing a significant amount to be absorbed in the oceans thus reducing the climate impact. If we stop emitting additional amounts, a balance will be reached in which the oceans will no longer remain a sink leaving behind a long-lasting residual.

In "cycling," cycles of concentrations can occur in the atmosphere. These may be parts of a larger process by which elements move and transform in the earth's environment over various time scales. It therefore intersects with climate when one of its branches controls the atmospheric abundance of a greenhouse gas. For the environmental sciences, it is most important for the dominant gas in the atmosphere that contains the element, such as CO_2,

Global Climate Change and Human Life, First Edition. M. A. K. Khalil.

Companion Website: www.wiley.com/go/khalil/Globalclimatechange

O_2, and N_2 are for the carbon, oxygen, and nitrogen cycles. As an example of a short time cycle, we can follow a single carbon atom initially tied up in atmospheric carbon dioxide through one of its shorter paths in the carbon cycle. The carbon dioxide is taken up by plants that use sunlight to produce sugars and grow, using our carbon atom which now becomes part of a larger organic molecule, but it is removed from the atmosphere. It can come back into the atmosphere by the plant's respiration as CO2, or the plant may be eaten by animals, or die and be consumed by bacteria (decay). The by-product of their metabolisms is carbon dioxide, or methane in some cases. So, our carbon atom may come back into the atmosphere as one of these two gases, but in time, the methane converts to carbon dioxide too. Such processes can create fast cycling of carbon dioxide as seen in the mid-latitude observed concentrations that vary seasonally in response to plant and microbial activities. This sort of cycle contains both a source and a sink arm in it. For carbon dioxide, if we look at the summer months, the mid-hemisphere biosphere is a net sink, and in winter, it is a net source, and if we look to time scales that are averaged over the whole year, it is neither a source nor a sink. Over the time scales of our interest, cycles shorter than a few years are awash as sources and sinks and can be neglected, while long cycles, say a century, can be treated like pure sources or sinks in the mass balance. The major reservoirs aside from the atmosphere are the oceans and land (soils, biota and the deep earth) where the ingredients of greenhouse gases, or the gases themselves, can be stored for long times acting as sinks. From our perspective, if it is out of the atmosphere, it will not affect the climate or cause global warming, so the end result is the same as if it was destroyed.

The "lifetime" represents mostly the first of these mechanisms but it includes transport processes within the atmosphere as we have discussed. The more general term "residence time" is sometimes used to distinguish the added aspect of how long a gas persists in the atmosphere when cross-media exchanges and cycling are present. For many trace gases of our interest, the residence time is dominated by the lifetime, or a destructive removal from the atmosphere as would happen with chemical reactions. However, almost all gases have some amount of storage in the land and oceans. The concept of residence time has to be invoked for carbon dioxide because its movement out of the atmosphere is dominated by shifts to the land and ocean reservoirs while there is virtually *no pure sink* in the atmosphere, which is our vantage point.

6.1 Reservoirs and Source-Sink Relationships

So far in our discussions we have relied on established laws of nature, mostly physics, that determine the characteristics of atmospheric structure and transport processes. The laws of nature governing the sources and sinks are founded in chemical and biological sciences making them more complex and less definite. You will recall that there are a few key gases that cause global change (Figure 3.1, Table 1.1). To begin the discussion let's look at Tables 6.1a,b. In these, sources and sinks are classified in two dimensions: four categories of mechanisms form one dimension, and where these gases are generated or destroyed is the other (land, oceans, and the atmosphere).

The locations are made up of reservoirs where the gases can be produced, destroyed, held for some time, and exchanged with other reservoirs. The several reservoirs increase the complexity of the mass balance models by extending dynamic processes outside the atmospheric component expressed in Eq. 4.1. Each reservoir acts similarly to the description in Eq. 4.1 as it balances between production, destruction, and transport internally and by

Table 6.1 Mechanisms of Sources and Sinks. These are expressed in two dimensions. The locations in the earth's system and the mechanism.

SOURCES				
	Location			
		Atmosphere	Land	Oceans
Mechanism	Microbial	x	CH_4, N_2O, N_2	CH_3x, DMS, OCS, N_2
	Plant Biochemical	x	O_2, NMHCs, VOCs,N_2	O_2, N_2
	Chemistry and Physics	O_3, CO, Particles	x	H_2O
	Industrial and Agricultural	Aircraft	CFCs, PFCs, HFCs, SF_6, CH_4, N_2O	x
SINKS				
		Location		
		Atmosphere	Land	Oceans
Mechanism	Microbial	x	CH_4, N_2O	x
	Plant Biochemical	x	CO_2	CO_2
	Chemistry and Physics	CH_4, N_2O, CFCs, HFCs, O_3, CO, HCs, Particles, ...	x	CO_2
	Industrial and Agricultural	x	x	x

exchanging gases with the other reservoirs, including the atmosphere. The latter will be discussed as *cross-media transport*. Ultimately our interest is focused on what happens in the atmosphere because that is what determines climate and climate change. As alluded to earlier, we can often circumvent the complexity of cross-media processes by treating the land and ocean reservoirs as either net sources or sinks of gases in the atmosphere without explicitly taking into account what happens within them (Endnote 6.1).

The mechanisms in Table 6.1 represent one of the most complex pieces of the global environment and habitability because they come about from eons of evolutionary processes that have resulted in the world as we see it today. Microbial processes primarily in the soils are fundamentally responsible for the production and emissions of methane and nitrogen oxides including nitrous oxide – all of which we see in the atmosphere. Plants emit non-methane hydrocarbons, which together with reactive nitrogen oxides from soil microbial processes can lead to ozone production in the natural troposphere where it acts as a greenhouse gas. Furthermore, this mix can generate fine particles that create a haze often seen over forests. Plants also free up the oxygen we breathe. On the sink side, plants remove carbon dioxide from the atmosphere, both natural and from man-made sources and store it in biomass. The last category of land-based sources is human activities which include emissions from technological, agricultural, and energy sources. In most cases, these surface emissions from human sources generally do not involve the complexities of cross-media processing.

Looking to the oceans we see that they are a source of water vapor and oxygen and an important net sink of carbon dioxide including the amounts emitted from human activities. Moreover, ocean biological activity produces and releases gases such as carbonyl sulfide (OCS) and dimethylsulfide (DMS) which produce particles that scatter sunlight and may thus exercise a modest direct cooling influence on the natural climate.

Ozone is produced entirely in the atmosphere. The fine particles come mostly from chemical processes that convert gases into particles. *But the remarkable role of the atmosphere in the global environment is its action as a sink.* It produces no stable gases, but it can remove them as we will see later in this chapter. Atmospheric chemistry is the main sink of the greenhouse gases such as methane, nitrous oxide, hydrofluorocarbons, chlorofluorocarbons, all gases with carbon–hydrogen bonds that cycle through the atmosphere and virtually every other gas that gets into the atmosphere, except carbon dioxide. This central role of atmospheric chemistry in removing trace gases and generating ozone both in the troposphere and the stratosphere is so important to global climate science that we will devote a large part of this chapter to laying out the relevant and fascinating characteristics.

Ultimately all sinks of atmospheric trace gases are chemical in nature. For the aerosol particles for instance, physics controls transport and settling but the formation and ultimate dissolution are chemical. Particles settle out of the air at time scales that are governed by gravity and air resistance. After reaching the soils, water, or other surfaces, further chemical degradation has to occur to re-cycle the components of the particles before their fate is complete. The presence of trace gases are parts of processes and cycles by which elements are shuttled from one molecule to another, which is the very essence of chemistry. The control of atmospheric composition by chemical processes is represented in Figure 6.1 where a number of mechanisms are introduced that will be discussed next.

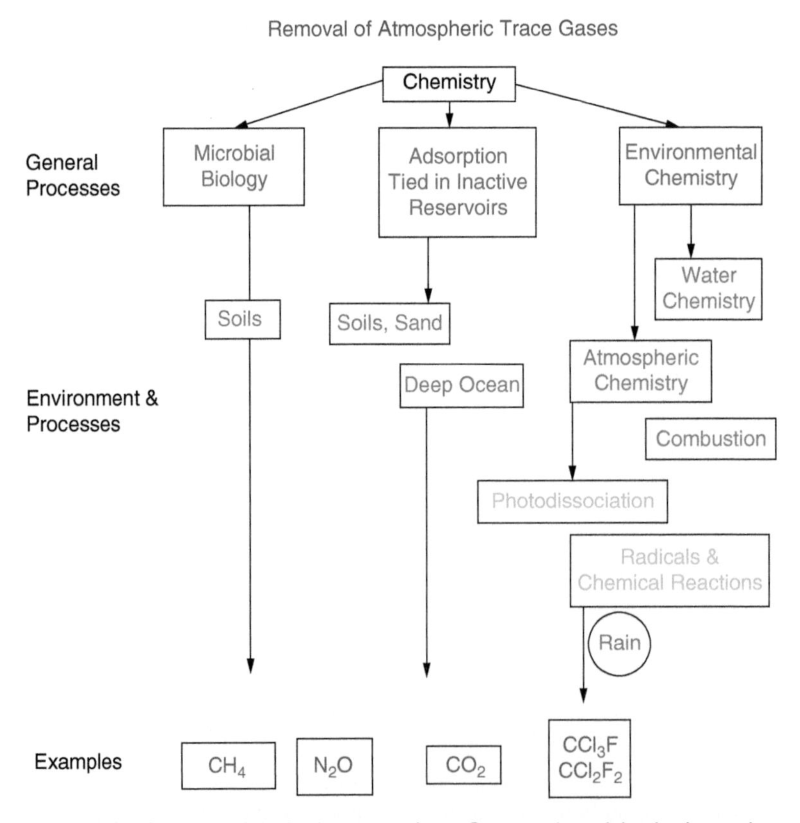

Figure 6.1 Chemical sinks in the atmosphere. Gases and particles in the environment are formed and removed continuously. The sink processes are almost entirely chemical or biochemical in nature. Aspects are shown that are connected to environmentally active gases.

6.2 Atmospheric Chemistry

6.2.1 Chemical Reactions and Photolytic Processes

Let's start by looking at two molecules A and B in a volume *dV* over a time dt during which they interact. As a result, there is a probability *p* that two new molecules C and D are produced conserving the atoms of each element but combining in different ways. The probability that they will stay the same is therefore (1–*p*); maybe they are moving in different directions and speeds than before, but they are the same chemicals. If instead of one B molecule we have two to start with, it will double the probability of making a C and D in the time *dt* because either of the B's is equally effective, and similarly if we have two A's. For N(A) = [A] *dV* molecules of type A, and one B, the probability that a C and D will be produced is *p* [A] *dV*. Here we have adopted the chemistry notation for concentration where [A] is the number of molecules per cm^3 of a gas (A); dV is the volume in cm^3. If we now consider that the volume contains many B molecules as well, N(B) = [B] *dV*, then the number of C molecules that will be produced during time dt will lead to a change of concentration *d*[C] and so, *d*[C] dV = (*p* [A] dV) × ([B] *dV*). Likewise for D. This is analogous to saying that if you flip a coin that has a probability of 0.5 or "50-50" that it will come up tails; then we expect 50 coins to come up tails if we flip 100 at the same time, that is, probability times the number of coins. Readers can convince themselves that the probability *p* itself is inversely proportional to the volume *dV* and directly proportional to *dt*, so we can write it as $p = k\ dt/dV$, where k is the proportionality constant called the reaction rate constant. The probability gets smaller if we make the volume small or shorten the time allowed. It is apparent that the units of k must be cm^3/molecule-s since *p* is the probability of a successful interaction per molecule of B that produces a C and a D. Putting this into the equation for the number of Cs formed and cancelling the *dV*'s we get:

$$\frac{d[C]}{dt} = \frac{d[D]}{dt} = k[A][B] \tag{6.1a}$$

$$\frac{d[A]}{dt} = \frac{d[B]}{dt} = -k[A][B] \tag{6.1b}$$

It shows that each occurrence of the reaction makes both a C and a D and results in the loss of one A and one B. It would be expected that the probability *p* must be a function of the temperature. If all else is the same, with higher temperature the molecules will have more kinetic energy when they collide and that should increase the chance that they will mash together and come out as the products C and D. From the relationship $p = k\ \delta t/\delta V$ this temperature dependence must be manifested in k, because the time and volume are of our own choosing. Both from observations and theoretical considerations, the dependence of k on temperature for most cases of our interest can be written as:

$$k = Ae^{-E/T} \tag{6.2}$$

This is called the *Arrhenius equation*, where A and E are constants, usually determined from laboratory experiments, and T is the absolute temperature. The E here is an activation temperature; however, it is often expressed as an activation energy $E_A = E/R$, where R is the gas constant encountered earlier. The activation energy embodies the idea that molecules must have enough kinetic energy before the reaction takes place. The values of A and E for various reactions of interest in atmospheric chemistry are tabulated in readily available compendia (Endnote 6.2).

The situation we have been discussing can be written symbolically as:

$$A+B\rightarrow C+D \tag{6.3}$$

There are two other possibilities that are subject to the same principles as already discussed but represent different processes. These are:

$$A+B\rightarrow C \tag{6.4}$$

$$A\rightarrow B+C \tag{6.5}$$

The first of these is an *association* and the second is a *dissociation* reaction. In the association reaction a third molecule has to be involved in order to conserve energy, which in the atmosphere, will be mostly nitrogen and oxygen because of their abundance, but the nature of this needed extra molecule is not important for the reaction. In the environment, this reaction is more completely written as: A + B + M → C + M, where M is an air molecule and [M] is the density of air "N" from before. The rate of formation of C in such a case is therefore:

$$\frac{d[C]}{dt}=k[A][B][M] \tag{6.6}$$

Similarly, the rates of destruction of A and B that occur simultaneously are represented by the minus on the right-hand side. It is a straightforward extension of Eq. 6.1. A point to note is that such reactions are favored near the earth's surface where the density of air [M] is large and can become hard to achieve high in the atmosphere.

The dissociation reaction often occurs when a molecule (A) absorbs a ray of sunlight, becomes unstable, and breaks apart in a process called photo-dissociation. In that case it is more completely written as A + hν → B + C indicating that a ray of light is needed. This leads us to an important aspect of atmospheric chemistry that involves sunlight.

Light can be represented by the quantum mechanical idea that it acts like a particle under appropriate circumstances, which includes colliding with atoms and molecules that exist in the realm of very small objects. A ball of mass m has kinetic energy $\frac{1}{2}mv^2$, some of which it can impart to another similar object upon collision. Light rays are electric and magnetic fields oscillating as sine waves and therefore have a frequency ν and a wavelength λ which are related by the speed of light c as $\nu\lambda = c$, as explained in Figure 6.2. When light waves act like particles, called photons, their energy is $E = h\nu$. Here h is Planck's constant, which, like the speed of light c, is a fundamental constant of nature. Some or all of the photon energy can be transferred to molecules upon collision. A ball can be given more kinetic energy ($= \frac{1}{2}mv^2$) by making it move faster. A photon cannot move faster or slower than the speed of light; however, since its energy is hν, more energy can only come about due to a higher frequency ν. That means it has a shorter wavelength if it contains more energy because $\nu\lambda = c$. Short wave radiation from the sun is like a bullet and can cause sun burn while the long-wave radiation from the earth is felt as warmth

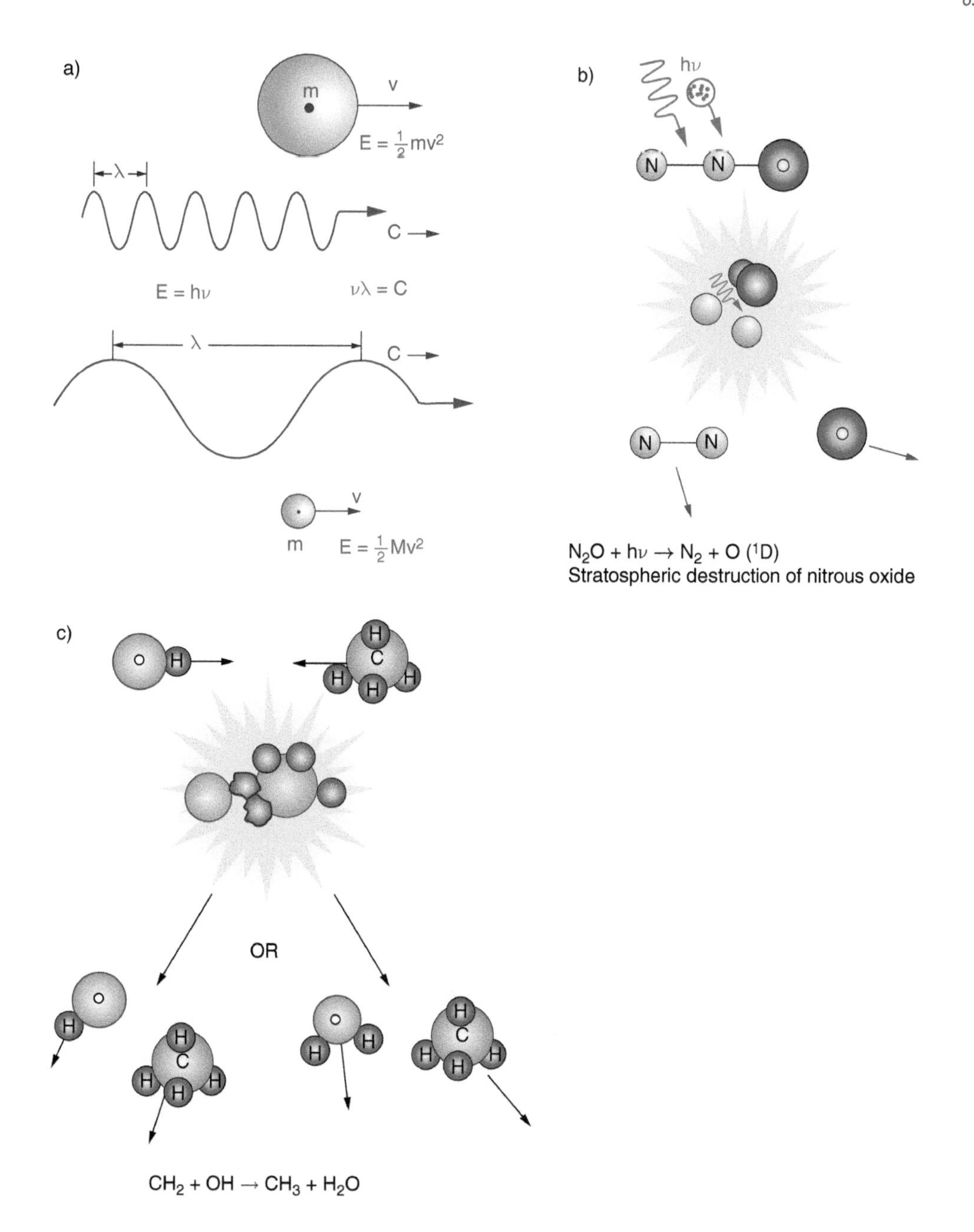

Figure 6.2 Photons, particles, and molecules and their interactions. (A) Light can be represented as a particle that can hit a molecule and break it apart just the same as another molecule can. But light is also a wave. These dual characteristics of light are shown and connected here. (B) Light hits a molecule and breaks it apart. (C) Radicals interact with atmospheric gases.

(Figure 6.3). There are several possible outcomes when a photon from the sun hits a molecule in the atmosphere. If the energy is high enough it can break the bond and result is Eq. 6.5.

Here are three important examples of these types of reactions in the atmosphere that we will use again: Reaction (Eq. 6.3): $CH_4 + OH \rightarrow CH_3 + H_2O$, which occurs everywhere but

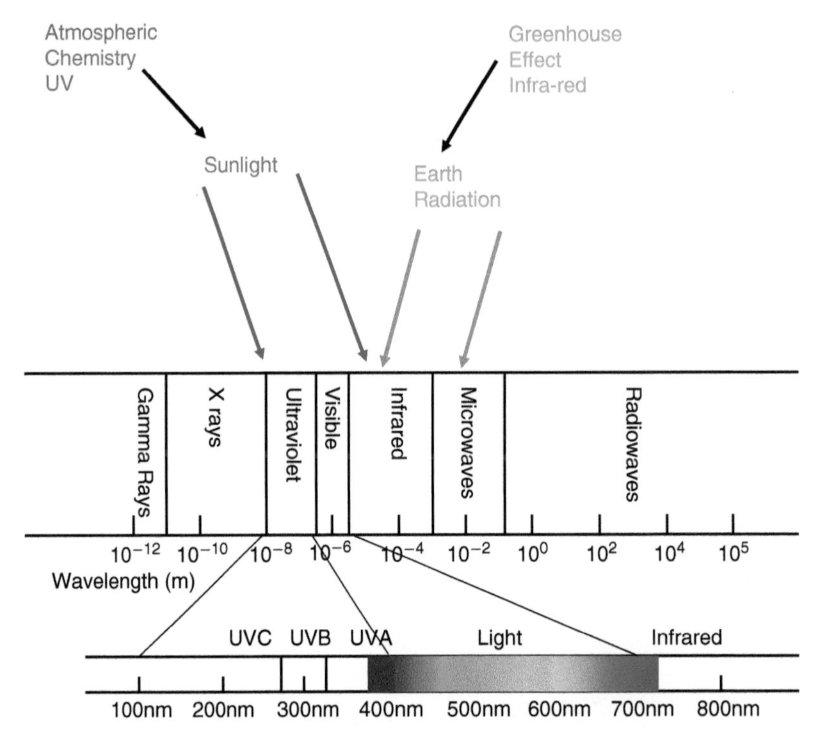

Figure 6.3 Electromagnetic radiation spectrum. About half of the solar energy is in the visible part of the spectrum with small amounts in the ultraviolet. The natural energy from the sun and earth is confined between the ultraviolet and infrared zones. Atmospheric chemistry is affected by the high-energy ultraviolet and violet radiation, whereas the greenhouse effect is mostly affected by infrared radiation. Outside these regions of radiation, we make and use micro-wave, radio-wave, and X-ray frequencies in everyday life.

is most influential in the troposphere. It constitutes the major mechanism for removing methane from the atmosphere. Association (Eq. 6.4): $O + O_2 + M \rightarrow O_3 + M$, which is the main mechanism for making ozone in both the troposphere and the stratosphere where it generates the ozone layer. Notice that the M appears on both sides of this reaction meaning that it is not involved in the chemistry which changes the nature of the molecules, but rather is required by the physics to conserve energy. But if M is not there, no reaction will take place. Photo-dissociation (Eq. 6.5): $N_2O + h\nu \rightarrow N_2 + O(^1D)$ which occurs in the stratosphere and represents the most important sink of N_2O, a greenhouse gas.

6.2.2 Chemical Lifetimes

When reactions represented by Eq. 6.3 occur, they are sinks of both the molecules A and B, and sources of C and D. If we choose to represent a stable atmospheric trace gas by A, then B is usually a short-lived reactive radical gas which serves as its sink. In such a case we can write the rate of chemical loss of A from Eq. 6.1b as $d[A]/dt = -(k[B])\,[A]$. Comparing this equation to our definition in Eq. 4.3 we can conclude that the lifetime of A due to its reaction with B must be:

$$\tau_B = \frac{1}{k[B]} \tag{6.7}$$

If we look at Eq. 6.7 as an average over times a day or longer, it will be a constant unless we fundamentally change the environment that affects the formation and destruction of B. This is a remarkable equation because it quantitatively connects the lifetime of a gas with a specific atmospheric chemical process. It also shows us that in this prevalent circumstance the lifetime gives a meaning to the abstract loss rate of Eqs. 4.2a and 4.3 (sink = C/τ) and supports the idea that the sink is proportional to the concentration of the gas being lost. It should be noted that the lifetime of A by reaction with B, τ_B, depends on two factors – the reaction rate constant k and the amount of the reactant B that is available. It often happens that radicals represented by B react rapidly with many atmospheric constituents and are therefore present in small quantities. So even though the reaction rate may be fast, the overall lifetime of a gas can still be long.

Readers are reminded that there is still a difference between the global average lifetime τ in Eq. 4.2 and the τ_B in Eq. 6.7. The global lifetime balances the concentration or burden of the gas in the atmosphere with its global sources and therefore represents the total lifetime due to all processes as in Eq. 4.7, whereas, the chemical lifetime in Eq. 6.7 is just one of the possibly many sinks; and there is also the matter of how transport processes can affect the global lifetime. If one sink is so dominant that the impact of all others together is less than 10%, and the lifetime is much longer than transport times, then we can approximate the global lifetime τ as $\approx \tau_B$, otherwise not. We have already seen that this is virtually impossible for the chlorofluorocarbons and similar gases (Eq. 4.16).

Example Calculate the lifetime of methane in years due to its reaction with hydroxyl radicals which have an average concentration in the atmosphere of 10^6 molecules/cm^3.

Solution: From the discussion earlier: $CH_4 + OH \rightarrow CH_3 + H_2O$. $k = 2.3 \times 10^{-12}\ e^{-1700/T}$ cm^3/molecule-s.

Step 1. At the average temperature of the atmosphere of 260°K the rate constant $k = 3.33 \times 10^{15}$ cm^3/molecule-s. Step 2. From Eq. 6.6, $\tau_{CH4/OH} = 1/k[OH] = 1/(3.33 \times 10^{-15}$ cm^3/molecule-s $\times\ 10^6$ molecules/cm^3) $= 3 \times 108$ s. Step 3. Convert to years: 3×108 s/(3600 s/h $\times$ 24 h/d $\times$ 365 d/y) says $\tau_{CH4/OH} = 9.5$ y.

There are two ideas that can be represented quantitatively at this point that will be useful when we apply atmospheric chemistry to key aspects of global change science. The first is to extend the concept of reaction rates to the case when photons are breaking molecules apart as in Eq. 6.5: $A + h\nu \rightarrow B + C$ where $h\nu$ represents the energy of a photon. Now:

$$\frac{d[A]}{dt} = -J[A] \tag{6.8a}$$

$$\tau = 1/J \tag{6.8b}$$

The interpretation of J is particularly simple as the inverse of the lifetime due to photodissociation, but figuring out the values is quite difficult. Since a bond is being broken in the reaction, only photons with energies high enough to do this would be effective. Additionally, there is an efficiency (j) with which a single photon can cause this reaction that is analogous to the rate constant k for molecular reactions. We can write J (1/seconds) = j (cm^3/photon-s) [$N_{photons}$] (photons/cm^3) to make it look similar to the usual reactions between molecules (Eq. 6.3) and note that the number of available photons [$N_{photons}$], called actinic flux, is more than the direct sunlight provides. This is because some of the light that is reflected or scattered from clouds, particles, air molecules or the surface features of the earth causes additional qualified photons to arrive at the location of interest. These factors are blended in calculating J. The values are available from existing resources (Endnote 6.2).

We have been discussing long-lived gases so far; however, atmospheric chemistry works on these gases through the creation and destruction of short-lived molecules and fragments that are behind the sinks. These short-lived molecules are also subject to the same mass balance theory we have discussed in Eqs. 4.1 and 4.2 and subsequent equations.

Before moving forward to see how this works in the real world, we need to derive a formula that will tell us how much of a short-lived gas we can expect in the atmosphere if it is produced and destroyed only by chemical reactions discussed earlier. Let's start with two reactions: one that creates the short-lived gas of interest, call it C, and one that takes it out. We are not interested in the other products of these reactions at this time, so:

$$k_s: \quad A + B \rightarrow C + products$$
$$k_L: \quad C + D \rightarrow products$$

Here, k_S is the reaction rate constant for the "source" and k_L is for the "loss" or sink. From our previous discussion:

$$S = k_s[A][B] \quad \text{and} \quad \tau = \frac{1}{k_L[D]}$$

Therefore, applying Eq. 4.4, $C = S\,\tau$, we get:

$$[C] = \frac{k_s[A][B]}{k_L[D]} \tag{6.9}$$

Recall that for short-lived gases $C = S\,\tau$ is a very good approximation even though the S and τ may be changing from one place to another or one time to another as discussed earlier in Endnote 4.2.

6.3 Global Environmental Applications

We are now ready to move on to two of the most profound manifestations of atmospheric chemistry in our natural world and its environment. The first is the making of hydroxyl radicals resulting in an oxidizing atmosphere that cycles many gases by creating sinks to maintain the atmospheric environment as we know it. Radicals are fragments of molecules that have a high affinity to react with many available compounds so as to return to completion. The second phenomenon is the creation of a prevailing ozone layer in the stratosphere that is believed to be necessary for life on the surface as we know it. We will see shortly that both these phenomena arise from the continual activity of short-lived compounds driven by sunlight and understandable by the processes that led to Eq. 6.9.

6.3.1 Hydroxyl Radicals – A Key Oxidant for the Global Environment

In the troposphere, reactions with hydroxyl radicals are the principal sink for hundreds of gases as can be seen in NASA's compendium of reaction rate constants (Endnote 6.2).

The presence of hydroxyl radicals rapidly takes out many complex molecules from the atmosphere leaving behind the stable trace gas composition we saw in Table 3.1. In the natural environment, plants emit a large number of organic compounds and non-methane hydrocarbons that are recycled by hydroxyl; and they are a principal sink of methane, one of the greenhouse gases of interest in global warming. Similarly human activities in urban and industrial areas emit organic compounds that are removed by hydroxyl. The man-made chlorofluorocarbons and many halogen containing technological gases do not react with hydroxyl so they accumulated in the atmosphere and threatened the ozone layer. One solution was replacement with compounds that can be removed by hydroxyl so they don't reach the stratosphere. The hydroxyl radicals take the hydrocarbons, carbon monoxide, and some other gases ultimately to carbon dioxide completing a loop in the carbon cycle and thus connect with the climate, greenhouse effect and global warming. Furthermore, hydroxyl starts a chain of chemical reactions that convert sulfur gases into fine particles which are involved in forming clouds and affect their albedos. In this way too they influence the earth's climate.

Here are the basic chemical reactions that make hydroxyl radicals in the atmosphere, OH to you and me or HO if you are a chemist:

$$O_3 + h\nu \xrightarrow{J} O\left(^1D\right) + O_2 \qquad J: \lambda < 310 \text{ nm} \tag{6.10a}$$

$$O\left(^1D\right) + M \xrightarrow{k} O + M \qquad k = k_1[N_2]/[M] + k_2[O_2]/[M] \tag{6.10b}$$

$$O\left(^1D\right) + H_2O \xrightarrow{kp} OH + OH \qquad k_p = 2.2\,x 10^{-10} \text{cm}^3/\text{molec-s} \tag{6.10c}$$

So, the source S from Eq. 6.10c is:

$$S_{OH} = 2k_p[H_2O][O(^1D)] \tag{6.11}$$

The $[O(^1D)]$ using Eq. 6.9 is:

$$[O(^1D)] = \frac{J[O_3]}{k[M] + k_p[H_2O]} \tag{6.12}$$

Substituting Eq. 6.12 into Eq. 6.11 gives the source of OH as:

$$S_{OH} = \frac{2k_p J[H_2O][O_3]}{k[M] + k_p[H_2O]} \tag{6.13}$$

Let's see what is happening in these equations. First an existing ozone molecule is dissociated by a photon that has wavelengths less than 310 nm (Eq. 6.5-type process). This requires the photon to have high energies in the ultraviolet range. It forms an excited oxygen atom called O singlet D or $O(^1D)$, which means that the outer electron in this atom is at a level higher than its lowest allowed quantum mechanical circuit. Excited atoms react faster than normal ones. The O singlet D's created by this process can be quenched to a normal state of ordinary O atoms, that is, they can transfer some energy to the molecules in the surrounding air (M, mostly nitrogen and oxygen). Most O singlet D go this way because there are a lot of available molecules M, but some react with water vapor (H_2O) to form two hydroxyl radicals (Eq. 6.10c). You need the excited versions of the oxygen atoms to split water into these two components; ordinary oxygen atoms won't do. Then we have taken the results from the previous discussion about how chemical reactions generate a source or a sink and applied it to figure out the source of OH as in Eq. 6.11. An extra step had to be undertaken to eliminate the $O(^1D)$ from the original source Eq. (6.11) because we want to represent the production process in terms of the stable ingredients that generate hydroxyl, namely, sunlight, ozone, and water vapor, and not an ephemeral intermediary such as $O(^1D)$. Without these precursors OH will not exist, but if they are present, then the $O(^1D)$ will always be there as an intermediate. Moreover, the precursor ingredients are readily measurable while the $O(^1D)$ is not. The factor of two arises because for each reaction 6.10 c, two hydroxyls are produced.

Here are the main reactions that remove OH from the atmosphere:

$$\mathrm{OH + CO \rightarrow CO_2 + H} \qquad k_{OH/CO} = 1.5\,x10^{-13}(1 + 0.6\ \mathrm{P\ atm}) \tag{6.12a}$$

$$\mathrm{OH + CH_4 \rightarrow CH_3 + H_2O} \qquad k_{OH/CH4} = 2.45\,x10^{-12} e^{-1775/T} \tag{6.12b}$$

So, the lifetime is:

$$\tau_{OH} = \frac{1}{k_{OH/CO}[CO] + k_{OH/CH_4}[CH_4]} \tag{6.13}$$

The sinks are normal reactions with methane and carbon monoxide and the lifetime is the application of the general Eq. 6.7. It was mentioned earlier that OH radicals react with hundreds of gases, but we don't include them in the sink Eqs. 6.12 because their combined effect on the lifetime of OH is very small and presumed to be within the 10% rule, especially when taken over the prevailing large portions of the non-urban atmosphere. According to Eq. 6.13 the average lifetime of OH is about one second! Based on these five reactions we can write the atmospheric concentration of OH as $C = S\,\tau$ or:

$$[OH] = \frac{2k_p J[H_2O][O_3]}{\left(k_{CO}[CO] + k_{CH_4}[CH_4]\right)\left(k_1[N_2] + k_2[O_2] + k_p[H_2O]\right)} \tag{6.14}$$

This daunting equation is still a simplified representation of the real world, but it produces results accurate enough for our interests. Average values of the precursors are shown in Figure 6.4 which are put into Eq. 6.14 to produce the resultant OH concentrations shown in Figure 6.5.

Nearly all the variables in this equation change with latitude and seasons. Averaged over a year, the major change with latitude is in water vapor (H_2O) and sunlight (J), both are in the numerator and both are much higher in the tropics than other latitudes leading to a commensurately higher OH in the tropics and therefore a much greater sink of trace gases. A noteworthy role of OH is that it cycles all hydrocarbons back to atmospheric CO_2 and in the process, hydroxyl generate about half the carbon monoxide seen in the atmosphere as an intermediary.

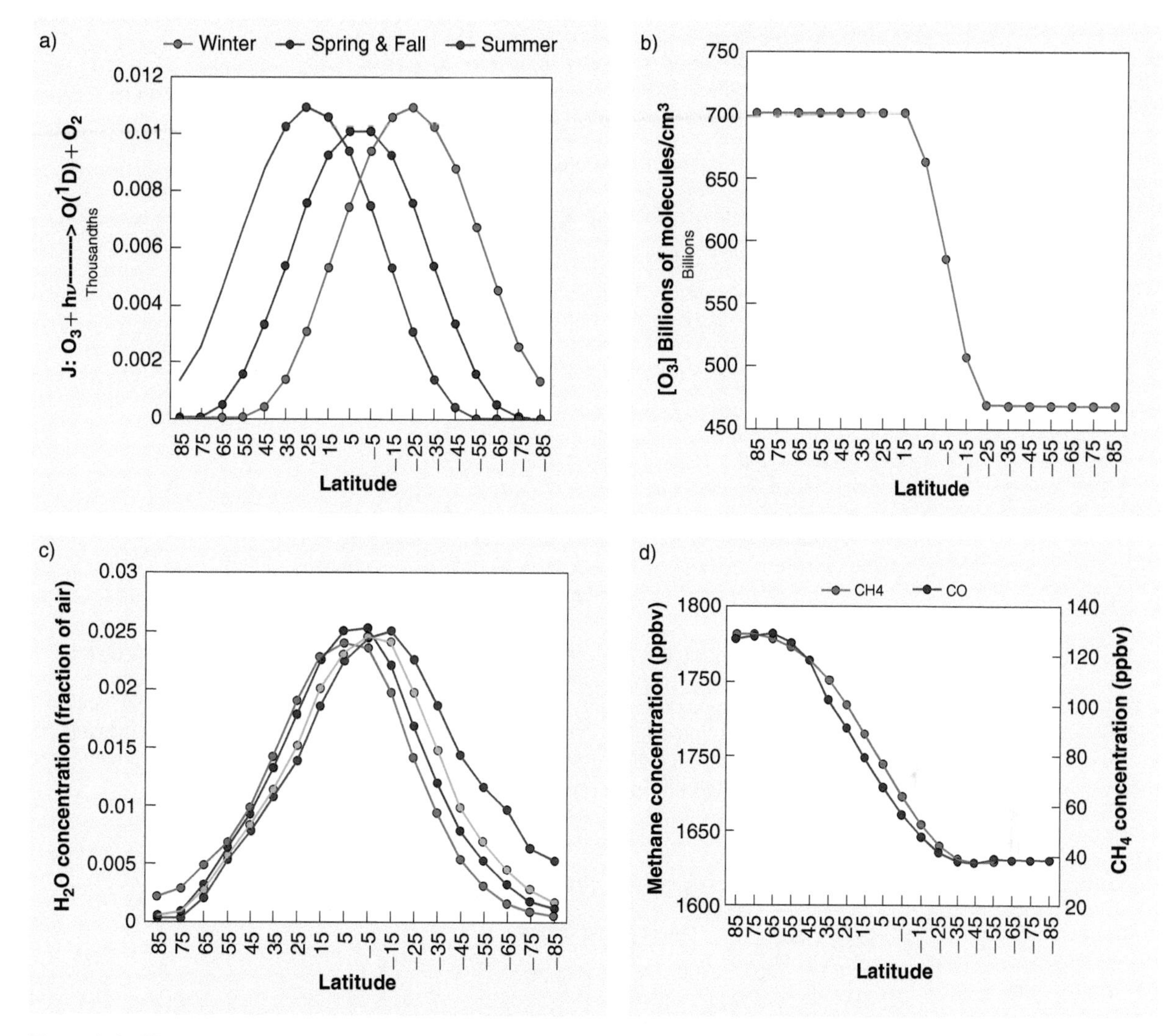

Figure 6.4 The precursors of the hydroxyl radical. (A) The action of sunlight on OH production – J for ozone photo-dissociation. (B) Water vapor concentrations. (C) Ozone. (D) The sinks of OH, methane, and carbon monoxide.

There is some disparity of concentration between the hemispheres because there is much less carbon monoxide in the southern hemisphere creating a smaller sink that would lead to higher OH, but this is compensated by the lower ozone concentrations which lead to a reduced production. According to the calculations based on Eq. 6.14, if we take an average over the latitudes and seasons the result is that there are approximately 10^6 molecules per cubic centimeter in the atmosphere. There are also variations of OH with altitude, cloudiness, and the presence of other chemicals such as nitrogen oxide and hydrogen interactions, and complexities added by cycling and regeneration processes, but further discussions about this fascinating subject will take us too far from our story about global change (Endnote 6.3).

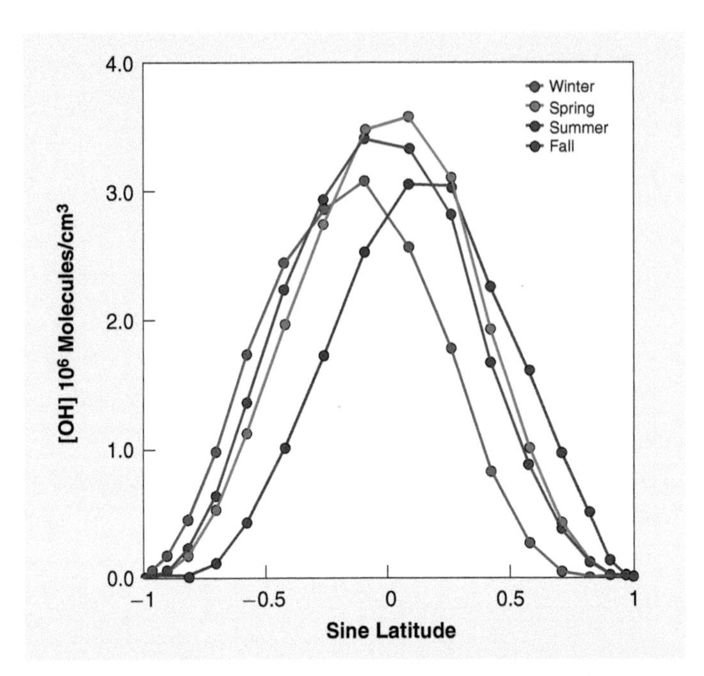

Figure 6.5 Tropospheric hydroxyl concentrations. Hydroxyl radicals are short-lived molecular fragments that exist in the atmosphere maintained by photochemical processes and readily available chemical precursors (sunlight, ozone, and water vapor). They are a major sink mechanism in the atmosphere and remove many atmospheric gases emitted by natural and human processes. The concentrations are much higher in the tropics because water vapor and sunlight are plentiful there compared with other locations. Seasons are for the northern hemisphere.

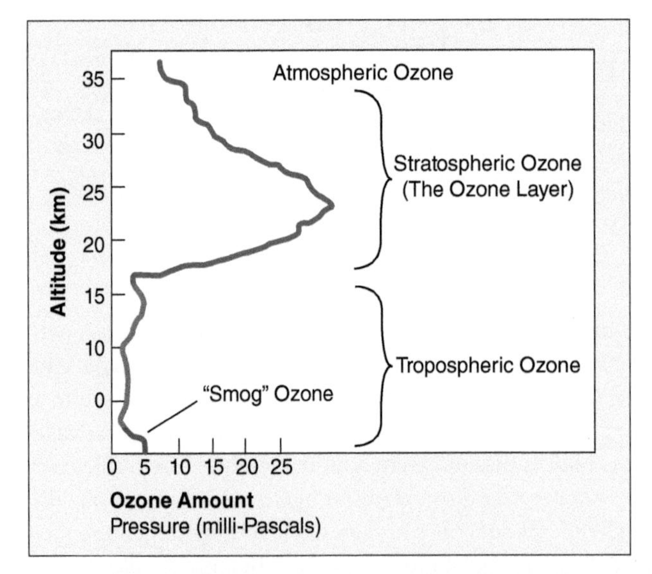

Figure 6.6 The stratospheric ozone layer. A layer of ozone is a permanent and most significant feature of the stratosphere. Ozone here defines the stratosphere by causing it to be warmer than the air below and separating it from the troposphere by creating a thermodynamic barrier to the free movement of air. It removes ultraviolet radiation which is harmful to living things on the surface (Endnote 6.4).

6.3.2 The Stratospheric Ozone Layer and Its Depletion

Ozone is generated by chemical processes in the atmosphere, both on a global scale and in urban pollution. It plays important roles in the environment and affects living things, including us. In the stratosphere, ozone concentrations reach a peak of about 5000–8000 ppb at 20–30 km (Figure 6.6). Here ozone takes out ultraviolet radiation of wavelengths less than 300 nm. Such radiation is considered harmful to humans and other life if it reaches the surface. It is thought that without the ozone layer, evolution of complex life on land may not have been possible because the ultraviolet radiation from the sun would cause mutations and damage early forms of life that are more fragile than the behemoths that exist now.

The ozone layer causes the temperature of the stratosphere to be warmer than the air below and thus explains the barrier to transport that we discussed earlier. Furthermore, it supplies some of the ozone found in the troposphere by air exchanges. As a contrast, in urban areas, ozone can build up to levels reaching 200 ppb and is considered a serious threat to human health. Standards are set at 70–80 ppb as maximum allowable concentrations averaged over 8 h periods. Since it is formed by chemical reactions involving sunlight, nitrogen oxides, and hydrocarbons, regulations have been imposed on the precursors so as to limit ozone concentrations in cities.

From the decades of the 1970s to recent times, it was recognized that human activities were depleting the ozone layer primarily from the emissions of chlorofluorocarbons and nitrous oxide. It has been known for some time that exposure to ultraviolet radiation leads to a greater incidence of skin cancer, formation of cataracts and possibly other health effects such as a compromised immune system. Given the important role of the ozone layer, it is understandable that its depletion would lead to a societal concern and eventual action to protect it.

Our interest here is to understand both the natural occurrence of the ozone layer and the impact of human activities. It is one of the permanent features of the global environment that affects human habitability in many ways and the climate in indirect ways.

Here are the Chapman reactions from 1930 AD that were the first chemical theory to explain the concentration of stratospheric ozone:

Production:

$$O_2 + h\nu \xrightarrow{J2} O + O \tag{6.15a}$$

$$O + O_2 + M \rightarrow O_3 + M \qquad k_{O/O2} = 6\times10^{-34}\left(T/300\right)^{-2.3} \tag{6.15b}$$

Destruction:

$$O + O_3 \rightarrow O_2 + O_2 \qquad k_{O/O3} = 8\times10^{-12}e^{-2060/T} \tag{6.15c}$$

$$O_3 + h\nu \rightarrow O + O_2 \tag{6.15d}$$

The last reaction (Eq. 6.15d) is of the most environmental importance because it represents the mechanism by which ozone absorbs solar ultraviolet radiation and protects life on the surface from its harmful effects. This absorption of solar radiation is also one reason the stratosphere is warmer than the tropospheric air below, the other being absorption of the earth's radiation that we will consider in the next chapter.

We can apply Eq. 6.9 to the Chapman reactions, assuming a pseudo-steady state so that $C = S\tau$ and write out the equations for $[O_3]$ just as we did for [OH]. But it is hardly worth the effort because you will find that the ozone concentrations calculated from these are about six times higher than the actual measured concentrations and the calculated peak itself is located higher than the altitude where it is seen. The resolution of this disconcerting situation makes it difficult to determine ozone concentrations accurately without including more complex chemistry and transport processes similar to our discussion in Section 5.2. We can, however, understand how the ozone mass balance works theoretically.

The first discoveries to fix this issue were that the destruction of ozone is greatly accelerated by catalytic cycles; in particular, Eq. 6.15c is sped up (Endnote 6.4). Here is the main cycle:

$$\begin{aligned}&\text{Type I}:\\ &X + O_3 \rightarrow XO + O_2\\ &XO + O \rightarrow X + O_2\\ &\text{Net}: O + O_3 \rightarrow 2\,O_2\end{aligned} \tag{6.16}$$

Here X is a chemical radical, which in this situation is likely to be one of NO, Cl, OH, Br, or H. What happens is that this radical first interacts with ozone and generates an intermediary XO and an oxygen molecule. The XO then reacts with an O making two oxygen molecules and recovering the original X. In the completion of one such cycle we see that an O atom and an ozone molecule are destroyed and replaced by two stable oxygen molecules. At the same time the X is free again to repeat this cycle. It keeps doing this, until it reacts with something else that is also present in the stratosphere and gets tied up into a more stable molecule. The Type I catalytic cycle is particularly effective, because it takes out an O atom that is needed to make ozone by Eq. 6.15b. So even if the catalyst X is not very effective at removing O_3, the XO is capable of taking out the O, preventing ozone from being formed in the first place and resulting in a decrease in ozone concentration. Depending on the catalyst, one X can take out tens of thousands of ozone molecules before it is terminated!

Another cycle, Type II, is equivalent to a reaction between two ozone molecules resulting in the creation of stable oxygen.

$$\begin{aligned}&\text{Type II}:\\ &X + O_3 \rightarrow XO + O_2\\ &XO + O_3 \rightarrow X + 2\,O_2\\ &\text{Net}: O_3 + O_3 \rightarrow 3\,O_2\end{aligned} \tag{6.17}$$

This type of reaction is an additional loss for ozone, not included in the original equation set 6.15. The general situation of Type II cycles can include more than the two reactions shown, but result in the same consequence, that is, $2\,O_3 \rightarrow 3\,O_2$ (Endnote 6.4). The catalysts X are as before, although each has its own effectiveness and some may be negligible.

Type I cycles are most effective at altitudes above about 30 km, and Type II are more effective below this altitude. The reason is that [O] increases as we get higher up.

We can see this from Eq. 6.15 for the production of O which is by solar ultraviolet. The amount of ultraviolet radiation is higher at the top of the atmosphere and decreases as we head toward the tropopause because it is continuously being used up along the way and absorbed by the ozone that builds up to a peak at 25–30 km. Below this level, the production of O will decrease commensurately. Moreover, the loss of O by 6.15b depends on the density of air [M] which decreases as we go higher. Both these effects cause [O] to be much more abundant at altitudes higher than ~ 30 km than below. This allows Type I processes to be more effective higher up because they depend on removing O atoms. Below this level Type I processes get limited by the availability of O atoms and Type II processes become more effective because they do not depend on it.

The catalytic cycles are terminated because X and XO react with molecules outside the cycles in Eqs. 6.16 and 6.17, usually forming a more stable molecule that gets transported to the troposphere by the air exchange processes discussed earlier. It eventually reaches its terminus at the surface, carried by rain or other mechanisms. The effectiveness of catalytic destruction of ozone allows small emissions, that produce the catalysts, to have big impacts demonstrating once again the reason why human activities can cause a significant global change; in this case a depletion of the ozone layer.

The matter of the lifetime of ozone in the stratosphere requires further discussion because on the surface it appears to be quite short at ~1 h from Eq. 6.15d near the ozone peak. This reaction, however, doesn't completely terminate the ozone because it produces an O atom at the same time. This O atom goes out by one of two ways – it either takes out ozone (Eqs. 6.15c and 6.16), or it re-generates ozone (Eq. 6.15b). Except for high in the stratosphere, the regeneration is much faster than the termination because there is a lot of O_2 present to make more ozone (Eq. 6.15d). This situation occurs when we have chemical cycles that couple production and loss processes. In an air mass that is moving or being mixed with other air, the ozone would persist through this constant regeneration. As far as the transport is concerned the lifetime that determines how far ozone will get is from processes that take it to O_2 and not the much faster instantaneous loss determined by action of sunlight in Eq. 6.15d. For processes that take ozone to O_2, the occurrence of the catalytic cycles makes this lifetime much shorter than would be calculated from the Chapman reactions 6.15 and turns out to be about a year near the tropopause, a few months in the lower stratosphere and a few days above the ozone peak. The consequence is that much more ozone gets transported to higher altitudes than its instantaneous lifetime of about 1 h would suggest due to the regeneration process. Because of this transport, ozone can go from below to higher up where it is destroyed faster further reducing the overall concentrations and shifting where the peak occurs. The coupling of chemistry and transport that we discussed earlier plays an important role here.

Where do the catalysts come from? The naturally occurring catalysts include NO from nitrous oxide (N_2O), OH and H (HOx) from chemical production with water vapor (H_2O) as a precursor and the chlorine (Cl) comes from methylchloride (CH_3Cl). The natural origins of N_2O and CH_3Cl are from biological activity in the soils and oceans respectively and released to the atmosphere by cross-media transport processes that will be discussed shortly. The chlorine and bromine comes directly from the photo-dissociation reactions (Eq. 6.5) in which the molecule converts by: $XCl + h\nu \rightarrow X + Cl$ and likewise for Br. The fragments are subjected to further dissociation generally requiring more energetic sunlight, until the molecules are completely broken apart and re-cycled. The production of NO from N_2O comes from a couple of reactions: $N_2O + h\nu \rightarrow N_2 + O(^1D)$ and $N_2O + O(^1D) \rightarrow$

2 NO. Some of the product in the latter equation goes as $N_2O + O(^1D) \rightarrow N_2 + O_2$ and is thus taken out of the ozone processes. Nonetheless, this $O(^1D)$ reaction with N_2O is a major source of NO in the stratosphere and underscores the significant catalytic effect of nitrous oxide on the ozone layer. Calculations show that the effect on natural ozone sinks is from NOx = ($NO + NO_2$) ~ 40%, while the Ox, Clx, and HOx have similar contributions (~20% each) (Dessler in Endnote 6.4). We can say that the observed characteristics of the ozone layer can be understood by combining basic Chapman chemistry with the natural catalytic cycles and transport processes. And the depletion of the ozone layer is explained by emissions of catalysts from human activities.

Human activities added significant amounts of NO, Cl, and Br that would lead to a depletion of the ozone layer. The use and release of chlorofluorocarbons, CCl_3F (CFC-11) and CCl_2F_2 (CFC-12) in particular, provided a large amount of chlorine to the stratosphere because they are hardly removed at all in tropospheric or surface processes. These CFCs were said to be "miracle compounds" and used widely in refrigeration, air conditioning, foam insulation, and many other consumer goods. A supply of Br came from the halons, CF_2BrCl and CF_3Br, used in fire extinguishers and CH_3Br used as an agricultural fumigant. The addition of all these catalysts, mostly Cl from human activities led to an observed globally averaged depletion of the ozone layer by about 4% at its maximum. The depletion of the ozone layer triggered an international agreement, the Montreal Protocol, which together with its amendments, led to phased ban on the use of these compounds and encouraged the production of alternatives. The alternatives either do not contain chlorine or bromine, or react with OH in the troposphere so that not much gets into the stratosphere. Nitrous oxide is not controlled and continues to increase; the ozone layer, however, is seen to be returning to higher levels (Endnote 6.4).

Two interactions between global warming and the ozone layer would interest readers. One is that methane, which is a greenhouse gas of concern, can protect the ozone layer to some extent from destruction by chlorine atoms, which were the largest contributors to its depletion. One of the most effective processes to terminate the chlorine catalytic cycle is its reaction with methane: $Cl + CH_4 \rightarrow CH_3 + HCl$. The HCl continuously moves to the troposphere, but some reacts with OH: $HCl + OH \rightarrow Cl + H_2O$, reforming Cl and reducing the effectiveness of methane somewhat. The second effect is that as the earth warms from a buildup of greenhouse gases, stratospheric ozone is increased! This is because global warming causes the stratosphere to cool, a phenomenon we will return to later. In looking at the stratospheric ozone reactions, it is apparent that the cooling of the stratosphere will slow the destruction processes and it happens that the production process in 6.15b is speeded up by cooling. Without global warming, the ozone would have been depleted more due to the man-made chlorofluorocarbons and nitrous oxide.

6.4 Cross-Media Transport: Oceans, Soils, and Biota

The machinery that manufactures stable gases in the earth's atmosphere lies outside it, in the land and oceans and connects with it across interfaces (Table 6.1). Cross-media transport looks at the complex states within the land, oceanic, and atmospheric systems to explain the natural composition of the earth's atmosphere, how it evolved, how it can change, and how that may affect global habitability. When the earth's environmental system is in a stable and steady state, the effect of cross-media transport can be interpreted as sources and sinks into the atmosphere that are more or less the same from year to year and represented by Eq. 4.1 or 4.3. If the state of the atmosphere is disturbed, as, for instance, by global warming, the disturbance will reverberate through the different media which respond in complex ways that can profoundly affect the sources and sinks of the gases that come from them.

The importance of cross-media transport process for climate is because exchanges between the atmosphere and the oceans are a major sink of carbon dioxide, as are exchanges with plants and vegetated land. A major source of nitrous oxide is from dry soils and the largest source of methane is from wet soils, often mediated by plants. This is not to imply that other gases have no major sources and sinks in the land and oceans; they may, but presently, their role in climate change is small and so we will not consider them further.

In cross-media transport, material is being moved across hundreds of meters from either side of the interface to be exchanged. This transport has to be fairly fast and should mix material in the bulk of the reservoirs. The interface then represents the barrier where most of the transport time is determined. While this is known to be the case for ocean-air exchanges, it is also likely to be so for all other cross-media transport processes such as exchanges of gases between the soils or plants and the atmosphere. The understanding of the interface transfer processes then becomes the gist of cross-media transport. At the cross-media interfaces molecular diffusion processes come into play. These are generally very slow and the impedance they create is only over a small part of the path through which the gases are moving from one medium to another. But it is enough to give transport times of days to years between media. This general idea is similar to our previous discussions about the 2-box atmospheric models. In them, the impedances to transport between the hemispheres, or the troposphere and stratosphere, were at the interfaces (equator and tropopause), while in the rest of the atmosphere, fast mean-transport prevailed, moving and mixing the gases rapidly. In the case of the atmosphere, the barriers are not across media, but the concept is the same.

6.4.1 Ocean-Air Exchange

The most common application of the cross-media theory in environmental science is the exchange of gases between the atmosphere and the oceans, but it works just the same for lakes, rivers, and fish aquariums. Our system is now represented by two regions, which we will take to be boxes for simplicity. We will extend Eq. 4.1 to include the oceans:

$$\frac{dC_A}{dt} = S_A - \frac{C_A}{\tau_A} - \frac{(C_A - H_O C_O)}{\tau_{A \to O}} \tag{6.18a}$$

$$\frac{dC_O}{dt} = S_O - \frac{C_O}{\tau_O} + \frac{(C_A - H_O C_O)}{\tau_{O \to A}} \tag{6.18b}$$

We see now that even with a 1-box model of the atmosphere we must include transport, making the situation more complex. Note that we still have T_{IN} and T_{OUT} in each box represented by the two terms $H_O C_O/\tau_T$ and C_A/τ_T. We are adopting the convention that if T_{NET} is negative, then net transport is from the atmosphere to the ocean as would happen if the atmosphere has a new source and the oceans acted as a sink. The work left is to figure out the transport times $\tau_{O->A}$ and $\tau_{A->O}$. The net transport from which the Eqs. 6.18 are derived can be written as:

$$T_{NET\,A\leftrightarrow O}(molecules/s) =$$
$$-(C_A - H_O C_O)K_v(cm/s)A_O(cm^2)N_0(molecules/cm^3) \tag{6.19}$$

This equation is based on the idea that the transfer is driven by a difference (or gradient) of the concentrations in the two media and the rapidity of the transfer is represented by the K_v. Since we are taking the *C*s to be mixing ratios, we have multiplied by N_0, the density of air at the surface to convert to molecules/cm^3 because the transport is in terms of molecules being transferred per second. A_O is the area of the oceans, or, more generally, the interface. The transport of molecules in Eq. 6.19 is translated into a change of mixing ratio by dividing it by the number of molecules (of air or equivalent) in each box which gives (Endnote 6.5):

$$\tau_{A\rightarrow O} = (A_E/A_O)H/K_v \tag{6.20a}$$

$$\tau_{O\rightarrow A} = D/K_v \tag{6.20b}$$

Here A_E = the area of the earth, since the atmosphere covers all of it, *H* is the scale height, and *D* = the depth in the ocean to which the gas is mixing; it could be the mixed layer and up to the entire ocean. For typical conditions, $\tau_{A\rightarrow O}$ ~6–8 years and $\tau_{O\rightarrow A}$ ~20 days for the mixed layer. Note that the ratio of transport times is an expression of the different sizes of the reservoirs, $\tau_{A\rightarrow O}/\tau_{O\rightarrow A}$ = Volume of Atmosphere/Volume of the Ocean of depth *D*.

Eqs. 6.18–6.20 are similar to the transport processes in the atmosphere for the 2-box models, but three new concepts have been introduced: K_V, H_O, and C_O; these are explained next. K_V is the *transfer velocity;* H_O is the *volatility constant* which is the inverse of the traditionally defined Henry's law constant or solubility and C_O is the concentration in the water.

The idea of concentration in the water C_O as it applies to these equations is understood as follows: Take a cubic meter of water. Remove all the gas in question from this cubic meter and put it in a cubic meter of air at the earth's surface, which is, let's say, devoid of this gas. The resulting concentration of the gas in the cubic meter of air will then be C_O. This is the concentration of the gas of our interest dissolved in sea-water.

In our use, H_O is a dimensionless quantity and its meaning is demonstrated by the following experiment. Take a closed container half filled with pure water and half "zero-air" in which there is only nitrogen and oxygen at atmospheric concentrations in equilibrium at atmospheric temperature and pressure. Put one mole of the selected gas of interest in the air side. After some time, perhaps speeded by agitation, the one mole of gas will be distributed between the water and the air in a fixed ratio $H_O = C_A/C_O$ where the Cs represent the fractions of the mole that ended up in the air (A) and water (O). It doesn't matter how many moles of the gas are added, the ratio of concentrations in the air to water will be the same and it will be H_O when equilibrium is reached. This is Henry's law. H_O is a characteristic of the gas and water into which it is dissolving. Soluble gases have low values of H_O (low volatility) and insoluble ones have large values (high volatility). More of a highly volatile gas evaporates into the atmosphere if placed in water.

Example If, in the experiment described above, $H_O = 0.2$ and there are 10^9 moles of air in the container. What will be the concentrations of the gas in the air and water parts of the container after one mole is added?

Solution: Adding one mole will initially give us an air side concentration of 1 ppb (1 mole/10^9 moles). After equilibrium is reached, $C_A = C_O H_O = 0.2 C_O$. Since there are no sinks, $C_A \, x \, 10^9$ moles + $C_O \times 10^9$ moles = 1 mole that we put in. Substituting for C_A (= 0.2 C_O) we get that $C_O = (1/1.2)$ ppb = 0.83 ppb and from this $C_A = 0.2 \times 0.83$ ppb = 0.17 ppb. Because the volumes of the water and air are the same, this also means that 0.83 moles will end up in the water leaving only 0.17 moles in the air. In this example, the gas is quite soluble in water as represented by a relatively small value of H_O which is the inverse of solubility.

Based on the example, a point to consider is that if the volumes of the water and air were not the same, the mixing ratios would still be 0.17 ppb in air and 0.83 ppb in water and the ratio will be 0.2, but in terms of moles in the air and water parts the ratio will not be the same and no longer 0.2. The volatility, or its inverse, the Henry's law constant are defined for mixing ratios, or concentrations and not the partitioning of the total number of moles between reservoirs. That depends also on the relative sizes of the reservoirs: C_A(moles)/C_O (moles) = H_O (V_A/V_O), where the V's are the volumes into which the gas partitions).

Thermodynamic theory tells us that the solubility, and hence its counterpart the volatility, are functions of temperature. Often, this dependence can be represented as $H_O = A \, e^{\alpha/T}$ where the constants (A and α here) are often determined from laboratory experiments and available as tabulated values for various gases (Endnote 6.6). This equation tells us that, all else being the same, more of the gas will dissolve in cold water than warm.

The ratio $H_O C_O/C_A$ is called the *saturation* S; it is 1 in equilibrium indicating that the water is "saturated," > 1 if the water is super-saturated and < 1 if it is under-saturated. Consistent with Eq. 6.18, no net exchange can take place in equilibrium; the oceans are a source if they are super-saturated and a sink if under-saturated. In the earth's ocean-atmosphere system, the super-saturated states usually arise when there is substantial net production of a gas in the oceans, as for instance, is the case for natural methyl halides (CH_3Cl, CH_3Br, CH_3I) which can be produced by living things. Under-saturation may represent a net sink in the ocean, or a transient effect of constantly increasing atmospheric concentration as the ocean adjusts. Both apply to CO_2.

The idea of a transfer velocity can be applied in any circumstance where there is a transport time. For instance, when the transport is by the blowing wind, the transfer velocity is just the wind speed (U), and the transport time is τ_T = L/U where L is the distance over which the wind transport is being considered, as we saw earlier. Comparing with Eqs. 6.20, K_v is like U and L is represented by the depth of the oceans (D) or the scale of the atmosphere (H). When it is extended to other aspects of environmental science, the physical meaning is obscured but the concept is still useful. For ocean-air exchange, while there are several theories, it is mostly determined from empirical observations. A simplified but useful approximation is the following equation:

$$K_v = 0.251 u^2_{Avg} (Sc / 660)^{-1/2} \qquad (6.21)$$

Here K_v is in cm/h, which are the traditional units, u_{avg} is the wind speed in meters/second averaged over some time of at least an hour, and Sc is a dimensionless Schmidt number that depends on the gas involved; it reflects the ability of a gas to diffuse into the water at the interface and the constant (0.251) is an empirical factor in cm-s^2/h-m^2 (Endnote 6.7). Sc is defined as the kinematic viscosity of water divided by the diffusion constant of the gas in water or Sc = v/D. D can be written as $D_o\, e^{-B/T}$ the D_o and B are properties of the gas and often obtained from measurements. These factors depend on the sizes of the diffusing molecules making heavier and larger molecules slower to diffuse than smaller ones. The annual average winds over the oceans are 8–9 m/s which can be used in Eq. 6.21 to produce a rough approximation: $K_v \approx 18 \times (Sc/660)^{-1/2}$ giving a range of transfer velocities from 14 to 20 cm/hr for most gases of environmental interest (Endnote 6.7). The equation is valid for gases of modest to low solubilities and low reactivity in water into which they are diffusing. These conditions hold for most of the gases of interest in climate and environmental sciences including those tabulated in Endnote 6.7.

Example Carbonyl Sulfide (OCS) is found to be produced in the oceans and released to the atmosphere. It can cool the earth by making fine particles in the atmosphere. Find the annual emissions from the oceans in Tg/year. The information available is: $D = 1.4 \times 10^{-5}$ cm^2/s at the effective temperature of the oceans; suitably averaged concentrations are $C_A = 500$ ppt, $C_O = 1100$ ppt, and $H_O = 1.7$.

Solution: Applicable equations are Eqs. 6.18, 6.19, and 6.21. Step 1: Obtain additional readily available but needed information: $N_0 = 2.5 \times 10^{19}$ molecules or air/cm^3 (calculated earlier), area of the oceans, $A_I = 3.6 \times 10^{18}$ cm^2, $M_W = 60$ g/mole, v = kinematic viscosity of seawater = 0.012 cm^2/s. Calculate the Schmidt number = v/D = 0.012 $cm^2/s/1.4 \times 10^{-5}$ $cm^2/s \approx 860$. Step 2: Use Eq. 6.21 with average winds speeds to calculate $K_v = 18 \times (860/660)^{-½}$ = 15.8 cm/h. Step 3: Use Eq. 6.19: $T_{NET} = (15.8$ cm/h$) \times (3.6 \times 10^{18}$ $cm^2) \times (1.7 \times 1100 - 500)$ $\times 10^{-12}$ molecules of OCS/molecule air $\times 2.5 \times 10^{19}$ molecules of air/$cm^3 = 2 \times 10^{30}$ molecules OCS/h. Step 4: Convert to Tg/year. S(ocean) = $(2 \times 10^{30}$ molecules/h $\times$ 24 h/d $\times$ 365 d/y)/(6 $\times 10^{23}$ molecules/mole) $\times$ 60 g/mole $\times 10^{-12}$ Tg/g = 1.7520 Tg/year. Wrong! S is about 2 Tg/yr. With all the approximations, we don't know it to three decimal point accuracy.

6.4.2 Effective Lifetime

The global balance of a gas can always be represented by Eq. 4.1 with a proper interpretation of the components including the lifetime, or residence time in which τ is replaced by an effective time τ_{eff} to reflect the general case when cross-media transport is present. Even though a gas may not be destroyed in the atmosphere, it can, and will, migrate to other reservoirs where it can be destroyed or stored for a long time. As these processes are occurring the observed concentrations will follow Eq. 4.1, but the simplicity of a constant lifetime may be lost. We saw before that the observed residence time is a combination of the transport times and the lifetime due to destruction processes in the atmosphere, especially if the sources and sinks are not in close proximity. If the dominant means of removing a gas from the atmosphere is by shifting it to other reservoirs, such as the case for CO_2, then the τ_{eff} must also include the cross-media transport times along with the destruction and the transport time within the media. In most such cases it cannot be represented as a constant until after the sources and sinks have not changed for a while or a steady state has been achieved.

Let us examine this concept further using the simplest model of cross-media transport as in Eq. 6.18 and assume that there is a source in the atmosphere, but no sinks and there is a sink in the oceans but no source – a situation similar to that of carbon dioxide ($S_O = 0$, $S_A \neq 0, \tau_A = \infty$, $\tau_O \neq \infty$). We will take some liberties with the values of the variables to illustrate the concepts. We can calculate the effective lifetime or residence time of a gas by comparing Eq. 6.18a with Eqs. 4.1 and 4.2a to obtain the following formula based on how the gas behaves as observed in the atmosphere alone. If you take measurements of the gas in the atmosphere (C_A), then this is the lifetime you will see, with a known source (S_A) and it is the one that affects climate because it reflects how much the gas will accumulate in the atmosphere:

$$\tau_{eff} = \frac{C_A}{S_A - dC_A / dt} \tag{6.22}$$

Now we are ready for experiments with this model (Eqs. 6.18 and 6.22). We will do two sets: in the first, we look at a pulse of a gas put into the atmosphere that instantly raises the concentrations to 100 ppm and we will assume for all cases that there was zero concentration in the atmosphere and the oceans when we started. We look at the situation when the lifetime of the gas in the oceans is $\tau_O = 10$ yrs and when it is infinite ($\tau_O = \infty$). In the second set of experiments, we add two cases – a constant source and one that increases exponentially, both with $\tau_O = 10$ yrs.

Let's look at the salient points of the results shown in Figure 6.7. In all cases the residence time increases as time goes on. Once a pulse of the gas is added to the atmosphere, you have no further control on what happens to it. If there is a loss in the oceans, at first the residence time looks like it is about 5 years reflecting the transport to the ocean, but after a while it rises to about 20 years (Figure 6.7b), neither is the same as the actual "chemical" lifetime of the gas in the oceans which is fixed at 10 years in this simulation. The concentration in the ocean rises at first because it takes time for the pulse to move in and the destruction rate is small because it depends on the concentration (C_O/τ_O). But after a time of buildup, the sink takes over and the concentration starts to fall. The concentration always falls in the atmosphere because it is leaving to go to the oceans from the very beginning. For the case when the lifetime in the oceans is infinite, after the pulse is added, the atmospheric lifetime is again around 5 years, but after a while the amount we had added partitions itself between the atmosphere and the oceans in the fixed ratio given by Henry's law (dashed lines in Figure 6.7a). The remaining fraction, 50% in the case of our example, stays in the atmosphere permanently because we have reached a new steady state between the atmosphere and the oceans with the added amount. The residence time for the remaining fraction becomes infinite. This situation illustrates how reservoirs can take up gases, but they "fill up" and can no longer take more. This is quite different from the situations when a chemical reaction or some other process can permanently remove the gas from the atmosphere assuming that the chemical process is maintained, such as the case here when we put the ocean lifetime at 10 years. In that case, the residence time becomes nearly constant and the concentrations continue to decline, until the gas added in the pulse is more or less all gone.

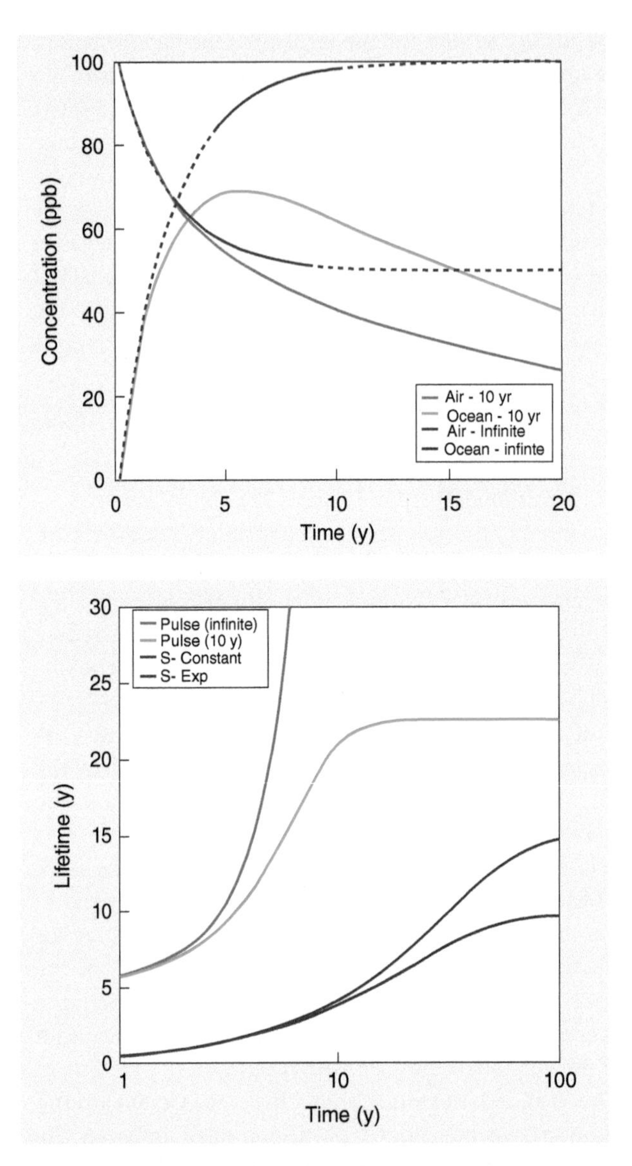

Figure 6.7 Illustration of cross-media transport effects on concentrations and residence times. (A) A pulse of a gas put into the atmosphere decays as it moves into the ocean. There are no atmospheric sinks. This is like the case of carbon dioxide. Two cases are considered – when the ocean lifetime is 10 years (solid lines) and when it is infinite (dashed lines). In both cases the gas concentration declines in the atmosphere as it would if were destroyed there, but unlike that case, the lifetime gets longer over time as the reservoir "saturates." (B) The change of atmospheric residence time with time is shown for three cases – pulse, constant emissions, and increasing emissions. The residence times are the longest for the pulse and get shorter with increasing sources.

For the second set of experiments (Figure 6.7b), the case of constant emissions illustrates again that the lifetime increases slowly, but as long as emissions are being put in, the ocean continues to take up its share and remains out of equilibrium longer, thus absorbing more of the gas. Similarly, if the emissions are increasing, the disparity between the equilibrium concentration in the ocean and the atmosphere is even higher every year than in the other cases and so a larger amount keeps flowing into the oceans keeping the atmospheric residence time lower than the other cases (Figure 6.7b).

6.4.3 Atmosphere–Land Exchanges

The exchanges of gases with the land are either with soils or the biota. The mechanisms of production and destruction in the soils are microbiological and chemical processes. The active zones where this happens is in the top few meters partly because of the difficulty of exchanging oxygen with deeper regions. The soils can be inundated, wet, or dry. Wet and inundated soils are the primary source of atmospheric methane, while dry soils release most of the nitrous oxide that we see in the atmosphere.

Let's look more closely at the wet or inundated soils first. It is rare though to have such places without plants. The plants, when present, are a part of what may be thought of as a master mechanism by which methane is released from wetland ecosystems, rice fields, or even suitable areas with trees. Deeper in the soil that is inundated with water, oxygen levels dwindle to near zero. This is because oxygen doesn't dissolve readily in water and must pass through it to get to the soil surface below. The wetland plants have to transfer oxygen to their roots. Many species have evolved special structures called aerenchyma that are conduits from the air above to the root zone. But as they supply oxygen to the roots, the reverse transport takes soil gases such as methane and nitrous oxide, which are produced by bacteria, to the atmosphere. The bacteria would not be able to live in soils without a food source. Roots exude sugars, which along with dead roots and even some flow of nutrients from the surface cycling of plant material can provide ample sustenance for many types of micro-organisms. A series of bacterial degradations produce the food source for methanogenic bacteria that thrive in the anoxic environments in the soil below or near the root zone (rhizosphere). The by-product of anaerobic bacterial processes is methane which can be produced in great abundance in the soil. Some of it bubbles out through the water and gets into the atmosphere. This process is supplemented, or even replaced, if mature wetland plants are present because of their well-developed aerenchyma. The plants are constantly transpiring which draws water from the soil to the roots creating a fast transport process within the soils and toward the roots. This water is supersaturated in methane which can seep into these structures and is readily transported out to the stems and from there to the atmosphere. Before it can get through, methanotrophic bacteria that live in the aerobic part of the root zone or perhaps even in the roots, capture it and use it for food. Their metabolic waste product is carbon dioxide, which is also emitted from wetland ecosystems whether natural or rice fields. What we get into the atmosphere is a relatively small residual methane between a large production and a large oxidation in the soils. Studies have shown that 50%–90% of the methane that is produced does not make it to the atmosphere. What we see is that there is nutrient rich carbon-based food available to bacteria in the soils below the plants and the outcome of the metabolism of the bacteria is a partitioning of carbon between methane and carbon dioxide. The carbon dioxide is not important from the climate change perspective because it represents a return of the same carbon that was in the atmosphere a few months back. We see that in the present environment, there is a complex and intimate relationship between living things and atmospheric emissions of methane.

In these circumstances, the cross-media transport processes have a steady-state relationship between production and emission to the atmosphere – it is:

$$S = \frac{PDA}{1 + \tau_T / \tau_R} \tag{6.23}$$

Here S is the cross-media source of a gas (molecules/s), P is the production in molecules/ cm^3 -s in the soil, mud or water, D (cm) is the depth from the surface throughout which this production occurs, A (cm^2) is the area over which the soils are releasing the gas. τ_T is the transport time from the reservoir to the atmosphere and τ_R is the lifetime of the gas in the reservoir due to the loss processes, perhaps oxidation, that prevent some of the gas from getting out into the atmosphere (Endnote 6.8) . The limits are instructive. If the transport time is short compared with the lifetime in the reservoir (τ_R), then everything that is produced gets out and in steady state the source becomes S (molecules/s) = P × Volume. In this situation, the cross-media transport does not have an effect and we get a pure source. If the lifetime is much shorter than the transport time, the flux in the most extreme case goes to zero and nothing gets out. It is the in-between cases that are of the most interest (Endnote 6.8). For trees and wet soil ecosystems, the soil below and around the root zone will have ample supply of nutrients as in the case of inundated soils, and methanogenic bacteria may be plentiful, but because the trees and plants in such environments do not have aerenchyma the transport can be slowed down and much of the methane will be consumed near the surface by methanotrophic bacteria so that only CO_2 will be emitted. This will result in a depleted level of methane in the top soils, which would allow methane from the atmosphere to seep in and that too will be consumed by methanotrophic bacteria that are there, but mostly living on the methane from below. Dry soils are known to be a sink of methane as this mechanism would suggest.

Example Consider methane production in an inundated wetland soil. Methanogenic bacteria are responsible for producing it in the soils and methanotrophic bacteria consume it. Take the present values of the transport time $\tau_T = 1$ h and that of the oxidation $\tau_R = 2$ h. With climate change consider three cases: (a) methanogens increase in population and the production goes up by 20%, but the oxidation remains the same, (b) methanotrophs increase utilization by 40 % but the production remains the same, and (c) both these changes happen; that is, more methane is produced and more is consumed, which is expected, although the numbers are not known very well. Calculate the percent change in the wetland source of methane for these situations.

Solution: (a) From Eq. 6.23, the source will be S(new)/S(before) = P(new)/P(before) = 1.2 or 20% more methane will be released to the atmosphere. (b) S (new)/S (before) = $[1 + \tau_T/\tau_R(\text{before})]/[1 + \tau_T/\tau_R(\text{new})] = (1 + 1/2)/[1 + 1/(2 \times 0.6)] = 0.8$. The flux will decrease by 20%. (c) S(new)/S(before) = $(0.8) \times (1.2) \approx 1$, or no change; the effect of climate change on production by methanogenic bacteria is cancelled by the greater oxidation by methanotrophs.

The land exchanges carbon with the atmosphere through the uptake of CO_2 by plants in another manifestation of global cross-media transport. Some of it is shuttled to the surface soils. Many microbial and biotic processes in the soils cause further transformations, storage, and a return to the atmosphere. For simplicity, the land carbon reservoir can be taken to be the storage in both the soils and the plants. We can write this land carbon as $C_L = H_L C_A$ that represents the equilibrium condition. These two reservoirs, land (soils and plants) and oceans, can store significant amounts of carbon together that would be in balance with lesser amounts in the atmosphere. The amount of carbon in the atmosphere is estimated to be about 800 PgC. If we add another 800 PgC, and there is no cross-media transport, then the atmospheric concentration of CO_2 will be doubled (because there is no chemical destruction and most of the carbon is CO_2). As it is, during the century before 2020, fossil fuel combustion has contributed some 440 PgC to the atmosphere and land use change may have added another 100 PgC making about 500–600 PgC from human activities. From the observed atmospheric increase, we find that only about half of this is in the atmosphere showing the powerful effect of cross-media transport processes and the role of the land and ocean reservoirs.

Another view of the role of the land in the cross-media transport processes with the atmosphere can be constructed as follows:

If we change the sources and sinks, the change in the land storage from one steady state to another, will be:

$$\Delta C_L = \tau_L \Delta S_L + S_L \Delta \tau_L \tag{6.24}$$

Here ΔC_L reflects the change in the land carbon storage and is the counterpart of the change of mixing ratio of a gas in the atmosphere (CO_2 in this case). The effect of the land processes on the fate of man-made CO_2 added to the atmosphere, including burning fossil fuels, can be seen as an increase of S_L, since we are taking the source of carbon to the land biotic system as proportional to the amount in the atmosphere. This increases the land storage by $+ \tau_L \Delta S_L$ (Eq. 6.25). The durability of the storage depends on the residence time τ_L which is different for different types of soil processes. Some of it is cycled back rather quickly, but some fraction persists from decades to centuries, or even longer, making the land an effective sink for this fraction of the extra carbon we put in. The mature forests are an example of long-term land storage reservoirs of carbon from the atmosphere. Trees take up carbon dioxide from the atmosphere at a characteristic rate. It is accumulated as the trees grow. But at the same time trees die or shed leaves and branches with some average lifetime, eventually reaching a near steady state with the prevailing environmental conditions, including the carbon stored in the forest soils. An amount of carbon is stored in the forest for as long at the forest lives.

Deforestation caused by human activities can be seen as a reduction in the land storage lifetime of carbon τ_L, which would then increase the atmospheric concentrations, and decrease the land storage ($- S_L \Delta \tau_L$). The biomass will readjust toward its original equilibrium if deforestation stops and land is allowed to grow a forest again. This reflects the idea that there is a canonical ratio of land to atmospheric carbon for prescribed environmental and climatic conditions. In the end the two processes, accumulation of carbon on land, both in the soils and plants, and deforestation, counteract each other and what we see in the real world is a residual of the two (ΔC_L (net) $\approx + \tau_L \Delta S_L - S_L \Delta \tau_L$, in steady state, or neglecting the transient factor). Over the last century, the net effect of human activities is that uptake and storage of the fossil derived CO_2 by land processes exceeds the losses by deforestation as discussed above. These aspects of the CO_2 budget will be considered further in the next chapter.

Review of the Main Points

1. The abstract discussion of sources and sinks was made concrete by looking at the major mechanisms by which gases are put into the atmosphere or removed from it. In particular the roles of biological and chemical processes were delineated and classified for further scrutiny.
2. The general principles of atmospheric chemistry were discussed and then applied to two important areas. The first is the hydroxyl radical concentration in the atmosphere. These fragments of water vapor are responsible for removing hundreds of gases and have been called the "detergent of the atmosphere." In particular they remove the greenhouse gas methane, and the hydrofluorocarbons. The second area we considered was the stratospheric ozone layer that is also determined entirely from chemical processes. The study of the ozone layer is an important part of understanding the natural environment as well as its depletion due to human activities. It is thought that the ozone layer was necessary for life to evolve, and remains so for it to flourish. Because of the catalytic chemical cycles that control the concentration of stratospheric ozone, we can see the point that in global change science, small changes can sometimes be amplified to create big effects.
3. The mass balance of gases in the atmosphere was extended by cross-media transport processes to include the effects of other parts of the global environment, namely, the oceans, soils, and biota. The interactions of gases with these other media lead to considerable complexities in their global atmospheric abundances and how human activities affect these concentrations. As long as a stable atmosphere is being maintained, many of these complexities are not triggered. But when emissions are added, or atmospheric chemistry modified, new cross-media and within media processes are activated. One of the consequences can be expressed as a non-constant atmospheric residence time.

Exercises

1. Calculate the lifetime of OH due to its reactions with CO, CH_4 and the total lifetime from both. Take the concentrations of methane and carbon monoxide to be 1800 ppb and 150 ppb, respectively.
2. Estimate the lifetimes of the following trace gases in the troposphere due to reactions with OH: CH_3Cl, CH_2Cl_2, $CHFCl_2$, CH_2ClF, C_2Cl_4, C_2HCl_3, $CHCl_3$, OCS, H_2S, CH_4. To simplify calculations, use the rate constants at the average effective temperature of the troposphere of about 270°K. Take OH concentrations to be 10^6 molecules/cm^3. The rate constants can be obtained from the NASA compendium.

3 Show that the concentrations of OH should be similar in the two hemispheres based on the budget of methyl chloride. Assume that it is produced in the tropical oceans and is lost only due to its interactions with OH. Measurements have shown that its abundance in the two hemispheres is nearly the same and follows a parabolic curve in concentration vs latitude with a peak in the tropics.

4 Calculate the annual air–sea exchange of the following trace gases: CH_4, N_2O, OCS, CH_3Cl, CCl_3F, CCl_2F_2, and SF_6. For the answer show direction of flux: (+) out of ocean and (-) into ocean), and flux (in Tg/yr). See Endnote 6.7.

Data:

	H	Ca ppb	Cw Ppb
CH_4	26	1700	67
N_2O	1.6	305	190
OCS	2.3	0.5	0.6
CH_3Cl	0.4	0.6	10
CCl_2F_2	25	0.45	0.02
CCl_3F	4.5	0.27	0.06
SF_6	220	0.0004	<.01 ppt

5 Listed below are the concentrations and emissions of methylchloroform in ppt and ppt/y. Use this information to calculate the global concentrations of OH for each year in molecules per cubic centimeter and plot the results. The rate constant of OH and methylchloroform (CH_3CCl_3) is $5 \times 10^{-12}\, e^{-1800/T}$ cm^3/molecule-sec. Use the average temperature of the atmosphere = 255 K. Assume that OH is the only sink of methylchloroform.

(Hint: Use a 1-box global model and do the calculation for each year for which you can calculate the rate of change of atmospheric concentration and note that $\tau = 1/K$ [OH]).

Year	C	S
1979	120	20.0
1980	125	21.0
1981	130	22.0
1982	135	23.0
1983	140	24.0
1984	145	25.0
1985	150	25.5
1986	155	25.0
1987	160	24.5
1988	165	24.0
1989	170	23.5
1990	175	23.0

6 Plot the concentration of ozone in the mid-latitude stratosphere as a density (molecules/cm^3) and as a mixing ratio (ppb).
(a) Where is the maximum of ozone concentration according to these calculations (mixing ratio and density)?
(b) Which maximum is more important for global problems? Why?
Use the following data. Assume that the density of air falls off exponentially with a scale height of 8 km.

Z (km)	O_3 (in 10^{12} molec/cm^3)	Z (km)	O_3 (in 10^{12} molec/cm^3)
2	0.68	35	1.40
6	0.57	40	0.61
10	1.13	45	0.22
15	2.65	50	0.066
20	4.77	60	0.0073
25	4.28	70	0.00054
30	2.52		

7 Fluorine atoms also destroy ozone catalytically (as do all halogens). The cycle is the same as for chlorine. The original chlorofluorocarbons and some replacement compounds contain fluorine. Why is fluorine not considered to be important for the ozone layer problem?
Useful reaction rate data at 250 K in units of cm^3/molec - s: $K(Cl + O_3) = 1.0 \times 10^{-11}$, $K(Cl+CH_4) = 4.3 \times 10^{-14}$; $K(F+O_3) = 1.1 \times 10^{-11}$; $K(F+CH_4) = 6.1 \times 10^{-11}$.

Endnotes

Endnote 6.1 The land and oceans have to be further subdivided to address more complex questions regarding transport phenomena, similar to our separation of the atmosphere into the troposphere and stratosphere. For the land, divisions can be: top soils and deeper land, or vegetated areas and deserts. Top soils (~0.3 m) and the root zone (~0.3–6 m) are the most biologically active parts that can affect atmospheric composition. For the oceans, the divisions are: the mixed layer (top 50–200 m, $T_M \approx 288°K$), the middle ocean, called the thermocline (200– ~ 1200 m, $T_{TC} \approx T_M - (10°C/km)\,Z$, where Z is the depth), and the deep (1.2–3.7 km = average depth, $T_{Deep} \approx 1–2°C$). These global average conditions vary by latitude and seasons. Biological activity declines with the depth in the oceans. These factors are mentioned here because some of them will be considered again in later chapters. Although these subdivisions can add to our understanding of climate and environmental change, separating them makes the mass balance equations much more complex. Solutions such as the ones we have considered in Chapter 4, even if they exist, may become too complex to illuminate the concepts and the mechanisms in action.

Endnote 6.2 The rate constants of immediate interest to us here are included in the chemical equations. The consensus values of hundreds of rate constants and chemical and photo-chemical data are tabulated in "Chemical Kinetics and Photochemical Data for Use in Atmospheric Studies" Evaluation Number 18, NASA, Jet Propulsion Laboratory, Publication 15-10, 2015.

Endnote 6.3 Two matters: The role of OH in atmospheric chemistry was first explained by H. Levy II in 1973: "Normal Atmosphere: Large Radical and Formaldehyde Concentrations Predicted," *Science*, 173, pp. 141–143. Second, there is another way to estimate globally averaged concentrations of hydroxyl radicals first proposed by J. Lovelock, that readers may find interesting. The idea is to use a tracer gas for which the only sink is OH. Further, if we know the sources well and have accurately measured concentrations in the atmosphere, we can calculate how much OH there must be in the atmosphere! It works by inverting our 1-box model Eq. 4.3 for the lifetime, and setting this mass balance lifetime equal to the chemical lifetime in Eq. 6.7. This is done as follows: $\tau_{MB} = C/[S - dC/dt]$ (the mass balance lifetime) (call the preceding equation, Eq.1), $\tau_{OH} = 1/k[OH]$ (the chemical lifetime) and assume $\tau_{MB} = \tau_{OH}$, therefore, $[OH] = 1/(k\ \tau_{OH}) \approx 1/(k\ \tau_{MB})$ (Eq. 2). The tracer of choice has been an industrial solvent methylchloroform (CH_3CCl_3) that was in the atmosphere for some decades. The source S was determined from industrial production records, C and dC/dt were calculated from observed concentrations in the atmosphere that gave a mass balance lifetime, from the equation above, of about 6 y (in Eq.1). Then using the rate constant from laboratory measurements, we get that $[OH] = 1/[(1.4 \times 10^{-7}\ cm^3/molecule\text{ -}y)\ (6\ y)] = 1.2 \times 10^6\ molecules/cm^3$ from (Eq. 2). It has many uncertainties, but it gives us more confidence that this is the level of OH in the present atmosphere.

Endnote 6.4 After the Chapman's reactions were shown to require modifications, D. Bates and M. Nicolet proposed the HOx catalytic cycle in 1950; in 1969, P.J. Crutzen added the NOx catalytic cycle and in 1973, R. Stolarski and R.J. Cicerone proposed the chlorine catalytic cycle. The importance of the latter in the depletion of the ozone layer was recognized when, in 1973, M.J. Molina and F.S. Rowland proposed that the man-made chlorofluorocarbons could deplete the ozone layer. Early work on ozone depletion that contributed to regulating the chlorofluorocarbons is discussed in "Causes and effects of stratospheric ozone reduction: An update," Chapter 5, National Academy Press, 1982. A review of health impacts can be found in L.A. Leaf, "Loss of stratospheric ozone and health effects of increased ultraviolet radiation," in *Critical Condition: Human Health and environment*, ed. E. Chivian, et al. 139–50. MIT Press, Cambridge, MA, 1993. The many subtleties and details of the ozone layer are discussed by A.E. Dessler, "The chemistry and physics of stratospheric ozone," Academic Press, San Diego, CA, 2000. Figure 6.6 representing a global ozone distribution in the atmosphere is adapted from World Meteorological Organization, 1994, Scientific assessment of ozone depletion (GORAMP Report 37).

Endnote 6.5 The derivation of the cross-media transport times, $\tau_{A->O}$ and $\tau_{O->A}$, is given here. The concept is similar to that for the stratospheric-tropospheric transport times in Endnote 4.4. A noteworthy difference is that in the atmospheric case, air was being exchanged and it carried a certain amount of our trace gas from one box to another. Here, the movement of the gas is not by air transport, but of the molecules of the gas itself. The net transport is given by the basic formula in Eq. 6.19 expressed in molecules per second. But the transport in the mass balances of Eqs. 6.18 is not in molecules/second, but in mixing ratio units/second. If T_{NET} molecules per second (from Eq. 6.19) go into the atmosphere, they will cause a change of mixing ratio in the atmosphere of $(C_A - H_O C_O)/\tau_{A\to O}$ (Eq. 6.18a), which is T_{NET} (mixing ratio/s) = T_{NET} (molecules per second)/($N_0 A_E H$ molecules), where the term in the denominator is the total number of molecules in the atmosphere found in Chapter 3 (A_E = area of the earth, H is the scale height, and N_0 is the number density of air at the surface, where the interface is). Then, T_{NET} (ppb/s) = $(C_A - H_O C_O)(K_V A_O/A_E H)$, assuming the Cs are in ppb. Comparing this last equation with Eq. 6.18a yields $\tau_{A\to O} = (A_E/A_O) H/K_V$. The same transport of molecules/second from the atmosphere to the oceans has a different effect on the mixing ratio of the gas in the oceans because the size of the ocean box is not the same. By the same reasoning, $(C_A - H_O C_O)/\tau_{O\to A}$, is T_{NET} (molecules per second)/($N_0 A_O D$ molecules), where the denominator represents the equivalent number of air molecules in the volume of the ocean box of depth D. This gives $\tau_{O\to A} = D/K_V$. D can be anywhere between the depth of the entire ocean of about 3500 m to that of the topmost mixed layer (50–200 m).

Endnote 6.6 Values of the Henry's law constant and volatility, used in this book, are tabulated as functions of temperature for all gases of current environmental interest by R. Sander (2015), "Compilation of Henry's law constants (version 4) for water as solvent" in *Atmos. Chem. Phys.* 15, 4399–4981 (582 pages). The paper also contains a detailed review of the various forms of solubility.

Endnote 6.7 The transfer velocity for ocean–air exchange in Eq. 6.20 is derived from the work of R. Wanninkhof in "Relationship between wind speed and gas exchange over the ocean," *J. Geophys. Res.* 97, 7373–7382, 1992; R. Wanninkhof with others, "Advances in quantifying air-sea exchanges and environmental forcing," *Annual Reviews of Marine Science*, 1, 213–244, 2009; and "Relationship between wind speed and gas exchange over the ocean revisited," *Limnology and Oceanography Methods*, 12, 351–362, 2014. In this work the relationship is written as: K_v (m/s) = $a\,u^2\,(Sc/660)^{-1/2}$ where a is an empirical constant and Sc is the dimensionless Schmidt number defined as = ν/D. The 660 is a normalization using the Sc for CO_2. It can be incorporated into the "a" and would then not appear in the K_V. As mentioned in the text, the Schmidt number can be approximated as $A\,e^{B/T}$ where A and B are constants and T is the absolute temperature. The values of these coefficients are derived from experiments and available from data compendia. A point to note is that this shows the explicit dependence of the transfer velocity on the diffusion constant of the gases in water, which is discussed in the text as one of the processes that impedes cross-media transport and is likely involved in all such interfaces, but not in the bulk of the media such as the deeper ocean away from the surface. Diffusion is active only near the interfaces because it is too slow to transfer molecules over the large distances involved in the bulk of the land, oceans, and the atmosphere of the earth. To estimate the fluxes, you need to calculate the Schmidt number at the prevailing temperatures in the oceans and lakes. The values of A and B (in °K) for gases of interest in climate science are given here as Gas (A, B): Major gases – N_2 (0.0001229, 4559), O_2 (0.0001229, 4557). Greenhouse gases – CH_4 (0.000285, 4310), CO_2 (0.000173, 4448), N_2O (0.0001252, 4460). Technological gases – CFC-11 (0.0002407, 4520), CFC-12 (0.0005471, 4277), CCl_4 (0.0002702, 4519). Other gases of environmental interest – dimethylsulfide (0.0004377, 4276), SF_6 (0.0003725, 4350), CH_3Br (0.0002289, 4380). These values are computed from the data in the paper by Wanninkhof (2014) mentioned above where the Schmidt number is represented as polynomials.

Endnote 6.8 The equation for the flux to the atmosphere from a reservoir where both production and destruction is taking place is derived here (Eq. 6.22). For concreteness let's take the reservoir to be the soils. The production is P (molecules/cm^3); the transport time is τ_T; the removal by biological processes in the soil is represented by the lifetime τ_O, although the nature of the sink does not matter; the concentration in the soil is C (molecules/cm^3) and the volume of the reservoir is V. Then, in steady state, the mass balance says that the production must be balanced by the sink plus the transport: P V (production) = C V/τ_O (sink) + C V/τ_T (transport). In steady state C V = P V τ and therefore $\tau = 1/(1/\tau_T + 1/\tau_O)$ from collecting terms. The source to the atmosphere is the amount transported out of the soil to the atmosphere or $S = C\,V/\tau_T = P\,\tau\,A\,D/\tau_T = P\,A\,D/(1 + \tau_T/\tau_O)$ as we aimed to prove. The source S (molecules/s) = Flux (molecules/cm^2-s) × Area (cm^2), so the flux can be easily calculated if desired.

7

Balance of Climate Gases and Aerosols

7.1 Anthropogenic vs Natural Components

It is customary to separate the origins of environmentally important atmospheric gases into "natural" and "anthropogenic" components. It can be argued that this separation sets humans apart from nature although they are not, and neither the most influential nor the center of the environmental world. Nonetheless, it is justified because it can be used to isolate one influence on our environment from all the others and is therefore consistent with the need to define the system on which we want to focus our scientific inquiry, especially since the observed global warming is man-made. Here we will see it in this light and consider the "natural" part merely to be anything other than caused by humans. Aside from semantics, the separation has important implications. One is that it carries a tacit understanding, that to manage climate or other global environmental issues, we must reduce anthropogenic emissions. It sets aside the possibility that we can leverage "natural" parts of the environment to accomplish the same goals. In particular, it is contrary to the idea of geo-engineering in which we can modify the natural system to reduce global warming without having to attend to the human causes. The geo-engineering concept leads to uneasiness with such solutions. A legitimate reason is that global change science is just emerging and has not reached the level of understanding that gives confidence in our ability to manipulate the earth's system predictably. Reducing the human impact seems safer for the environment, although it will not serve the economic aspirations of our societies. Moreover, the experience from dealing with urban pollution has shown that reducing emissions from human activities leads to cleaner air in the cities. Likewise, stopping the manufacture and use of chlorofluorocarbons is showing a clearly measurable recovery of the ozone layer. These situations give us confidence that it is workable. We will return to these ideas later but for now, we will separate the sources of greenhouse gases according to whether they are from human activities or not, as a means of continuing our inquiry into global change.

In Figure 3.1 we saw that the gases that cause the greenhouse effect are water vapor and carbon dioxide; and to a lesser extent: methane, nitrous oxide, and ozone. Moreover, aerosols or fine particles in the atmosphere may offset warming influences and are therefore an important part of climate change. Here we will look at the budgets of these gases and aerosols as described by the mass balance theory of Chapters 4 and 6. A tabulation of the global emissions from each source and loss from each sink constitutes "the budget" and any

Global Climate Change and Human Life, First Edition. M. A. K. Khalil.

Companion Website: www.wiley.com/go/khalil/Globalclimatechange

disparity should be within the observed trend (dC/dt) which is usually much smaller than the sources and sinks. For use in climate science, the budget is usually defined over a whole year, but it varies during the year particularly with seasons.

How the annual emissions and sinks are determined depends on the gas under study. For biogenic gases such as methane and nitrous oxide, direct flux measurements are undertaken from land-based sources and extrapolated to annual and global scales. For carbon dioxide and industrial gases such as the chlorofluorocarbons, and technological gases in general, industrial production, energy, or commodity use data and sales records may be used to estimate the emissions. For these gases, direct measurements of fluxes are usually not taken and may be unreliable if done. Another method is to rely heavily on the atmospheric measurements of concentrations and trends and back track how much each source has contributed to cause the observed features. Such inverse methods were mentioned earlier and rely on independent estimates of the sinks and a model to take transport into account (Section 4.5). Either class of methods has uncertainties, which can be daunting sometimes, but they are the only measure we have of where these gases are coming from and why they are increasing. The observed atmospheric concentrations are shown in Figure 7.1 for the main global warming gases (Endnote 7.1) .

The small mass balance models we have learned so far can be used to advance our knowledge of the budgets of the global change gases. The budgets that explain the current levels and the history of these sources are a major piece of information necessary to understand global warming and where there may be leverage to manage it. These budgets, therefore, play a key role in our plans to manage global change effectively. We will return to this subject in the last section of the book. For now, let's see if we can apply the mass balance models to learn more about how much of the greenhouse gases come from human activities and how much are natural.

From Figure 7.1, we can separate the natural and man-made emissions for methane and nitrous oxide. Let's assume that the only change that has taken place since pre-industrial times, about 200 years ago, is that human activities have added new emissions. Starting with methane and taking the lifetime to be 8.5 years, which can be calculated independently, we estimate the natural source by Eq. 4.4: $S_n = C\,(\text{pre-industrial})/\tau = 700\ \text{ppb}/8.5\ \text{y} \times 2.6\ \text{Tg/ppb} \approx 220\ \text{Tg/y}$. The data show a constant concentration for centuries before that time. Then we can calculate the recent source by Eq. 4.4: $S = S_n + S_a = dC/dt + C/\tau = (7\ \text{ppb/y} + 1850\ \text{ppb}/8.5\ \text{y}) \times 2.6\ \text{Tg/ppb} \approx 580\ \text{Tg/y}$. Therefore, $S_a = (580 - 220) = 360\ \text{Tg/y}$. We see that the man-made source significantly exceeds the total natural source. This calculation puts a constraint on the evaluation of methane sources and sinks in the sense that the sum of all the individual man-made sources should not exceed about 320 Tg/y. Applying the method to nitrous oxide reveals $S_n = C(\text{pre-industrial})/\tau = 280\ \text{ppb}/120\ \text{y} \times 7.2\ \text{Tg/ppb} = 17\ \text{Tg/y}$ and now, $S = dC/dt + C/\tau = (0.8\ \text{ppb/y} + 330\ \text{ppb}/120\ \text{y}) \times 7.2\ \text{Tg/ppb} \approx 26\ \text{Tg/y}$ and the difference representing the man-made source is about 9 Tg/y. We see that the natural sources are still much larger than anthropogenic. The idea can be applied to other gases as long as pre-industrial concentrations can be measured or re-constructed by proxy methods, and the lifetime does not change much. It is not useful for the technological gases because most of them don't have natural sources and therefore they did not exist in the pre-industrial atmosphere. All of what we see today is man-made. This is so for the chlorofluorocarbons, sulfur hexafluoride, and many other gases. Readers may be curious to know whether it can be applied to carbon dioxide. It can be, but as we will see, there is some

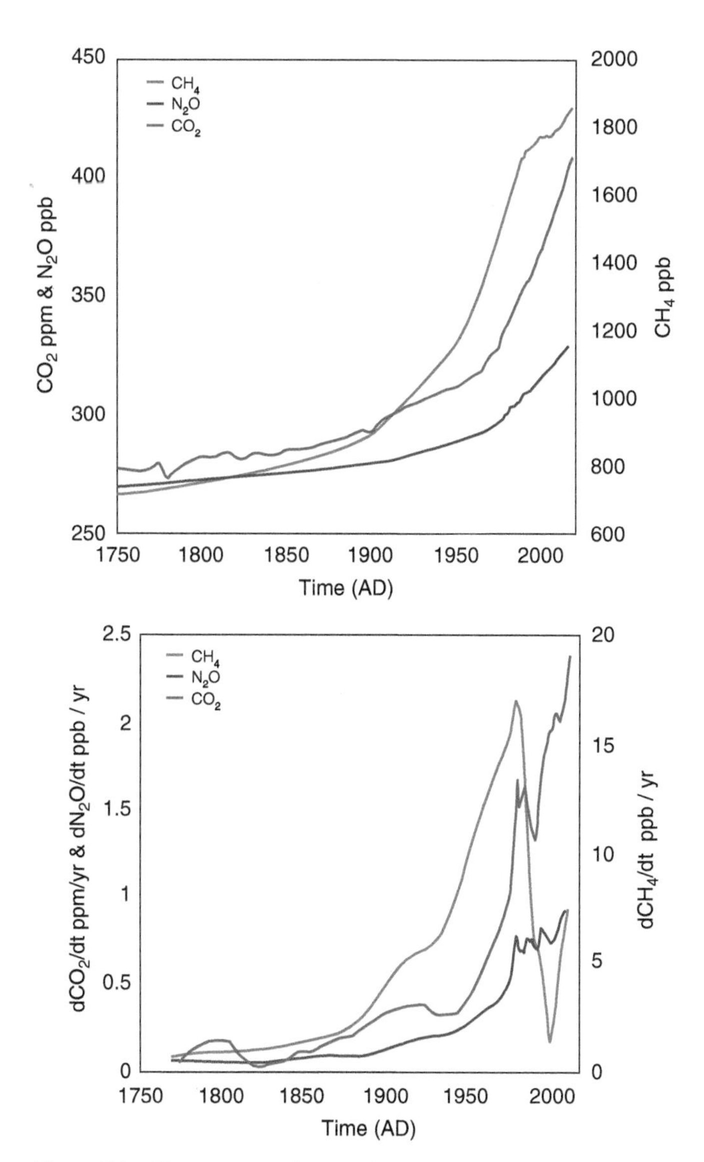

Figure 7.1 The concentrations and trends of greenhouse gases. (a) The mixing ratios of carbon dioxide, methane, and nitrous oxide. These gases are most significant for causing global warming. Concentrations taken in ice cores and the atmosphere are averaged over 5 year increments for earlier years and annual increments in recent decades. The figure shows the increase of these gases from pre-industrial to present times. (b) The rates of change. These are *dC/dt* in the mass balance equations. The CO_2 observations show dips, some of which are related to global human and natural events. Methane had a 20-year hiatus in the late-twentieth-century trends when the concentrations became more or less constant. It reflects shifting human sources. N_2O trends have picked up in recent years as the use of nitrogen fertilizer has increased significantly. *Source*: Data from IPCC AR5.

disarray about the sinks which prevents us from making such estimates reliably; moreover, the man-made emissions are well constrained by global energy use data and so this method may not add much to our knowledge.

7.2 Greenhouse Gases

Let's look at the salient characteristics of the budgets of gases that determine the greenhouse effect and climate change.

7.2.1 Water Vapor

Water vapor is the most important greenhouse gas causing the earth to be warmer by some 20°C. At the same time, it also cools the surface by evaporation and transpiration by plants. Its abundance in the atmosphere is closely related to the temperature and represented by the humidity, which is the ratio of the actual concentration of water (g water vapor/kg air) to what would be expected if the air was saturated (also g/kg). The ratio is multiplied by 100 to express it as a percentage.

Consider a container, such as an aquarium, filled half way with water and the other half with air. It is closed at the top. Water molecules will evaporate from the surface and move to the air. At the same time, some molecules from the air would move back to the water. After a while, the concentration of water molecules in the air part will become stable, or reach equilibrium. It is a remarkable scientific fact that what that concentration is, depends only on the temperature of the water and its fixed thermodynamic properties. If we increase the temperature, more molecules will end up in the air than under cooler conditions. This makes sense because we think of temperature as reflecting the average motive or kinetic energy of the molecules. The speeds of the molecules have a distribution with some molecules moving faster than others. Increasing the average temperature increases the energies causing more molecules at the higher end of the distribution to escape the liquid. Cooling does the opposite. The Clausius-Claperyon equation, applied to water, relates the amount of vapor that can exist in the air as a function of the temperature expressed as the saturation water vapor concentration or vapor pressure (Endnote 7.2). For the range of temperatures and conditions prevalent in the atmosphere, this relationship can be written as:

$$q_s = q_{s0} \exp[(L_v / R)(1/T_0 - 1/T)] \approx q_{s0} \exp[(L_v T_0^2 / R)T'] \tag{7.1}$$

qs is the (mass) mixing ratio at equilibrium or saturation in g water vapor/kg air above the water surface; qs_0 is the mixing ratio at the reference temperature taken to be the freezing point of water (273°K) and may be obtained from observations; it is 3.7 g/kg. R is the gas constant for water (0.461 j/g – °K); Lv is the latent heat of fusion at 0°C (2500 J/g). The second equation on the right is a further approximation, left as an exercise, in which T' is the temperature in degrees centigrade. Putting these numbers into Eq. 7.1 gives us a more convenient version:

$$q_s = 1.58 \times 10^9 (g/kg) e^{-5423/T} \approx 3.7 (g/kg) e^{0.07T'} \tag{7.2}$$

This equation shows clearly that the amount of water vapor in the air is determined by the surface temperature. More water vapor can exist in warm air compared with cold.

Example: Calculate the saturation water vapor concentration in the atmosphere at the current average temperature of 288°K. If the temperature of the earth warmed by 1°C, how much would the water vapor concentration change if it is proportional to the saturation levels?

Solution: $qs\,(288°K) = 1.58 \times 10^9\ \text{g/kg}\ e^{-5423\ K/288\ K} \approx 10.5\ \text{g/kg}$. $qs\,(289°K) = 1.58 \times 10^9\ \text{g/kg}$. $e^{-5423\ K/289\ K} \approx 11.2\ \text{g/kg}$. Change = $(0.7/10.5) \times 100\%$ = **6.7% per 1°C**. The result is apparent by inspection of the simplified equation on the right-hand side in Eq. 7.2.

We can go a step further with the aquarium model by putting some ice on the lid causing it to cool. Then, we will see that water will condense on the inside surface and fall back. The condensation occurs because the surface of the cover is cooler than the water below and it makes the air in contact cooler too. The amount of vapor arriving there was saturated with respect to temperature at the water surface and is now cooler so it must adjust to have less vapor consistent with the Clausius-Claperyon equation causing some of it to condense. What does all this have to do with the earth's environment and climate? With the cooled top, we have a model of the water cycle in the global atmosphere and the role of the Clausius-Claperyon equation in explaining it. The parts of the earth, represented mostly by the oceans, are a large source of water vapor to the atmosphere up to the amount that causes saturation or near saturation content. Since there is no top over the oceans, the vapor rises but as the saturated air gets higher up, it cools by expansion as discussed earlier, and the water condenses back to liquid and falls as precipitation, mostly as rain to complete the water cycle.

If we assume that the water vapor content in various parts of the world is proportional to the saturation concentration, we can explain the significant variations of its distribution in the atmosphere. The major variation therefore follows the temperature trends in the atmosphere. Water vapor concentration falls rapidly with altitude, leading to a declining mixing ratio as in Figure 7.2a. It shows that most of the water vapor exists close to the earth's surface. Compared with the mixing ratio at the surface, it is only 10% at 5–6 km and less than 1% at the tropopause. Because of this feature, the greenhouse effect caused by water vapor occurs near the surface. The other major spatial variation is with latitude – high levels of water vapor exist around the equator and very low levels are seen at the poles which are partly explained by the Clausius-Claperyon equation. As we saw earlier, this determines the latitudinal distribution of hydroxyl radicals making the tropical atmosphere a major sink of many gases, while the higher latitudes are not (Figure 6.5). The average amount of water is about 0.5% ranging from almost 0% up to 4% ≈ 13 exagrams in the whole atmosphere (Figure 7.2b). The total flux into the atmosphere, constituting the source, is about 500 Exagrams/y which gives us an approximate lifetime $\tau \approx 9$ days (1 Exagram = 10^{18} g and using $\tau = C/S$ from the mass balance model).

Although human activities can directly put tons of water vapor into the atmosphere, the natural flux and the atmospheric concentrations are so large that this direct contribution is truly negligible. Yet, indirectly, human activities are increasing water vapor by causing global warming since warmer temperatures increase it by evaporation from the oceans. We will come back to this situation later. For now, note that this phenomenon is supported by observations showing an increase in water vapor of about 0.4 mm/decade as the earth warms in recent times.

As a reminder, some water vapor is removed by the chemical reactions or photo-dissociation high in the mesosphere. Even larger amounts are removed by $O(^1D)$ reactions that make OH radicals, but these sinks are negligible when compared with rain and snow. Yet these reactions and processes have profound environmental consequences caused by hydroxyl radicals as we have seen earlier.

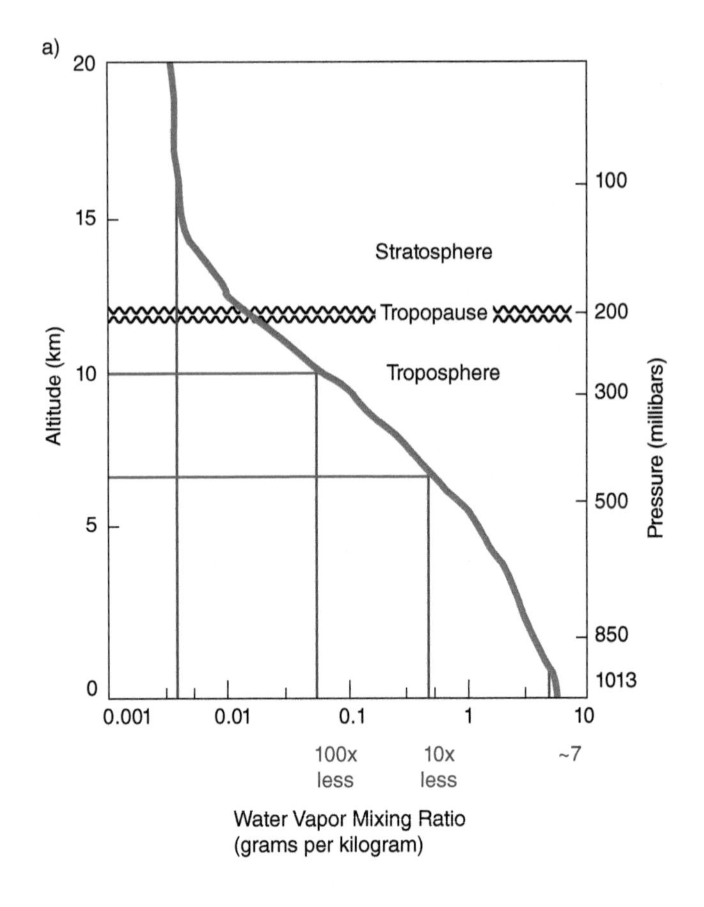

Figure 7.2 The global balance of water. (a) The vertical concentration is marked by a rapid decline with altitude. The effect of water on climate is therefore mostly from near the surface. (b) Atmospheric water is supplied by evaporation and transpiration processes and lost by precipitation, mostly rain and by runoff from rivers. A budget is shown for the three main parts of the global environment. Chemical sources and sinks also exist but are not shown here because they are negligible in the budget, but have major roles in atmospheric chemistry (Endnote 7.3). The units used are Eg = Exagrams = 10^{18} g.

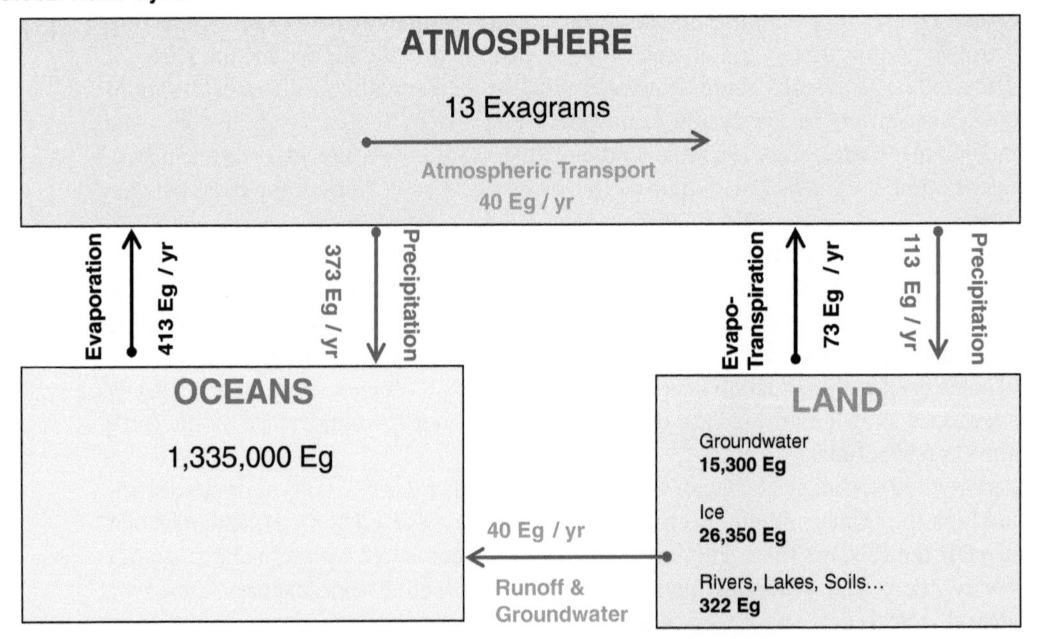

7.2.2 Carbon Dioxide

Carbon dioxide contributes to the natural greenhouse effect, although not as much as water vapor. This is in part due to the much lower concentrations of about 400 ppm (or 0.04% in 2020). Which, as previously mentioned, also makes it easier for human activities to significantly increase its concentrations. There is little doubt that increases of carbon dioxide are the main cause of the global warming we are seeing, and that this increase is driven by burning fossil fuels for energy. *In other words, carbon dioxide is not the largest contributor to the natural greenhouse effect, but it is to global warming.* The two are not the same thing. This has resulted in a significant societal investment to understand the mechanisms that drive the carbon cycle of which carbon dioxide is the major part in the atmosphere. The carbon cycle is represented in Figure 7.3 showing the large reservoirs and the fluxes between them. From this figure we see that the budget of man-made part is $dC/dt = S$ – Losses: 4 PgC/y (Increase) = 7.8 (Fossil Fuels) + 1.1 (Land Use Change) – 2.3 (Net Ocean Sink) – 2.6 (Net Land Sink). It is noteworthy that the land and ocean sinks are of comparable magnitudes.

Example: An effect of the buildup of CO_2 from fossil sources of carbon is that it is tying up atmospheric oxygen. That is, the oxygen is declining due to this process. By how much? What is the lifetime of O_2 due to this process?

Solution: $dC(O_2)/dt = -dC(CO_2)/dt = -4$ PgC/y (Figure 7.3) $= -(4 \times 10^{15}$ gC/y)/12 g/mole $= 3.33 \times 10^{14}$ moles/y. Therefore, $dC(O_2)/dt = -3.33 \times 10^{14}$ moles/y $= -3.33 \times 10^{14}$ moles/y/1.667×10^{20} moles air in the atmosphere $= -2$ ppm/y. There is about 21% in the atmosphere, so in percentage per year, the decline is: $(d(O_2)/dt)/C(O_2) = -$ [$(2 \times 10^{-6}$/y)/0.21] $\times$ 100% $= -0.001$%/y. $\tau(O_2)$ due to this process is $\approx C/(dC/dt) = 1/0.00001 = 100{,}000$ y from the previous step. The residence time of O_2 is estimated to be 3000–4000 years, so this will not affect it, and we will not notice the decrease even if it occurs. Many ongoing processes can move the oxygen concentration by these or larger amounts for some lengths of time.

Next, we will consider the salient mechanisms that lead to the impacts of the man-made sources and sinks on the atmospheric concentration of carbon dioxide and its increase.

The carbon cycle is a machinery with many gears and sprockets. When the cycle is disturbed, it sets into motion buffers and feedbacks that make predicting the consequences complicated as well as unreliable. Let's look at two key differences of carbon dioxide when compared with other greenhouse gases: it is not removed from the atmosphere by chemical processes, and when it does leave, some of it inevitably comes back into the atmosphere to continue its role as a greenhouse gas. Man-made gases such as the chlorofluorocarbons, or the natural greenhouse gases such as methane and nitrous oxide, have well-defined pure sources and sinks that create the known balance as discussed earlier. For instance, once the methane molecule reacts with hydroxyl radicals, it is presumed to be gone forever. It is not expected that the fragments will regenerate into another methane molecule and return to cause climatic influences. It is safe to say that all man-made technological greenhouse gases share this characteristic with methane except for carbon dioxide. Since it does not react chemically, it is removed from the atmosphere by cross-media transport processes – specifically, by dissolving into the oceans and the soils, and incorporation into the vast expanse of plant, microbial, and animal life.

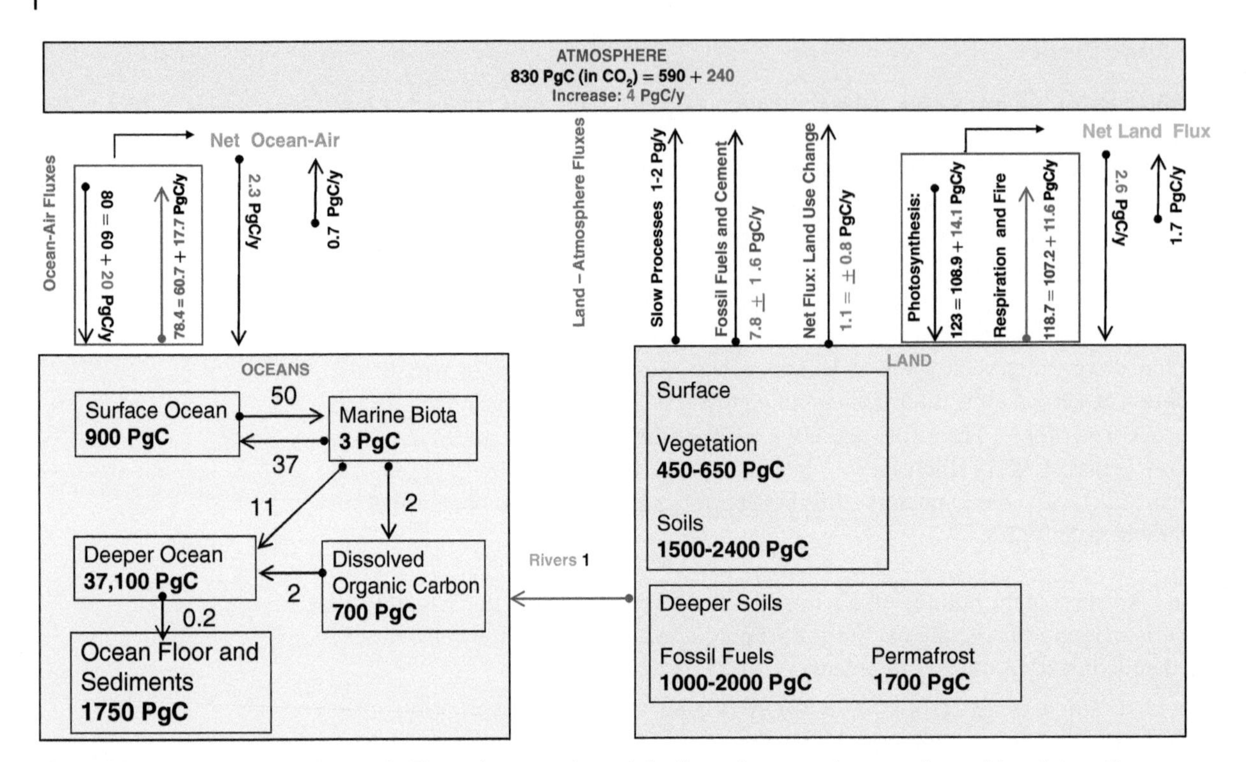

Figure 7.3 A view of the carbon cycle. The main reservoirs and the fluxes between them are shown. Many interactions occur within these reservoirs that affect atmospheric concentrations. Emissions of organic compounds by living things usually end up as carbon dioxide in the atmosphere by atmospheric chemical processes and the carbon cycles back to the other reservoirs. The balance for man-made emissions, which are increasing, is for early twenty-first century (adapted from IPCC AR5, Endnote 1.1).

The carbon in the atmospheric CO_2 moves into other molecules that revert back to CO_2 resulting in an equilibrium that can shift with changing environmental conditions. Carbon dioxide cycles can be divided into slow and fast for further analysis. Carbon atoms may be eternal, but not all the carbon on earth can cycle itself to atmospheric CO_2. The slow cycles are essentially a burial of the carbon that was once carbon dioxide into chemical species that are stable and located far enough that the carbon will stay there for thousands or even millions of years before it can get back to the atmosphere as CO_2. For carbon dioxide that goes into the oceans, this is the result of slow falling of carbonaceous material to the bottom of the sea and subsequent accumulation in sediments of organic biological material that has no way to decompose further. This includes shells, dead animals, and waste from various living things which sinks to the bottom and is not dissolved or recycled along the way. On land, it includes the slow burial of organic material from dead plants and animals, especially under water-logged conditions when bacteria cannot access it. This deep carbon returns eventually by weathering of rocks, volcanism, and related geological processes to complete the cycle. The carbon we use as fossil fuels accumulates this way. We will not consider these cycles further because they have little effect on atmospheric carbon dioxide over the time scales of our interest. On the other side, there are many fast cycles that are clearly observable in atmospheric measurements, as mentioned earlier, repeating seasonally or even shorter times. These too do not affect the systematic buildup of carbon dioxide that is responsible for climate change, and so we will set these aside as well. The burning of fossil fuels is taking carbon that would not have affected the atmosphere over the time scales of decades to centuries, and releasing it rapidly, adding to the atmospheric burden. What affects the budget of man-made emissions on decadal time scales is the cross-media transport of CO_2 to the oceans and the land ecosystems such as the surface plant life and the soil microcosms, followed by the subsequent processes that it undergoes in these reservoirs, including re-emission.

Carbon dioxide is exchanged with the oceans according to the theory we discussed earlier, proportional to the disparity between the atmospheric and oceanic concentrations. If that was all there was to it, the capacity of the oceans to take up man-made CO_2 would be much less than it is. The exchange reflects the levels of saturation in the water at the interface between the oceans and the atmosphere. The waters mix rapidly to a depth of 50–200 m in the mixed layer, but carbon dioxide is not very soluble, so this can saturate the oceans quite quickly with the atmospheric CO_2 and therefore the exchange would stop as indicated by Eq. 6.18. Two phenomena intervene to greatly increase the ocean's capacity. One is mixing with the deeper waters, but it is slow, taking decades and still not highly efficient because of the low solubility. The most important process is the shift of the carbon in the oceanic CO_2 to carbonate and bicarbonate ions which greatly increases the ocean's capacity to take up carbon dioxide and hold it for very long times, tied up in these ions, constituting an oceanic sink. Let's look at this process.

7.2.3 Ocean Sink and Acidification

Some 20–30 % of the CO_2 emitted by human activities over the last century has been taken up by the oceans, making it a crucial sink over the decade to centuries time scale (additional amounts of the emitted CO_2 are tied up in the land reservoirs). A deeper look at this process is useful to understand the role of the oceans in its atmospheric residence time. When CO_2 is added to the atmosphere it dissolves by the ocean–air exchange process. Three key chemical reactions occur:

$$K_H : CO_2(g) + H_2O \leftrightarrow CO_2.H_2O \tag{7.5a}$$

$$K_1 : CO_2.H_2O \leftrightarrow H^+ + HCO_3^-(\text{bicarbonate}) \tag{7.5b}$$

$$K_2 : HCO_3^- \leftrightarrow H^+ + CO_3^{2-}(\text{carbonate}) \tag{7.5c}$$

In the first, the added CO_2 attaches to a water molecule to form carbonic acid, $CO_2.H_2O$ (hydrolysis). The "↔" symbol indicates that the process goes back and forth until an equilibrium is established determined by a rate constant, which for CO_2 (Eq. 7.5a) is the Henry's law constant K_H (or $1/H_O$). It should be noted that these reactions go both ways, unlike atmospheric gas phase chemistry. After a while, in a time determined by rate constants, the $CO_2.H_2O$ breaks apart, or dissociates to form bicarbonate and then carbonate ions represented by the reactions in Eqs. 7.5b and 7.5c. Using Eq. 6.9, we can write the concentrations of these species expected when equilibrium is reached:

$$[CO_2(in\,water)] = K_H[CO_2(g)] \tag{7.6a}$$

$$[HCO_3^-] = \frac{K_1 K_H}{[H^+]}[CO_2(g)] \tag{7.6b}$$

$$[CO_3^{2-}] = \frac{K_1 K_2 K_H}{[H^+]^2}[CO_2(g)] \tag{7.6c}$$

The units of concentrations are moles per liter of water, and for the gas phase, designated by (g), they are often expressed as pCO_2 or the partial pressure of CO_2, that is the pressure of CO_2/pressure of air at the interface.

In this picture, the CO_2 added to the oceans is taken out and the carbon is put into the bicarbonate and carbonate ions reducing the concentration of dissolved CO_2, which then makes it under-saturated and the oceans continue to draw it out of the atmosphere at a steady rate. This draw-down is only related to the CO_2 concentrations in the oceanic and atmospheric phases as represented in Eqs. 6.18 and 6.19 and not the total carbon-containing molecules in sea water that may have come from the dissolution of CO_2. From our concept of sinks, this transfer to carbonate and bicarbonate ions gives CO_2 a short lifetime in the oceans. But the carbon can shuttle back because these reactions go both ways.

The process of CO_2 dissolving in the oceans generates hydrogen ions or protons (Eq. 7.5). The concentration of H^+ in moles per liter is the very definition of pH and acidity according to the formula:

$$pH = -Log_{10}[H^+] \tag{7.8}$$

Solving for the hydrogen ion concentration gives the equivalent representation $[H^+] = 10^{-pH}$. Accordingly, the more H^+ ions there are the lower the value of pH. Even in pure water, hydrogen ions form giving it a pH of 7 at 25°C which is taken to be neutral; lower pH is acidic and higher is basic. Under natural conditions, the atmosphere causes all waters that are in contact with it, including rain drops, lakes and the oceans, to become slightly more acidic than without the CO_2. The natural rain water, for instance, has a pH of 5.5–6 since the CO_2 in the atmosphere exercises a controlling influence on this pH. In large reservoirs, including the oceans, several processes can contribute to the hydrogen ion concentrations in addition to the CO_2 effect. From Eqs. 7.5b and c we can conclude that if we add CO_2 to the atmosphere, as from burning fossil fuels, it will cause "ocean acidification" because more H^+ ions will be generated (Endnote 7.4).

We can go one step further from Eq. 7.6c by considering the fate of the carbonate ions. These can associate with the abundant calcium in the ocean water by: $Ca^{2+} + CO_3^{2-} \leftrightarrow CaCO_3$ or calcium carbonate. The calcium carbonate is used by living things to make shells, limestone, coral reefs, and such. With increased acidification, we see that the ratio (Eq. 7.6b and c) $[CO_3^{2-}]/[HCO_3^-] = K_2/[H^+]$ decreases as the hydrogen ion concentration increases which means that carbonate ions become less abundant, thus reducing calcium carbonate availability. This can cause one class of ecological changes associated with ocean acidification including thin shells and coral reef damage. It should be noted that the calcium carbonate formation connects the faster cycles of carbonate with the longer parts of the carbon cycle discussed earlier that causes a slow burial of carbon in the ocean sediments removing some of the carbon dioxide we have put into the atmosphere almost permanently.

The second major reservoir that affects atmospheric CO_2 is the land storage in plants that can further shift to the soils where more carbon can be stored as shown in Figure 7.3. The cycling times from this reservoir are months to centuries for various types of carbon molecules. One major difference between the ocean and biotic reservoirs is that the existing size of the former is fixed by the physical extent of the oceans, as it is for the atmosphere as well. But the sizes of the biotic and soil reservoirs can also change in response to the amount of carbon dioxide available from the atmosphere (see Section 6.4). Once the fossil source of carbon dioxide stops, it means that the re-growth of the forests, increase of ocean plant life and shuttling of carbon to soil microbial processes can continue to take up more carbon than at present and could therefore return the atmosphere to more natural conditions than the action of the oceans alone.

Residence Time: The amount of CO_2 we put in the atmosphere will partition itself into the land and ocean reservoirs by cross-media transport, and then into the many sub-reservoirs. The principles that control the residence time were illustrated using a simplified model in Section 6.4. The amount of CO_2 put into the atmosphere moves rapidly to the biota and somewhat more slowly to the oceans. This may include existing or new forests where the increased CO_2 provides a fertilization effect causing a more robust growth and storage. In time the pulse of CO_2 added to the atmosphere will come to equilibrium with the oceanic and land biotic reservoirs described by following equation based on carbon cycle models (see Chapter 6 and Endnote 7.5).

$$C_P(t) = C_{P0}(a_0 + a_1 e^{-t/\tau_1} + a_2 e^{-t/\tau 2} + a_3 e^{-t/\tau 3}) \tag{7.9}$$

Here C_{P0} is the amount put into the atmosphere at time zero and P is for pulse. If it is in kilograms, then the kilograms of this initial input that remains in the atmosphere at a later time is given by Eq. 7.9. The a_1 and τ_1 are constants determined from the models. The a_0 term represents the fraction that has an "infinite" atmospheric residence time, so its e $^{-t/\tau}$ term is just 1. The other terms have the following values (τ's in years): $a_0 = 0.2173$, $a_1 = 0.224$, $a_2 = 0.2824$, $a_3 = 0.2763$, $\tau_1 = 394.4$, $\tau_2 = 36.54$, $\tau_3 = 4.3$. The τ's represent combinations of the times due to the fast- and slow-acting reservoirs along with transport times. If you plot Eq. 7.9, it will be apparent that it cannot be approximated by a single exponential as we were able to do for other gases in our mass balance models. This equation has a number of important implications. It says that if you put 1 kg into the atmosphere at time zero, then about 30% will be gone in less than 10 years and half of it would be gone in 40 years but the rest will stay for a long time including about 22% that will remain for even longer times ($a_0 \approx 0.22$). As in the example of Section 6.4, the residence time increases with time after the pulse is put there. At first it will appear close to the shortest time scale of τ_3 representing oceanic and biotic exchanges. But this term goes to zero quite fast in about 2 $\times \tau_3 = 8$ years because of the saturation effect, leaving the other terms to continue to cause a decline of the pulse at ever-slowing rates. Similarly in about 70 years the τ_2 term drops out and what is left now follows the very long lifetime of τ_1 and after about a thousand years, you are left with the 22% of the a_0 term.

Although a single pulse decays according to Eq. 7.9, it is not the same as how sustained constant or increasing releases behave, as we already saw in the simplified model of Section 6.4. For such situations, which can be represented as a series of pulses one for each year, the residence time in the atmosphere is lower and remains lower for long times compared with a single pulse. Indeed if we apply Eq. 4.3 to the observed concentrations of CO_2 during the last 60 years we find that $\tau_{eff} = (C_{avg} - C_0)/[S_{avg} - (C_F - C_I)/\Delta t] = (355\text{ ppm} - 280\text{ ppm})/[3.8\text{ ppm/year} - (408.5\text{ ppm} - 316.5\text{ ppm})/60\text{ years}] \approx 34$ yrs (Endnote 7.6). Here C_I and C_F

are the concentrations (ppb) at the beginning and end of the period of observations Δt (yrs), S_{avg} is the average emissions rate (ppb/year), C_{avg} is the average concentration during this time, and C_0 is the pre-industrial concentration at time 0. This residence time for CO_2 is consistent with the observations over the period Δt (here, 60 yrs), but it cannot be extended confidently into the future because that depends on how the future emissions will evolve, which is not known, and ultimately on the saturation effect, which is certain to occur and extend the residence time of the remaining fraction. The short residence time of 30–40 years for the present budget represents the effects of the fast processes in Eq. 7.9 and is in fact the time scale on which our present environment and global warming are functioning for CO_2. The reason for this disparity between the pulse and sustained release residence times is explained as follows, similar to the explanation discussed earlier for a hypothetical and simpler situation. The exchange with the reservoirs, such as the oceans and land, is proportional to the near surface concentrations of CO_2 in the air and water. Every year that we add a larger amount than the previous year, this imbalance is re-established and a large transfer occurs. As long as we keep adding CO_2 the imbalance remains high and we see a sustained residence time dominated by the shorter lifetimes (τ_3 and τ_2). In other words, the terms they modify, the a's, are constantly being added to. If we stop adding CO_2, then the concentration gradient is not renewed and each year it becomes a little less, causing a slowdown of the uptake until the point of equilibrium. At that time the remaining CO_2 will not go into the water at all. It may still go into the biotic and soil reservoirs if they are not saturated and the carbonate in the oceans could go to deeper storage keeping the ocean sink active but slow.

It is noteworthy that while atmospheric chemistry does not remove CO_2, it does make some. When living things utilize the complex carbon molecules for metabolism, they emit a number of carbon-containing gases including methane, many non-methane hydrocarbons, and carbon monoxide (CO) into the atmosphere. Similarly burning carbon-containing materials, whether trees in forests or fossil fuels for energy, all produce non-CO_2 carbon-containing gases. These gases play their own roles in the environment during the time they are in the atmosphere, which is generally not long. By atmospheric chemical processes discussed earlier, often controlled by OH, they end up as CO_2 within a few hours to a few years. CO_2 is a fully oxidized state, meaning that it will not react any further. So, in the end, knowing how much fossil fuels are being burned globally in a year, gives us an accurate estimate of how much CO_2 is generated despite the fact that other gases are also created when the fuel is burned that are the bane of air pollution scientists and city dwellers. They are merely intermediates before converting to CO_2 in the atmosphere which is safe for us in urban environments.

The preceding discussion has implications for what causes long-term climate changing increases of CO_2 in the atmosphere and what doesn't. Burning wood for energy as long as you plant as many trees as you harvest would not lead to a long-term increase. We take out the CO_2 as our planted trees grow and we put it back when we burn them. Burning fossil fuels and land use change add CO_2 because it is being released from more stable reservoirs. The fossil fuel reservoirs store the carbon almost forever from our perspective. As discussed earlier, forests are also a long-term reservoir of carbon that will become CO_2 if the trees are cut down and not replaced. At the same time, they saturate when a balance is reached between new growth and loss due to dying trees, as well as the supply and processing of carbon in the forest soils.

Carbon dioxide from human sources can be seen as the major by-product of the industrial revolution that was stimulated by the benefits of a large ready source of energy. As a complement to this, methane and nitrous oxide are two other greenhouse gases that have come mostly from food production, although methane is also being released from energy use in increasing amounts. These gases have contributed to current global warming and are likely to continue to do so in the future. Let's take a closer look.

7.2.4 Methane

Methane is regarded as the most important non-CO_2 gas for causing global warming, but its environmental budget and behavior is very different from carbon dioxide as shown in Figure 7.4. Most of it is of biogenic origin produced by anaerobic methanogenic bacteria. The environments without oxygen, or anoxic, in which they can exist, are in the soils, especially under waterlogged conditions and in the guts of ruminants and wood eating insects such as termites. Although deeper regions of the earth are also anoxic, the bacteria that produce methane need a food supply that exists only near the surface. Looking at the budget in Figure 7.4 the main categories of emissions are biological, geological, and combustion. We can summarize the early-twenty-first-century budget of Figure 7.4 in Tg/y as $dC/dt = S - C/\tau$; $dC/dt = 17$ Tg/yr (increase) = 220 Tg/yr (natural) + 360 Tg/yr (man-made) - 500 (tropospheric OH) - 60 Tg/yr (stratosphere, chlorine, and soils). It should be noted that the trend is small compared with the sources and sinks and can therefore be taken as approximately zero in evaluating the mass balance representing a pseudo-steady state.

Despite the uncertainties, the importance of the biological source is apparent. On the natural side it comes from wetlands which also represent most of what is needed to explain the pre-industrial levels. The "man-made" biogenic source is a byproduct of agricultural

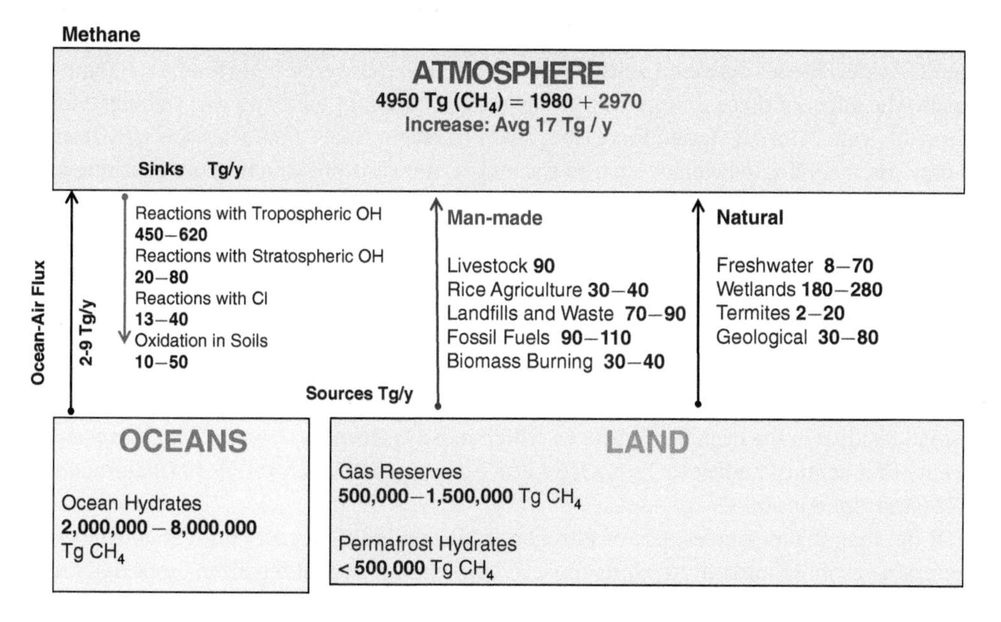

Figure 7.4 The global sources and sinks of methane (adapted from IPCC AR5, Endnote 1.1).

activities, mostly rice agriculture and domestic cattle, and to a lesser extent, waste management including wastewater treatment and landfills. The combustion sources are from biomass burning and there is a smaller natural source from forest or grassland fires. The geological or fossil sources include the seepage from deeper in the earth, which is small compared with the man-made component which includes drilling for oil, coal mining, and leakages from the use of natural gas. Deep in the earth methane is not as easily removed as in the atmosphere and therefore can persist for millions of years.

Since the atmospheric lifetime of methane is relatively short at around eight years a sizable annual emission has to be maintained even before industrial times for it to persist in the atmosphere at the levels we see. An increasing supply is needed to explain the trends assuming the lifetime is constant. The lifetime is dominated by reaction with hydroxyl radicals but dry soils, the stratosphere, and reactions with chlorine in the air above the oceans may together remove up to 25% annually.

We saw from our calculations earlier that there is now a large contribution of human activities to the methane concentrations which explains the increase by some 2.5 times over the pre-industrial values seen in Figure 7.1. Many of the details of the increase are also consistent with our knowledge of how the man-made sources increased over time during the last century. Earlier in the time from around the 1800s to the later decades of the twentieth century, there was a major increase in the hectares of rice harvested and the cattle populations, in keeping with the growing human population. In time, this connection was broken. It happened because of both social and economic changes and most importantly, as the agricultural sources reached their limits of growth. Rice agriculture extended to all the land where it could be readily grown, new varieties were developed that required less time to produce as much rice as before. In China, shortages of irrigation water led to reductions in the time the rice fields were inundated and the use of organic wastes were replaced with nitrogen fertilizers. These processes are ongoing and all lead to a reduction in emissions of methane even as rice production increases. For cattle, the populations in India, where there were the largest numbers, reached limits of sustainability. And in western countries, new breeds delivered dairy products and meat more efficiently than free ranging herds. The effect of these limits was that the concentration of methane was stabilized for about 20 years before it started increasing again in recent times. Sources, especially from energy use including new ones, such as fracking as a means of extracting oil, continue to rise and are now contributing to increasing levels of methane. They have plenty of potential left.

7.2.5 Nitrous Oxide

As we saw, even now, most of the annual emissions are from natural sources which we have identified in the budget as soils. It is produced there by bacterial processes and is only a small residual in the cycling of nitrogen which mostly returns N_2 (Figure 7.5). The early-twenty-first-century budget in $Tg(N_2O)/yr$ is $dC/dt = 5.7 \approx 17$ (natural) + 10 (man-made) – 22 (stratospheric sink).

Of the man-made sources, use of nitrogen fertilizers is the largest and it is constantly increasing as more agriculture shifts toward using manufactured fertilizers, especially in countries such as China. In addition, there are quite a few smaller man-made sources. Each one may be considered negligible, but taken together, they add up to a sizable annual emission rate comparable to the fertilizer source. This aspect of the budget makes it

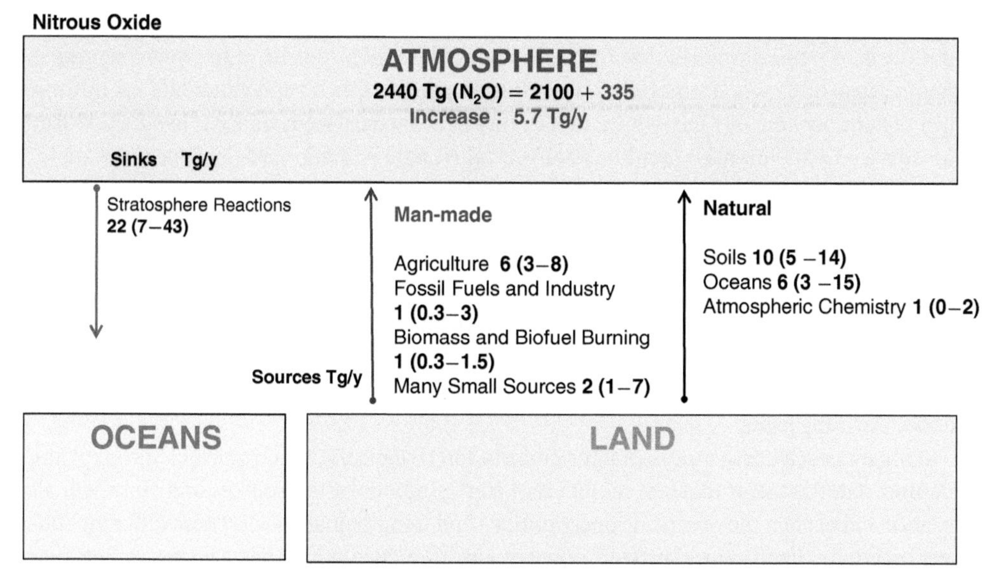

Figure 7.5 The nitrogen balance and nitrous oxide. Nitrous oxide is virtually the only nitrogen gas with a direct effect on the earth's climate and particularly on global warming. It is seen as a residual in the biological processes that occur mostly in the soils (adapted from IPCC AR5, Endnote 1.1).

difficult to reduce the global emissions of N_2O effectively. It is noteworthy that nitrous oxide is also an ozone depleting compound. It is likely that in the future its contributions to both global warming and ozone depletion will be of concern.

7.2.6 Many Other Greenhouse Gases

Aside from the gases already mentioned, there are dozens of man-made technological compounds that are very efficient at causing global warming, but their concentrations are so low as to have only a modest effect on the current climate, and never expected to get high enough to match the global warming effects of CO_2, CH_4, and N_2O. Chief among these are classes of perfluorocarbons (PFCs), hydrochlorofluorocarbons (HCFCs), hydrofluorocarbons (HFCs), and sulfur hexafluoride. They would be set aside to watch, except that they can have a role in a comprehensive plan to limit or manage future climate change. In current agreements to reduce global warming, controlling smaller amounts of these super greenhouse gases can be traded for emitting larger amounts of CO_2. We will return to this matter later.

Although each of these gases has a story of its own, perhaps that of sulfur hexafluoride will interest readers. It is present in the atmosphere at a few parts per trillion, but it is included in the target gases under the Kyoto Protocol. It is estimated to be 20,000 times as potent as CO_2 for causing global warming and has a lifetime perhaps around or longer than 3000 years. It is possible to measure its concentrations down to extremely low levels of around a tenth of a part per trillion. For that reason, it was used in stacks of power plants and other burners to study how the effluents spread out into the environment and cause city-wide or regional air pollution. So, some of the SF_6 in the atmosphere is there because of air pollution research. It was used in consumer goods such as tennis balls and air soles of athletic shoes. It is heavy, so it doesn't leak easily from these applications, and it makes the balls bounce better. So, some of it is in the atmosphere from sporting goods. Since the lifetime is so long, almost all that has been released is still in the air. It has been suggested that because of its wondrous global warming properties, it is a good candidate for putting into the atmosphere of Mars where it may stay near the surface because it is heavy, warm the surface by its intense greenhouse effect, rejuvenate a water cycle, and perhaps make Mars livable for us (Endnote 7.7).

7.2.7 Balanced Budgets and Uncertainties

A closer examination of the budget figures shows that there are considerable uncertainties in the estimates of present sources and sinks. These uncertainties reduce our ability to predict future concentrations, even though the causes of emissions are well defined for all the global warming gases.

In some cases, it doesn't mean that the budgets don't balance. For methane and nitrous oxide, the uncertainties mean that several different configurations of the sources and sinks will all balance and explain the observed concentrations and trends equally well. These different budgets may have significant effects on how we aim to reduce emissions. The separation into "natural" and "anthropogenic" components based on ice-core data can eliminate many, but not all the possibilities (Endnote 7.8). For carbon dioxide, the global burden from measured concentrations is precisely known; there are robust estimates of the emissions from known energy production; and extensive ocean concentration data are available that constrains the amount that has gone there. When these three items are taken together with the observed increase, they do not balance. The effect of land response is the remaining component and it is not well constrained. In the end, the land has to act as a net sink to balance the budget. It is apparent that deforestation is not the only major process on land affecting CO_2, since it is a net source, although hard to estimate accurately. Once that is taken into account, we need a sizable 2–3 PgC/year sink on land which may come about as the earth's natural system re-balances the added atmospheric carbon with the land reservoir as discussed previously. Changes in the storage of the carbon on land are more difficult to observe than for the oceans. This is partly because the effect of land not only includes changes of above ground long-lived biomass (natural regrowth, afforestation, etc.), but also changes in below ground biomass and increase in soil carbon stores by various biotic processes (see Ch. 6.4 Atmosphere-Land Exchanges).

7.2.8 Ozone

Tropospheric ozone is a powerful greenhouse gas that makes a contribution to the natural warming effect. Some of it comes by the air exchanges with the stratosphere but most of it is formed and destroyed in the troposphere by chemical processes similar to what we discussed in Chapter 6 for the stratosphere. Soils emit NO which is itself rather short-lived. It destroys any ozone that is present by the reaction: $O_3 + NO \rightarrow NO_2 + O_2$. But in the process, it generates NO_2 which participates in two reactions, $NO_2 + h\nu \rightarrow NO + O$ and then $O + O_2 + M \rightarrow O_3 + M$, that makes ozone. The latter is familiar from our discussion of the ozone layer (Eq. 6.15b) and the former supplies the oxygen atoms needed to make ozone in the troposphere. Here the direct photolysis of oxygen (Eq. 6.15a) cannot supply the O atoms because there isn't enough high-energy sunlight available. If we add carbon monoxide (CO) and hydrocarbon gases or volatile organic compounds in general (VOCs) emitted from plants and soils, these will breakdown by reacting with hydroxyl radicals by: $VOC_i + OH \rightarrow RO_2 +$ Other Molecules. The RO_2 can be various fragments that react rapidly with NO to make NO_2 as in: $RO_2 + NO \rightarrow NO_2$ + Other Molecules. The effect of this process is to remove NO, which would have destroyed the ozone that may be present, and generate NO_2 that will make more ozone. When this happens, sizable amounts of ozone can be added to the troposphere. Similar processes are responsible for causing air pollution in cities where the NO and the hydrocarbons are supplied by the exhaust from cars and other vehicles. As human activities supply increasing amounts of NO, CH_4, and other hydrocarbons, the background concentrations of ozone rise, adding to global warming. Various observational data sets have shown that tropospheric ozone has been increasing for some 50 years now.

The sinks are chemical processes as shown above, and a deposition to the surface. The latter represents a loss that arises when ozone encounters a surface. In the interaction ozone is lost as in a pure sink and cannot be regenerated. Ozone interactions degrade various materials in the human environment, especially rubbers and plastics, by this deposition process. A budget of ozone, with wide ranges, is approximately stated in Tg/yr as: dC/dt (small) $= 4500\ (= S_{\text{Chem}}) + 500\ (= S_{\text{Strat}}) - 4000\ (= C/\tau_{\text{Chem}}) - 1000\ (= C/\tau_{\text{Deposition}})$. Here the global tropospheric burden is ~ 300 Tg from observations, giving an average lifetime at ~20 days (see IPCC, TAR, Endnote 1.1).

The role of ozone in the environment can be seen in four space scales. In large urban areas it is a major component of photo-chemical smog. It causes damage to materials and human health. Its precursors as mentioned earlier are NO, NO_2, CO, and hydrocarbons. Since it is not emitted directly, to reduce air pollution the emissions of the precursors are controlled by regulations in most countries including emissions from automobiles. The chemical processes that make and destroy it play out over large regions downwind of urban areas creating high concentrations over rural agricultural fields and forests, supplemented by natural emissions of the precursors. Sufficient concentrations can occur that can harm crops and reduce yield. Further regulations are sometimes needed to manage these regional scale concentrations. Even further out, in the larger troposphere, its increases add to global warming. At the same time, it is a key ingredient in making OH radicals that remove and cycle gases. It can, by direct chemical reactions, remove some hydrocarbon and other gases. And finally, high up in the stratosphere, it is a good thing. International agreements such as the Montreal Protocol have been signed to limit the processes that can destroy it there. The story of ozone is enigmatic therefore – it is regarded as generally bad for human life in the troposphere and good in the stratosphere, rather jocularly referred to as "bad ozone" and "good ozone." Ironically, human activities tend to increase bad ozone and decrease the good.

7.3 Aerosols

Aerosols are fine particles in the atmosphere that are the bane of city dwellers where their concentrations surge, causing both short-term and long-term health problems and requiring environmental regulations to control their prevalence. It is the only form of air pollution that is visible to the eye as a haze. In the global environment however, fine particles generally act to cool the temperatures either directly by scattering sunlight, or by their complex interactions with clouds, including more cloud formation and increasing their albedos. Because of this role, aerosols are used for seeding clouds to make rain and have a potential to reduce global warming. Light-absorbing carbonaceous aerosols including soot and biogenic carbon absorb radiation and lead to warming, but their emissions and abundances are much lower that the light-scattering aerosols. For climate change as it is happening at this time, the effect of aerosols is to cool the earth (Figure 7.6).

Aerosols are distinguished as primary or secondary. Primary aerosols start out as particles that get suspended in air. These include dust and sea-spray which tend to be fairly large in size with diameters of 5–20 microns and usually settle out close to the sources due to the earth's considerable gravity (1 micron = 1μ = 1 micrometer = 10^{-6} m). Finer dust can be suspended in air for some times and travel over long distances. A well-known case is Sahara Dust. It can have particles that range in size from about 1 to 50 μ when it lands, but most of the particles are on the smaller side of the range. It starts out in the Sahara and can travel for up to ten thousand kilometers and spread out over large regions. It is a seasonal phenomenon that is generated by unique environmental conditions of land and winds. The Yellow Dust that originates in China is a similarly large-scale phenomenon.

For climate change, secondary aerosols, that are fine particles, are more effective. Most of these start out as sulfur- or nitrogen-containing gases, or organic compounds, and convert to particles over a time, often while still close to the source. Their atmospheric behavior is quite different from the larger primary particles. For one thing, the gas precursors can get their sources higher and spread out before they are even formed. These gases undergo chemical reactions often initiated by hydroxyl radicals (OH) and convert to sulfate and nitrate aerosols. At that point they are very small, of sub-micron sizes (0.01–0.2 μ), and can continue to spread

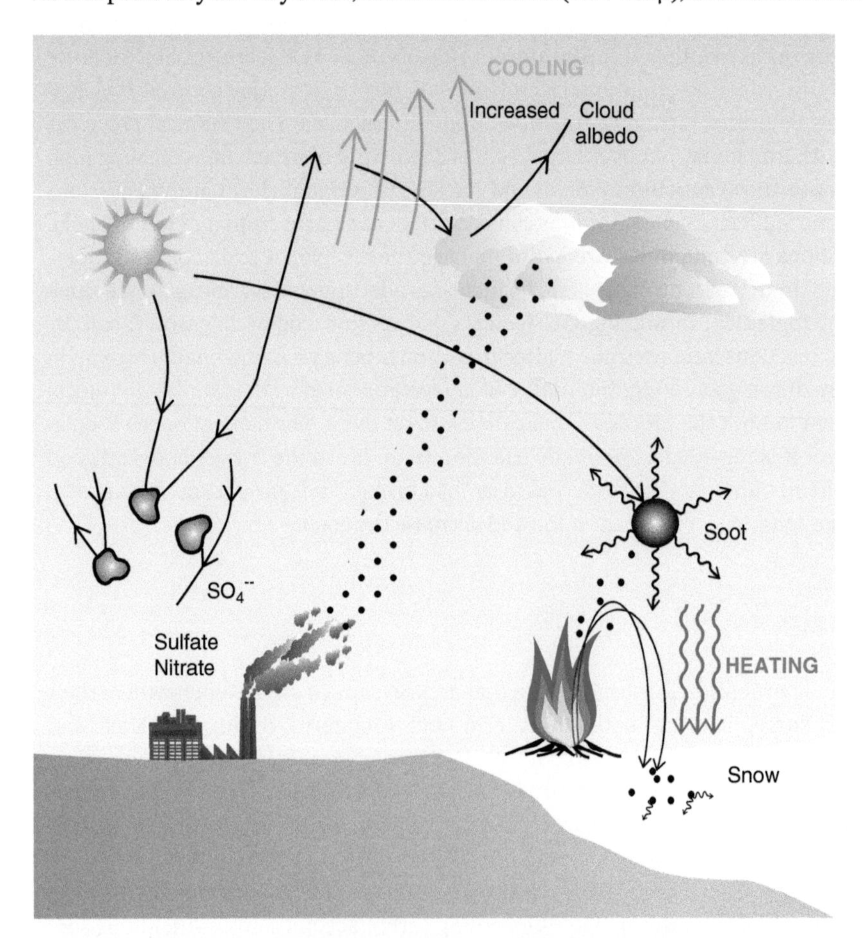

Figure 7.6 Aerosols and climate. Fine aerosols of diameters 1–5 micrometers are generally formed in the atmosphere and persist for some time. Such particles have large-scale or global effects on the climate that is mostly cooling by sulfate and nitrate aerosols, somewhat offset by light-absorbing aerosols. Larger particles have more localized influences.

because they can act more or less like a gas, and therefore resistant to settling due to gravity. They grow and pick up water reaching stable sizes which are still in the range of a few microns (0.1–5 μ) and stimulate the formation of clouds. Their residence times are extended by their small sizes and the effect of turbulent mixing processes. Their size also makes them good at scattering light, thus giving them a role in the climate. Moreover, they can contribute to the formation of clouds and to cloud albedos. Natural sources of such aerosols include the emission of sulfur gases such as SO_2, H_2S, OCS, CS_2, and DMS (CH_3SCH_3); nitrogen gases such as ammonia (NH_3) and nitrogen oxides, and biogenic hydrocarbons from plants and forests. The latter include isoprene (C_5H_8), a dominant plant emission, as well as other hydrocarbon species of increasing molecular complexity. These react quite rapidly with oxidants and generate natural ozone and fine particles. The often observed haze over forested areas is attributed to these natural emissions (Figure 7.7).

Human activities have been adding to these precursor gases. The sulfur gases come mostly from fossil fuel combustion and some industrial processes. The nitrogen gases from human activities include combustion, excreta, and fertilizer use. The addition of fine particles from human activities can scatter more light, make more clouds, and increase cloud albedos. By these processes, they offset global warming (Figure 7.7, Endnote 7.9). We will see later that the impact on climate is observable, in the sense that the increase of greenhouse gases would otherwise over-predict global warming at this time, compared with the observations.

This completes the consideration of how atmospheric composition is affected by the three main parts of the earth system we originally delineated – the atmosphere, oceans, and land. We will move to looking at how the earth's climate is formed incorporating what we have learned here.

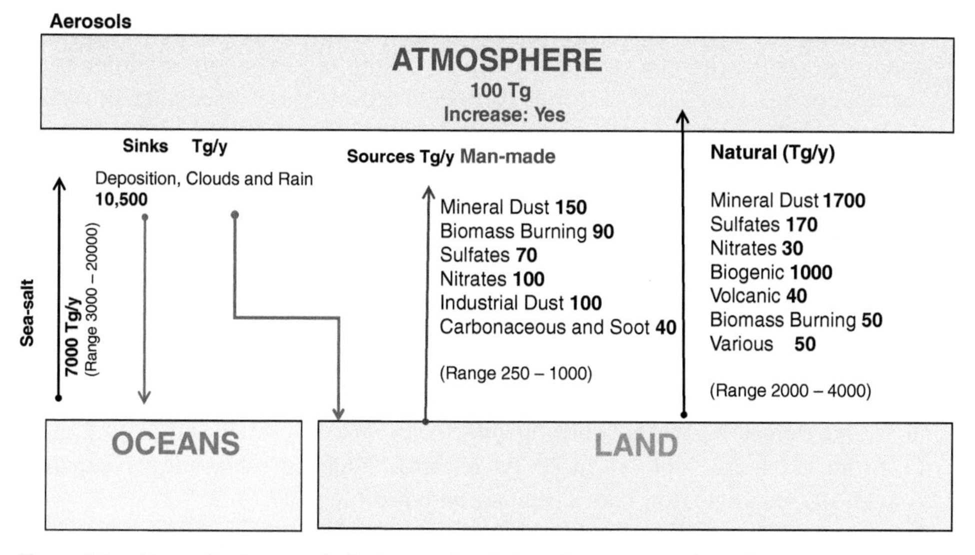

Figure 7.7 Atmospheric aerosols. Estimate of emissions from man-made and natural sources are shown. The fine aerosols that are generated from gas-to-particle conversion, such as from sulfate and nitrate, carbonaceous, biogenic or from biomass burning, persist in the atmosphere for some weeks in the troposphere. Larger aerosols, such as sea-salt or dust, have shorter residence times in the atmosphere. The uncertainties in the estimates are very large.

Review of the Main Points

1. An important topic of the discussion was the division between natural and man-made sources, which is artificial because humans are indeed part of nature, but it provides a tried and tested means to manage the environment and remedy undesirable developments.
2. Trace greenhouse gases, carbon dioxide, methane, and nitrous oxide have increased considerably over the last century as a byproduct of supplying energy and food. In addition, long-lived technological gases such as the chlorofluorocarbons, perfluorocarbons, and others have become a new component of the atmosphere and their use has increased social affluence. The precursors of ozone have added to its concentration in the troposphere, all contributing to a greater total abundance of climate changing gases in the atmosphere.
3. Both the black carbon aerosols that cause warming and the rest of the atmospheric aerosols that cause a cooling influence have been constantly added to the atmosphere. The sources of these are burning fossil fuels and biomass for land clearing and other purposes. They are short-lived, however, and their influences are more regional than global.

Exercises

1. A farmer plants a row of trees every year for ten years. Then he harvests the first row of trees (10 years old) for fire wood and re-plants the row. Next year he harvests row #2 and re-plants it, and so on. After year 10, the first row is ready for harvesting. This way the farmer has a perpetual supply of fire wood.

   Let's say that in the first year the trees take up C_1 tons of carbon from the atmosphere, and in the second year the same row takes up C_2 tons of carbon and so on for each year that passes. How much carbon is there in the row that is burned after 10 years? What is the effect of this practice on amount of carbon in the atmosphere? Plot the loss or increase of carbon in the atmosphere over 12 years from the time the first row is planted. Can this type of wood farming, slash-burn-replant, cause a buildup of CO_2 in the atmosphere? Why or why not?
2. The human population has increased greatly during the last 100 years going from 1.3 billion to 7 billion. It is claimed that the CO_2 exhaled by human beings has increased proportionately over this time and is now of significant magnitude compared with the amount of CO_2 put into the atmosphere by burning fossil fuels which is 7 PgC/yr. Assume that human breath consists of 5% CO_2, the average breath volume is 700 cm^3, and average breathing rate is 14 breaths/min.
   (a) Estimate the excess of CO_2 in PgC/yr released by human beings today compared with 100 years ago. How does it compare with fossil fuel CO_2?
   (b) Does this CO_2 contribute to the observed increase of CO_2 in the atmosphere? Why or why not?
3. The mass mixing ratio of water (Figure 7.2a) can be represented by this equation: w (g/kg) $= 7 \times 10^{-0.21\,z(\mathrm{km})}$. What are the heights below which we have 50%, 80%, and 90% of the water vapor in the atmosphere? Comment on how this affects the greenhouse effect and global warming.

4 It is claimed that in the process of producing OH, a lot of water vapor is lost from the atmosphere, making it an important sink.
(a) Evaluate this proposition. Start by assuming that the lifetime of OH is about 1 s and the concentration throughout the troposphere is 10^6 molecules/cm^3. From this calculate the annual production rate, or source of OH in molecules/year. Assume also that $O(^1D) + H_2O \rightarrow 2\ OH$ is the only way OH is made, which is consistent with the text. Then note that for every two molecules of OH produced, one water molecule is lost. Proceed.
(b) Compare your results with the budget shown in Figure 7.2. How important a sink is it? What is the lifetime of water vapor relative to just this sink?

Endnotes

Endnote 7.1 Precise and reliable measurements of atmospheric composition have been taken only in recent times, starting in the twentieth century. Concentrations of environmentally important gases for earlier times come from ice cores and more recently from measurements in the firn layer of polar ice. This is how it works. There is very little snowfall in polar regions annually. The snow doesn't melt, however, and accumulates for hundreds of thousands of years. As it is building up, air is exchanged with the atmosphere. As the weight of the snow pushes down, it compresses the lower layers causing them to turn to ice. The fluffier layers above the ice line are the firn. As the ice forms, it traps the air that was in the void fraction and seals it in bubbles. That air represents the atmospheric concentrations at the time the bubble closed off. The ice keeps building but at the bottom of the ice sheet the pressure is so high that it melts again and this water moves on. So the ice sheet doesn't grow any more, but it stores the air from various time periods. This time can be from 50–100 years when the firn closes to 100,000–600,000 years when the bottom ice melts! Ice is drilled and a core is extracted. The atmospheric composition is measured by high-precision modern techniques to get the long-term atmospheric concentrations. Despite the availability of the ice, there is not enough air to be able to measure most gases, but methane, carbon dioxide, and nitrous oxide, the main global warming gases, can be reliably measured and are reported in the literature. Still, the most recent times cannot be recovered from the bubbles because the earliest ones that closed are some 100 years old, more or less. In recent times, the techniques have been extended to measure the air in the firn layer. Large quantities of air can be drawn from various layers and many gases can be measured. Both the bubble and firn measurements require models to determine the age of the ice or the air and this creates uncertainties in the results. Other complications can arise due to slow chemical processes that can compromise the data over such long periods. And reactive gases such as OH cannot be so measured. Nonetheless, there are no better means to know the longer-term changes of atmospheric composition. The same ice cores are also used to get a proxy measurement of the earth's temperature over the last 400,000–800,000 years. This temperature record, as interesting as it is, does not tell us much about the recent times when man-made climate change started. Reliable direct measurements have been assembled for the last 170 years. These temperature measurements become increasingly complete and accurate as we consider more recent times. The discoveries about the increases of greenhouse gases, their pre-industrial and long-term concentrations were first reported by the following authors: CO_2 increase: C.D. Keeling. Tellus, 12, 200–203, 1960 and Proceedings of the American Philosophical Society, 114, 10–17, 1970. Increases and pre-industrial levels of CH_4 and N_2O: M.A.K. Khalil and R.A. Rasmussen in, (increase of methane): Atmospheric Environment 15, 883–886, 1981; J. Geophys. Res., 86, 9826–9832, 1981 (increase of N_2O): Tellus, 35B, 161–169, 1983; (pre-industrial CH_4): J. Geophys. Res., 89, 11,599–11,605, 1984; (pre-industrial N_2O): Annals of Glaciology, 10, 73–79, 1988. Pre-industrial CO_2: J.M. Barnola et al. Nature, 303, 410–415, 1983. Firn layer concentrations of gases: M. Battle et al., Nature, 383, 231–235, 1996 and James H. Butler et al. Nature, 399, 749–755, 1999. Long-term CO_2 record in ice cores: J.R. Petit et al., Nature, 399, 429–436, 1999.

Endnote 7.2 Detailed information and derivations of the Clausius-Claperyon equation can be found in J.M. Wallace and P.V. Hobbs Atmospheric Science, 2nd Edition, Academic Press, 2006. Readers seeking a definitive textbook on the atmospheric sciences may find this classic work of considerable value.

Endnote 7.3 Figure 7.2a is adapted from D. Geffen in: "Water Vapor in the Climate System," AGU Special Report (Washington D.C., 1995). Figure 7.2b is adapted from K.E. Trenberth et al. (2007): "Estimates of the Global Water Budget and Its Annual Cycle Using Observational and Model Data," *J. Hydrometeorology*, 8, pp. 758–769.

Endnote 7.4 In pure water, some H_2O dissociates into OH^- and H^+ giving it a pH of 7 based on the rate constant for this reaction, which is 1.82×10^{-16} M. M is for "molar" – the units of $[H^+]$ or $[OH^-]$. 1 M = 1 mole per liter of water. Since the ocean has a pH of about 8.189 it says that the ocean is basic – not acidic. Ocean acidification should be seen as a trend toward becoming less basic and not actually resulting in an ocean that is acidic. "Ocean less-bacification" doesn't sound as good as "ocean acidification." Useful rate constants are: $K_H = 0.0034$ M/atm, $K_1 = 4.3 \times 10^{-7}$ M, $K_2 = 4.7 \times 10^{-11}$ M from J.H.Seinfeld and S.N. Pandis, Atmospheric Physics and Chemistry, J.Wiley & Sons, 1998. Mathematically advanced readers may find additional information on many subjects in this 1326-page compendium of atmospheric sciences subtitled "From Air Pollution to Climate Change".

Endnote 7.5 Pulse decay formulas, including Eq. 7.3, are a result of a general principle of box models in which the different τ's reflect combinations of transport and lifetimes in the boxes. The one used here is from F. Joos et al. *Atmospheric Chemistry and Physics*, vol. 13, p. 2793 (2013). Such a formula was adopted by the IPCC and reported in AR4, Chapter 2, which was from F. Joos et al. GBC, vol. 15, p. 891, 2001 (Endnote 1.1).

Endnote 7.6 Starting with Eq. 5.3, $dC/dt = S - C/\tau$ a simpler equation can be derived that is often useful in examining budgets of atmospheric gases. We average both sides to get: $(CF - CI)/\Delta t = S_{avg} - C_{avg/}\tau_{eff}$. Here C_I and C_F are the concentrations (ppb) at the beginning and end of the period of observations Δt (yrs), S_{avg} is the average emissions rate (ppb/year), C_{avg} is the average concentration during this time τ_{eff} is an effective average lifetime. In applying it to CO_2, the S represents the man-made sources, so the pre-industrial concentration has to be subtracted from the measured concentrations to connect this source to its impact. If concentrations are known from measurements and the sources can be evaluated independently, the effective lifetime can be calculated from this equation that would represent the period over which the average was taken. The τ_{eff} depends on the averaging period, Δt, if it is not constant.

Endnote 7.7 Some speculations on how to make Mars habitable may be found in: C. McKay, et al. "Making Mars Habitable," *Nature*, vol. 352, pp. 489–496, 1991. M. F. Gerstell et al. "Keeping Mars warm with new supergreenhouse gases," *PNAS*, vol. 98, pp. 2154–2157, 2001.

Endnote 7.8 The issue of uncertainties in the estimate the of global sources for gases such as methane and nitrous oxide arises in part from having to extrapolate observations taken in localized environments to the whole world. For instance, we can put boxes on the soil and measure how much methane is coming off the soil in a wetland, but then extrapolating that to the whole earth with so many wetlands that function under varying environmental conditions, will have inherent uncertainties. The more places we can sample, the less the uncertainty may become, but that has its own practical limitations. This is the upscaling problem. It may be partially resolved by global space based remote sensing measurements. The constraints from pre-industrial concentrations compared with present can be very informative and must be satisfied for any budget to be considered valid: M.A.K. Khalil and R.A. Rasmussen, Constraints on the Global Sources of Methane and an Analysis of Recent Budgets, Tellus 42B, 229–236, 1990.

Endnote 7.9 Data for Figure 7.7 is taken from: C. Tomasi and A. Lupi, in *Atmospheric Aerosols: Life Cycle and Effects on Air quality and Climate*, pp. 1–86 (Wiley-VCH Verlag GmbH & Co., 2017). For more detailed information about atmospheric aerosols, readers may consult: S. Ramachandaran, *Atmospheric Aerosols* (CRC Press, Boca Raton, FL, 2018).

8

The Science of Climate

A plan was stated in Figure 2.8 consisting of three main components for our inquiry into climate, global change, and the human dimensions involved. In it the atmospheric composition was a major component that determines the earth's climate. Since the expected climate change is driven by increasing greenhouse gases from human activities and the slower natural cycles, we have spent considerable effort to understand atmospheric composition in the last seven chapters. Our discussion so far completes this component. The other two ingredients, whether to understand the natural climate or its changes, are solar radiation arriving at the top of the atmosphere and the albedo. We will describe these next, followed by the complex task of combining the three pieces to describe the earth's climate.

8.1 Solar Radiation

The solar radiation at the top of the atmosphere is the major external influence on the earth's climate as it is for every other planet. How this radiation interacts with the earth's system determines our environment and distinguishes it from what happens on other planets. The sun supplies all the energy that drives life on earth and has done so over its entire evolution. At this time, and for many millennia before, solar radiation has been about 1360 watts/m^2 with small, and at times meaningful, variations. Let's see how this comes about.

The sun, earth, and most objects in our experience can be approximated as black bodies. A black body is a hypothetical object that absorbs all electromagnetic radiation that falls on it regardless of the frequency and amount. It then emits energy at all frequencies with a spectrum that is described precisely by Planck's black body radiation law (Endnote 8.1). The gist of the law is to tell us how much energy is radiated at each wavelength of the electromagnetic spectrum and that *it depends only on the temperature of the body*. Hotter bodies will radiate more energy in the ultraviolet and cooler bodies radiate more in the infrared. This law can be simplified in two ways. One is to evaluate the total emission of energy at all wavelengths from a black body which results in the Stefan-Boltzmann law (Eq. 8.1) and the other is to find the wavelength at which the maximum radiation occurs, which is Wein's displacement law (Eq. 8.2):

$$E = \sigma T^4 \tag{8.1}$$

Global Climate Change and Human Life, First Edition. M. A. K. Khalil.

Companion Website: www.wiley.com/go/khalil/Globalclimatechange

$$\lambda_{Max} = \frac{2898}{T} \tag{8.2}$$

In Eq. 8.1, $\sigma = 5.67 \times 10^{-8}$ w/m^2-°K^4. Although not a fundamental constant of nature, it is made of them and is therefore an absolute constant to the extent that the fundamental constants are. E therefore has the units of w/m^2 and represents the energy flux of the black body that has the temperature T (in °K). Equation 8.2 gives the wavelength in microns (1 μm = 1μ = 10^{-6} m) at which the black body radiates the most energy. The constant in Eq. 8.2 is in μm-°K, and is also absolute. These results are valid for perfect black bodies; real objects often depart from the ideal. We can extend the idea to "gray bodies" by defining an absorptivity and an emissivity. The absorptivity α_λ is the fraction of the radiation the object actually absorbs at wavelength λ rather than all of it. It can be proved by thermodynamic arguments that if the body only absorbs a fraction α_λ of the radiation at some wavelength λ, then it will also emit exactly that fraction compared with a perfect black body emission at its body temperature. This is Kirchhoff's law, that the emissivity $\varepsilon_\lambda = \alpha_\lambda$, the absorptivity (Endnote 8.2). We can modify Eq. 8.1 to add an averaged emissivity making it $E = F\sigma T^4$ for objects that are not perfect black bodies.

We can now calculate the amount of solar radiation that arrives at the top of the atmosphere based on astronomical variables and the black body law. Let T_O be the surface temperature of the sun, R_O its radius, and D the distance between the centers of the earth and the sun (Figure 8.1). The total energy (joules) radiated by the sun per second (watts) is: $E_O = \sigma T_O^4 (4\pi R_O^2)$. Conservation of energy tells us that the solar energy must all go through any sphere we draw around the sun from its center including one that has the radius = D, the distance of the earth from the sun. The sun's emitted energy will be distributed evenly everywhere on the surface of this sphere, which we can designate as S_O. Then, $E_O = S_O 4\pi D^2$ which gives:

$$S_O = \sigma T_O^4 \left(\frac{R_O^2}{D^2}\right) \tag{8.3}$$

S_O is the *solar constant* representing the sun's energy flux arriving at the top of the earth's atmosphere. Putting numbers into this equation, with a solar temperature of 5760°K, gives 1358 watts/m^2. The solar constant varies cyclically a little (~0.15%) due to the 11-year solar cycle and even less by other cycles.

If we put a large screen some distance behind the earth – the side facing away from the sun, we will see a circular shadow of the same radius as the earth's, establishing that the earth has removed the energy that falls on an area πR_e^2, where R_e is it the radiu of the earth (Figure 8.2). The surface area of the earth however is four times larger at $4\pi R_e^2$; but at any time only half is lit, so the average radiation on the whole surface is ½ S_O. This amount occurs for only ½ the day, so if we average the flux over the day and night in addition to averaging it over the surface, we get $1/4\ S_O = S = 342$ w/m^2 representing the average amount of radiation that comes to the top of the atmosphere for every square meter of the earth's surface during one day. It is common practice to use this diurnally and globally averaged solar constant to represent the energy the earth receives. Since we will not be concerned with periods as short as a day, this will suit our calculations.

Figure 8.3 shows the spectrum of solar radiation at the top of the atmosphere and what happens to it along the way to the surface. In the figure the total solar radiation reaching the top of the atmosphere is closely approximated by the Planck black body radiation law applied to the sun (Endnote 8.1). Moreover, it shows that the peak of the spectrum, where

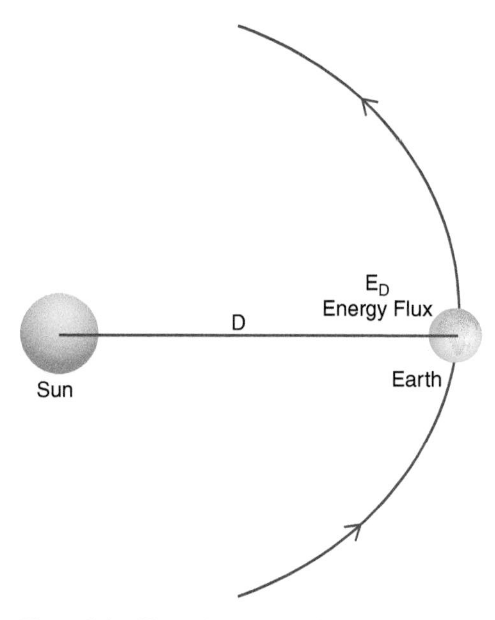

Figure 8.1 The solar constant. It is the solar energy flux (j/m^2-s = w/m^2), or the amount of energy per unit area arriving at the top of the earth's atmosphere every second. It will be the same on a sphere surrounding the sun with a radius equal to that of the distance between the sun and the earth (*D*).

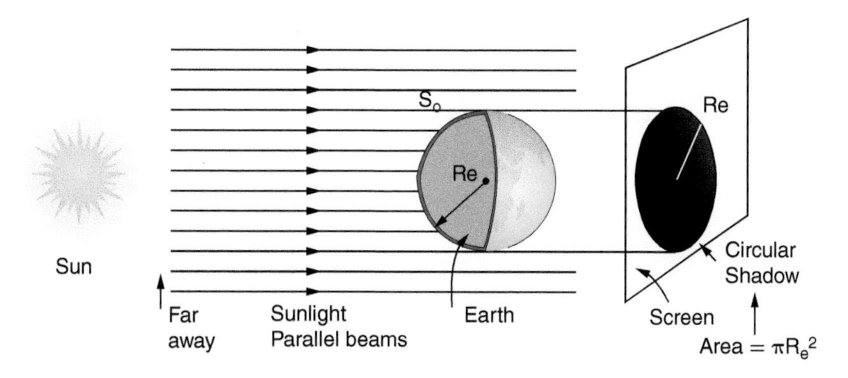

Figure 8.2 Interception of solar radiation by the earth. The sun radiates energy in all directions. The earth is far enough away that a small amount comes to the earth in beams that are nearly parallel. The earth absorbs as much radiation as a disk of the same radius as illustrated by putting a (hypothetical) large sheet of paper behind the earth.

the sunlight is most intense, is in the visible region, which we can also find from putting numbers into Eq. 8.2: $\lambda_{Max} = 2898/5760 = 0.5\ \mu m = 500$ nanometers. It is no accident that our eyes have evolved to see things in this "visible region" of the spectrum, because these are the wavelengths where most of the sunlight is. Finally, the figure shows that a significant amount of the sunlight is reflected and a lesser amount is absorbed. The absorption includes the ultraviolet that is taken out by the ozone layer, as we discussed earlier, leaving very little of it to reach the surface and almost none below 300 nm. The remaining absorption is in the troposphere, due mostly to water vapor. The major components of the atmosphere N_2, O_2, and even CO_2 absorb almost no sunlight; we can say that dry clear air is mostly transparent to solar radiation. Why this is so will become clear later.

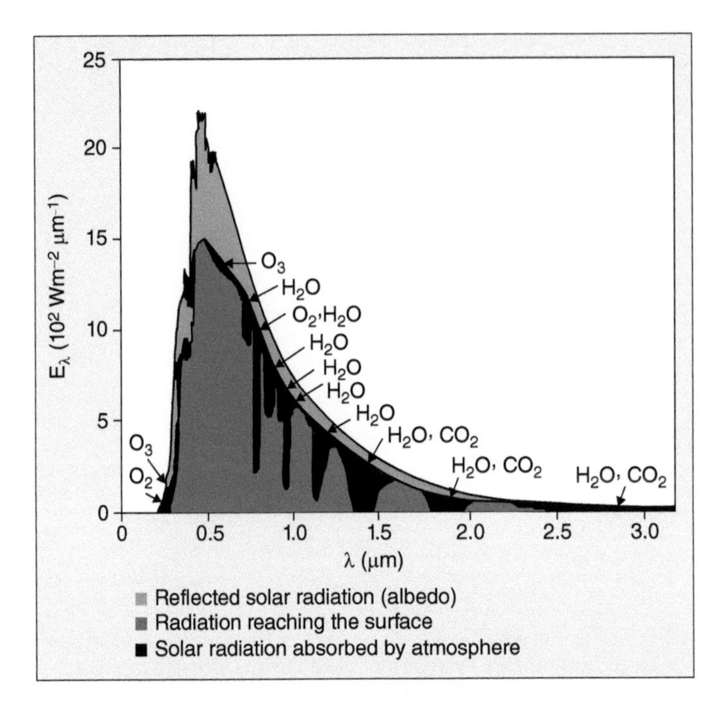

Figure 8.3 Solar spectrum. The figure shows the distribution of solar energy as a function of wavelength. The Planck black body curve (not shown) is close to the observed energy distribution of the sun. Absorption by water vapor is the largest effect. Absorption of the ultraviolet in the stratosphere by the ozone layer appears on the left. The albedo is the difference between the light at the top of the atmosphere and that absorbed at the surface and the atmosphere. It is the fraction of sunlight reflected to space.

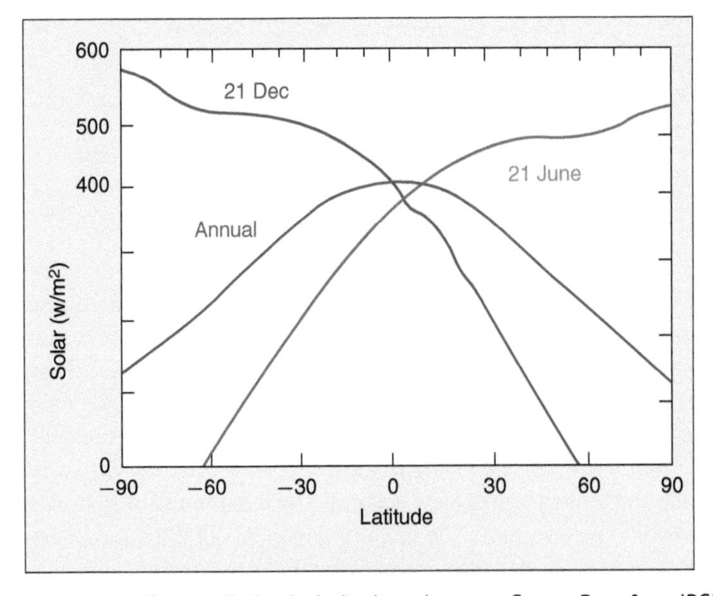

Figure 8.4 Solar radiation by latitude and season. *Source:* Data from IPCC AR5.

As we know from experience, solar radiation varies significantly by season and by latitude. Average conditions are shown in Figure 8.4 (Endnote 8.6). In it we see that near the equator, the radiation is the highest and varies little by season. At the polar latitudes it varies the most and is on average the least.

8.2 Albedo

Global albedo is the reflection or scattering of sunlight leaving at the top of the earth's atmosphere without having been absorbed. It has no role in the greenhouse effect (but can affect atmospheric photochemical processes). The albedo can be separated by the components previously defined as land, oceans, and the atmosphere. Between 22% and 25% of the incoming radiation at the top of the atmosphere is reflected, mostly by clouds with lesser contributions from scattering by air molecules and aerosol particles. Another 5%–7% is reflected from the oceans and land making a total reflection of about 30%. The contribution of a component to the global albedo depends on its specific reflectivity, surface extent, and location, both vertical and horizontal. Figure 8.5 shows the values and ranges of albedos of various elements of the earth's surface from which we can estimate the global importance of each. For comparison the albedos of the moon and the inner planets are also included.

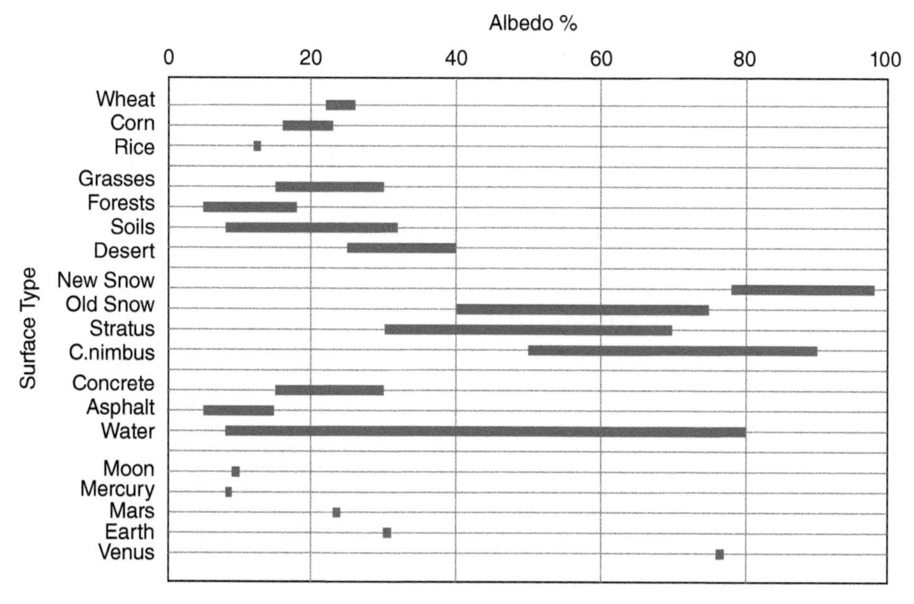

Figure 8.5 Albedos of features on the earth. The reflection of light by the various objects and regions on the earth is quite variable. Clouds have the largest effect on the global albedo both because they are highly reflective and there are lots of them at all times (~60% cover). Surface features tend to have less reflection, especially forests and plants because they absorb radiation to store and utilize solar energy.

Plants that represent a large fraction of life on land tend to have very low albedos, while dead zones such as deserts, snow, and clouds have high albedos. This makes sense because plants need solar radiation to survive and grow and would therefore evolve to absorb most of it rather than reflect it away.

There are some noteworthy nuances about the way the albedo affects the earth and its climate. The location of a reflecting component has an important effect. For instance, the same area of ice on a glacier at mid latitudes will reflect more light than if was at the poles because the solar radiation per square meter is less there (see Figure 8.4). The effect of the vertical location is more subtle. Let's consider a hypothetical situation in which clouds and surface snow are both present and assume that the cloud albedo is 60% and the snow albedo is 80%. What is the actual albedo of this region assuming no absorption by clouds? Clearly we cannot add the two albedos because the total is more than 100%. Since the clouds are on top, they will intersect the sunlight first and reflect 60% of it. That will only leave 40% to reach the surface, which will reflect 80% of that, or 32% of the incoming radiation: $A_{Land} = A_{Snow}\,(1 - A_{Cloud})$. If this 32% gets through the clouds, then the total reflection is 92% of the incoming radiation which has nearly twice as much contribution from clouds as from snow even though clouds are less reflective. If none of the surface reflection gets through the clouds, then the albedo will be 60% from just the clouds. On a day when there are no clouds the surface would revert to reflecting 80% from the snow. This situation makes the land reflectivity less effective on average than atmospheric, and even more so because whatever the surface does reflect will not all get through the clouds. Some of it will be returned to the earth by reflecting from the undersides of clouds, maybe back and forth a few times. The 30% global albedo is after these effects have been taken into account.

Figure 8.5 also shows the major effect of clouds on the planetary scale. Thick clouds are the most reflective and we see that the albedo of Venus is 78%! Its high reflectivity is evident in how bright it appears from the earth to the naked eye. Mars has no clouds so its albedo is only 22% despite the desert surface. The moon perhaps is even more surprising because it appears so bright to us, but has an albedo of just 12%. Seen from a far distance, it will appear considerably darker than the earth. Since the moon is thought to be a part of the ancient earth, we may surmise that the earth's albedo would also be about 12% if there were no clouds, ice, or life.

Observations of the moon were used to first measure the earth's global albedo and have been used in recent years to evaluate possible trends. When the moon is partial, we see the bright sunlit part and a dimmer but clearly visible remainder which is caused by the reflection of earthlight from the moon. Readers will intuitively see that the difference in brightness of these two parts is determined by the earth's albedo. The brighter the earth is the less dim the dim side will be. The bright part gives us the moon albedo since the solar constant there is nearly the same as for the earth. Using the moon albedo, the amount of light coming from the dim side is used to calculate how much earthlight must be reaching it. And that can be used to figure out how much light the earth was putting out toward the moon by reflection, which is the earth's albedo times the solar constant. This idea was put to use in the early part of the twentieth century by F.W. Very and later in the 1920s and 1930s by A-L. Danjon who arrived at an estimate of 0.29–0.39, which is close to the present value of 0.3, now also validated by satellite measurements.

Changes of albedo over the times of our interest can affect climate just as increasing greenhouse gases can. Human activities are affecting the albedo. Direct changes occur from increased levels of aerosol particles emitted by burning sulfur containing fossil fuels. As mentioned earlier, these particles can also interact with clouds and make them more reflective. Urbanization and deforestation can replace vegetation with lighter surfaces. All these would increase the global albedo. At the same time, albedo can decrease if global warming melts ice and reduces its extent near the poles or in glaciers (Figure 1.1). These are competing effects that may compensate in the global average, but can cause significant changes of regional climates at various latitudes.

8.3 Radiative Transfer

In the context of climate, *radiative transfer* is a term used to describe how electromagnetic radiation interacts with atmospheric constituents. In particular, how the atmosphere absorbs the radiation from the sun and the earth and what happens to the energy after absorption. This is the foundation of the greenhouse effect and, by extension, of global warming and climate change. It follows a two-step process: first an absorption and then a radiation. A balance is achieved that represents the state of the earth's surface and atmospheric temperatures and hence the climate.

As we discussed earlier, some 20% of the incoming solar radiation is absorbed by the atmosphere, mostly by tropospheric water vapor and clouds and to a lesser extent by stratospheric ozone (partly shown in Figure 8.3). Added to the 30% albedo leaves us with only about 50% of the solar energy reaching the surface. This energy is absorbed by the land and oceans which act as nearly perfect black bodies. The consequence is that the surface heats up. This must reach a point when the earth radiates as much energy as it gets, otherwise it would never stop getting warmer. The heat of the surface can be radiated as electromagnetic energy or transferred to the atmosphere by conduction and convection. All these processes have a role, but radiation is the most significant and described by Eq. 8.1. The average temperature of the earth is about 288°K (59°F). According to Eq. 8.2 the wavelength at which the earth radiates the most energy is $\lambda_{Max} = 2898°K\text{-}\mu/288°K = 10.1\ \mu$. This is in the infrared region of the electromagnetic spectrum (Figure 6.3). If we look at this as photons-bundles of energy, we see that these are much less energetic than the photons that come from the sun.

Example Calculate the energies of solar and earth photons at peak intensities of their spectra.

Solution: Basic information: Energy of a photon: $E(\text{joules}) = h\,\nu = hc/\lambda = (6.63 \times 10^{-34}\ m^2\text{-}kg/s) \times (3 \times 10^8\ m/s) = [2 \times 10^{-25}\ m^3\text{-}kg/s^2]/\lambda\ (m)$.

Solar photon at peak intensity (Eq. 8.2): $(2 \times 10^{-25}\ m^3\text{-}kg/s^2)/(0.52 \times 10^{-6}\ m) = \mathbf{3.8 \times 10^{-19}\ j}$.

Earth's photon at peak intensity: $(2 \times 10^{-25}\ m^3\text{-}kg/s^2)/(10.4 \times 10^{-6}\ m) = \mathbf{2 \times 10^{-20}\ j}$, or about 20 times less energetic.

Atmospheric greenhouse gases have properties by which they absorb the earth's low energy radiation even though they do not absorb much sunlight. The outcome is that 90% of the heat emitted from the surface as electromagnetic black body radiation is absorbed in the atmosphere. The absorption of both the solar and earth radiations can be described by common principles and by the specific absorption properties of the gases in the atmosphere.

Absorption of radiation by molecules can have a number of possible outcomes. Suppose we have a molecule AB made up of atoms A and B. If it absorbs a photon, it will temporarily go into an activated state AB* (absorption: $AB + h\nu \rightarrow AB^*$). This state is usually short lived, after which it can fly apart if the absorbed photon has enough energy to break the bond (dissociation: $AB^* + M \rightarrow A + B + M$); it can favor a chemical reaction with another molecule (reaction: $AB^* + C \rightarrow A + BC$); it can release the photon (scattering: $AB^* \rightarrow AB + h\nu$) and perhaps most importantly, it can transfer the excess energy to the atmospheric heat reservoir (thermalization: $AB^* + M \rightarrow AB + M +$ "heat"). All of these will take the original photon out of the stream. We have encountered the first two of these processes previously in the discussion of chemical reactions. However, as important as they are for atmospheric chemistry, they have a negligible effect on the stream of radiation going through the atmosphere. The reason is that the activated AB* has to find a C molecule by random collisions with all the molecules in the air. Readers are reminded that the concentrations of a viable reactive C are minuscule and the major gases, N_2 and O_2, don't react. For dissociation there are plenty of M's around, but the energy of the photon has to be enough to break a bond. That is not probable for the molecules in the troposphere, all of which have strong bonds or they would not be there. So most of the molecules that have absorbed a photon end up going to the remaining two cases that make up radiative transfer theory. The *scattering* reaction is also called *radiative decay* which is more evocative because it implies a lifetime over which a molecule decays back to its lackluster state. A photon comes in and a photon goes out, but not necessarily in the same direction; most of the time it has the same frequency. The decay lifetime is very short and different for different atmospheric molecules. All molecules scatter solar radiation with a relatively low efficiency. We will not consider the details here except to note that it contributes to the albedo and is included in our estimates (Endnote 8.3).

Example What is the wavelength of radiation needed to dissociate carbon dioxide molecules?

Solution: The bond energy in the carbon to oxygen bond in CO_2 is 8.1 × 105 j. It will take a photon of this energy or more to break the bond by dissociation. Keeping in mind that frequency times the wavelength is the speed of light ($\nu\lambda = c$), the wavelength required in each photon is $E = h\nu = hc/\lambda$ which implies that: $\lambda = hc/E = (6.62 \times 10^{-34}\ m^2\ kg/s) \times (3 \times 10^6\ m/s)/8.1 \times 10^5\ j = 0.15$ microns. This is in the extreme ultraviolet (Figure 6.3). It is more energetic than most of our sun's radiation and if a rare such photon does arrive it would not make it to the troposphere.

8.3.1 Thermalization

While the AB* is waiting to decay, it is colliding with all the air molecules around. If it collides with one and transfers the energy of the photon it had absorbed, before it can decay, we get thermalization. It stands to reason that in the troposphere where the air pressure is high, thermalization will be highly favored and when it is low, scattering will have a better chance. The rest depends on the characteristics of the molecule doing the absorbing.

Absorbing the earth's energy causes the greenhouse gas molecules to vibrate or rotate leading to warming of the atmosphere by thermalization ($AB^* + M \rightarrow AB + M +$ "heat").

That is, the energy is transferred to the other surrounding molecules increasing their kinetic energies that we interpret as a warming (Endnote 8.4). An important part of the greenhouse effect is that a lot of the energy is transferred to N_2 and O_2 by thermalization. These molecules are not radiators so the energy is kept as heat until radiated more efficiently by greenhouse gases both back to the earth and into space or lost by conduction at the surface or with clouds.

Thermalization can be visualized with this thought experiment that explains the concept for the greenhouse effect. Suppose we have a transparent box in which we can confine 10,000 molecules of nitrogen (N_2) and 10 molecules of methane – a greenhouse gas. Next, we irradiate this box with ample amounts of infrared radiation at a single wavelength of 6.52 μm which is one of the wavelengths at which methane is known to absorb radiation in a vibrational motion. The 10 methane molecules will absorb these photons right away and will start to vibrate. When they are busy vibrating, the rest of the radiation will go through and when they have released a photon, they will take up another. The result in steady state will be that the 10 methane molecules will be vibrating more or less all the time and therefore have more vibrational energy than before, storing some of the incoming energy. At the same time, they will be emitting radiation according to the scattering formula. If this was all there was to it, the temperature of gases in our box would not rise at all. But you will note that the molecules in the box are not stationary – they are moving all the time because they have some temperature, which is the average kinetic energy of all the molecules in the box (Endnote 8.5). The moving molecules will collide with each other since they are in a box. The collisions will transfer energy with every event according to the rules of mechanics and Newton's laws of motion. So now, a methane molecule can get to vibrating by absorbing a photon from our irradiation and maybe before it can get rid of the energy it collides with another molecule, most likely a nitrogen since there are 10,000 of them, and it can transfer the vibrational energy to it. If it did that, it would no longer be able to radiate a photon of the same energy it had absorbed because it no longer has it. The nitrogen molecule speeds up and adds to the temperature. This goes on with all the molecules, nitrogen or methane, until they are all moving much faster than before so that there is an equilibrium of energy exchanges. This is the same as having achieved a constant temperature. The infrared energy intercepted by the methane molecules is now invested in all the molecules in the box and mostly in the nitrogen molecules because there are so many of them. All molecules can acquire energy by collisions including the nitrogen in our box because it is an exchange of mechanical energy – not of electromagnetic radiation. Similarly, they can also transfer energy by collisions. The collisional exchange of energy is a form of conduction. So even the 10 methane molecules are enough to store a lot of energy in the box by transferring it to nitrogen. When the temperature is high enough, the collisions no longer result in an energy transfer because the methane molecules collectively get back as much as they lose. Then, no further net absorption can take place and we have a steady state. It is interesting to note that if we pull the methane molecules out of the box at some point, the temperature of the remaining gas (nitrogen) will then remain constant forever because the nitrogen cannot radiate the energy, as explained by Kirchhoff's law. The gas (nitrogen) will not warm or cool, unless there is conduction with the walls. The walls may act as a black body and radiate energy away to the outside. Then the box, and its nitrogen, will cool. That is how the greenhouse effect works. (Convection is the other transport process, but that requires pressure gradients, usually by an acceleration, whether by gravity or inertia; such a process is not a part of our situation here).

At the earth's surface too, the transfer of energy occurs by conduction involving all the molecules in the air and not just the greenhouse gases, so even though conduction is not very efficient, it is effective because 100% of the atmosphere participates in it. Patches of the earth that are warmed by sunlight can transfer energy to the air in contact this way. Similarly, energy at the surface can be conducted to deeper soils where it is stored for some time. When you feel warm air on your face, it is not from the greenhouse gases in the air – there are too few molecules of them; what you feel is the nitrogen and oxygen that were heated up by the greenhouse gases or conduction. And what you feel is also by conduction since the air molecules are touching your face. You can feel the radiant infrared energy if you sit near a fire, and even in the outside air, but it is always combined with the warm or cold air that is touching your body.

8.3.2 Mechanics of Radiation Absorption

Gases: When electromagnetic radiation interacts with atmospheric constituents including gases, particles, and clouds, some of it will be absorbed and distributed by thermalization as we have discussed, and the rest will either go through (transmission) or move on in various directions (scattering). The amount of radiation that goes through a layer of the atmosphere of thickness dz, relative to the amount that is coming in, is expressed by Beer's law:

$$\delta I_{v} = -\rho_{g} k_{v} I_{ov} \delta z \tag{8.4}$$

It is applicable to the absorption of solar radiation or the earth's radiation by the atmosphere, and the absorption of the infrared radiation from within the atmosphere as it makes its way

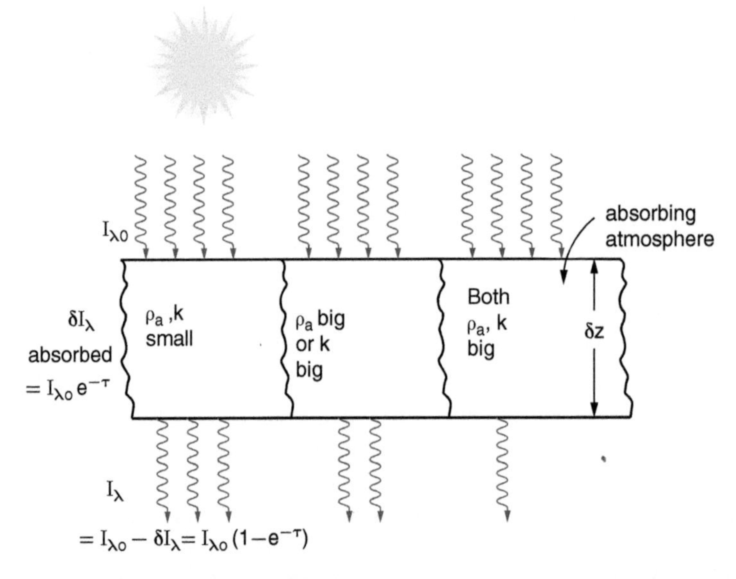

Figure 8.6 Absorption of radiation in the atmosphere. Appropriate absorbers in the atmosphere follow Beer's law for sunlight, the earth's radiation, and atmospheric radiation.

toward the surface or toward outer space. Eq. 8.4 says that if I_0 is the amount of electromagnetic radiation coming into a layer of the atmosphere, as it passes through a distance δz, the amount coming out the other end will be reduced by $\delta I_\nu = I_\nu - I_{0\nu}$ (Figure 8.6). In Eq. 8.4, k_ν is the mass absorption coefficient (m^2/kg) for radiation that has a frequency ν and ρ_g (kg/m^3) is the mass density or amount of the "absorber," which in our case will be a greenhouse gas (subscript g) although in general, it could be particles, air molecules or clouds. The units shown are commonly used in this context but you can convert to others that may be more suitable. We will soon see that the "k" is not the same for all frequencies. In fact, each gas only absorbs radiation at and around specific frequencies ν with the efficiency k_ν that depends on the properties of the gas. Readers will note that k is conceptually similar to the rate constant we discussed earlier for chemical reactions, however, here it reflects the likelihood that electromagnetic radiation entering the atmosphere will be absorbed along the path. Eq. 8.4 can be written as $dI_\nu/dz = -(k_\nu \rho_g) I_\nu$ and solved, as we did with Eqs. 4.3 and 4.5 earlier. Then we will get the fraction of radiation transmitted out of the atmosphere, $T_\nu = (I_\nu/I_{0\nu}) = \exp(-\tau_\nu)$ and consequently the fraction absorbed is $F_\nu = 1 - T_\nu$ or:

$$F_\nu = 1 - e^{-\tau_\nu} \tag{8.5a}$$

$$\tau_\nu = k_\nu M \tag{8.5b}$$

τ_ν is called the "optical depth," although it is a dimensionless quantity and hence not a depth in the sense of a length. M is the total mass per square meter in the column through which the radiation is passing (Endnote 8.5).

For greenhouse gases other than water vapor, we can assume that the mixing ratio is constant, at least in the troposphere where most of the absorption will take place. In that case, $\rho_g = \rho_{air} C_m$ where ρ_{air} is the density of air (kg/m^3) and C_{mg} is the mass mixing ratio of an absorbing gas ($kg\text{-}gas/m^3/kg\text{-}air/m^3$). Then, after some algebra, the optical depth for a given gas of our interest will be:

$$\tau_\nu = \rho_0 H k_\nu C_{mg} = \rho_0 H k_\nu \left(\frac{MW_g}{MW_a}\right) C \tag{8.6}$$

Here ρ_0 is the density of air at the surface (in kg/m^3), H is the scale height, and C_{mg} is the mass mixing ratio of the gas. It is converted to C (number mixing ratio) that we have used in this book by recognizing that C_{mg} (mass mixing ratio) = $(MW_g/MW_a) \times C$ (number mixing ratio) where MW_g is the molecular weight of the gas in question and MW_a is the molecular weight of air.

The importance of Eqs. 8.5 and 8.6 is to show explicitly the connection between atmospheric absorption of electromagnetic radiation and the mixing ratios of greenhouse gases. This determines the greenhouse effect and global warming when applied to the interaction of the earth's infrared radiation with the absorbing gases. If you increase their concentrations, you increase the absorption of the earth's radiation. Equations 8.5 and 8.6 tell us that the absorption represented by F_ν is exponentially related to the efficiency k_ν and the mixing ratio C. If $\tau_\nu \ll 1$, then, since $\exp(-\tau_\nu) \approx 1 - \tau_\nu$, $F_\nu \approx \tau_\nu = \rho_0 (MW_g/MW_a) H k_\nu C$ which means that the absorption by this gas is proportional to the concentration and that the absorption is the optical depth. If we double the concentration the absorption doubles. Often however, this limit does not apply because there is too much greenhouse gas in the atmosphere so $\exp(-\tau_\nu) \approx 1 - \tau_\nu$ is not a valid approximation. Then, changing the concentration of the gas has a more complex relationship to absorption.

In the discussion so far, the connection between the absorption coefficient k and the properties of gases remains unspecified. Clearly it is the most important characteristic of a specific gas that determines whether it will absorb radiation effectively or not – whether it will be a greenhouse gas or not. Let's start with a classical understanding of absorption of energy. Suppose you take a spring of spring constant k that reflects its stiffness, attach it to a ball of mass m, and hang it from the ceiling. Now pull the ball down a little and let go. We all know that this system will oscillate, that is, it will move up and down with some frequency. Physics tells us that the frequency will be $\omega = (k/m)^{1/2}$. It is determined entirely by the properties of the system embodied in the mass (m) and the stiffness of the spring (k) and unaffected by what pulled it down or by how much. Let's consider another scenario. Suppose we apply a periodic force to this system, that is, we push it with a regular frequency. Then, you will find that the system will vibrate most vigorously if the frequency of your pushes is close to the "natural frequency" or "natural mode of vibration," $\omega = (k/m)^{1/2}$, otherwise it may not move much. This is called resonance and it is an intuitively understandable phenomenon that is encountered in everyday life. It can be seen from pushing a child, or even an adult, on a swing. If you just move your arms back and forth, sometimes you will connect with the swing and it may move a little. It will also take almost no work from you. To make the swing work, you have to give it a push every time it comes back to you. That happens at a characteristic frequency determined by the length of ropes making up the swing. Then, the swing will swing higher and higher, and you will have to work harder to keep it going, since it is taking up energy from you. Atmospheric molecules are not balls and springs, but they are homologous systems. Instead of springs, there are bonds which act in much the same way, and instead of balls we have the atoms that make up the molecule, which we often visualize as balls. When you pull the atoms slightly and let go they will oscillate just like the spring and mass system depending on the strength of the bond (k) and the masses of the atoms (m). The number of different modes of vibration or frequencies at which a system vibrates in resonance increases with the number of atoms, N, up to 3 N-6 and each mode is stimulated by its own special frequency that can be determined from the established principles of classical and quantum mechanics. The modes represent all the different ways in which the vibration can occur. The vibrational characteristics of atmospheric molecules amenable to interacting with solar or earth radiation are shown in Figure 8.7.

Electromagnetic radiation is represented as sinusoidal oscillations of electric and magnetic fields. When they encounter charges, they exert forces as: $F = qE$ where q is the charge and E is the electric field in the wave. The charge q will then move according to Newton's laws $F = ma = qE$, with an acceleration $a = (q/m)\,E$, in the direction of the applied electric field. Air molecules are neutral in charge, so under ordinary circumstances electric fields will not be able to move the molecule. The molecules are however, made of positive charges tightly bound in the nucleus and negative in the surrounding clouds of electrons that are quite far away. A means of interaction arises when the charges are separated enough in the molecule to create a dipole. This is when there is an excess of positive charge on one end of the molecule and a consequent excess of negative charge on the other end. The strength of the dipole is expressed as $q \times D$, where q is the charge at each end and D is the distance between charges. Under these circumstances the forces imposed by the

Molecule	Arrangement	Permanent Dipole Moment
N_2	N—N	No
O_2	O—O	No
CO	C—O	Yes
CO_2	O—C—O	No
N_2O	N—N—O	Yes
H_2O	H, O, H	Yes
O_3	O, O, O	Yes
CH_4	H, H, C, H, H	No

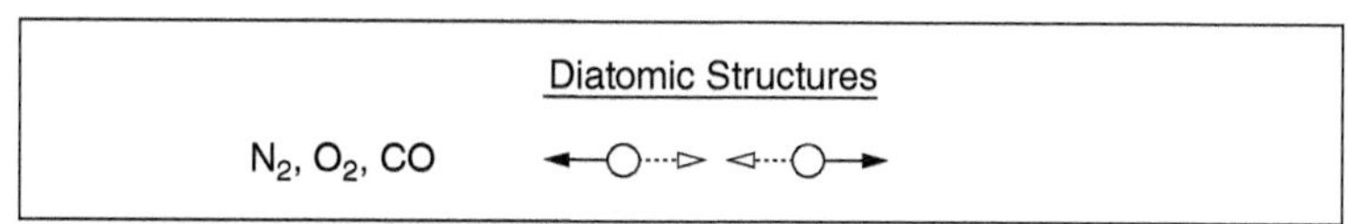

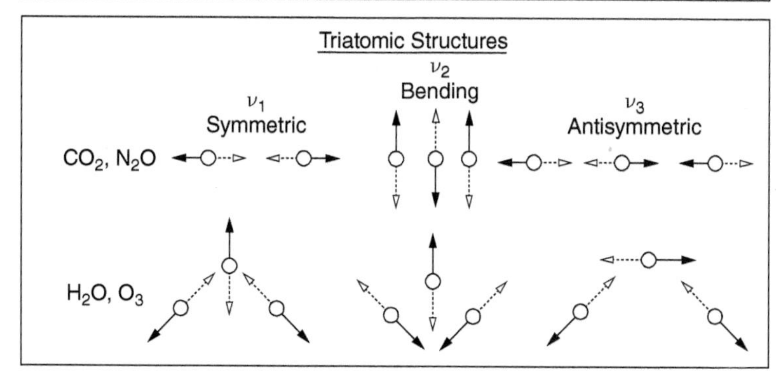

Figure 8.7 Absorption mechanics of infrared radiation by air molecules. Molecules that can vibrate or rotate absorb infrared radiation. The vibrational modes of the main greenhouse gases are shown as stretching and bending motions (Endnote 8.6). *Source:* Adapted from McCartney, 1983.

radiation will not be uniform and the molecule will vibrate as long as the wavelength of the radiation is less than the separation of charges D. The molecule will be pulled apart when the applied field is pointing one way and pushed together when the sine wave reverses. Now the molecule will act very much like the ball and spring system we discussed. It will pick up energy of the frequency of the radiation that is close to its natural frequencies and not otherwise. These oscillations of molecules are one of the primary mechanisms by which the atmosphere captures the earth's radiation leading to the greenhouse effect. The properties of the molecule determine at what frequencies it can vibrate (see Figure 8.7). Strong bonds are like large values of the spring constant since it is hard to pull apart the atoms. These will require higher frequencies to stimulate vibrations, and likewise, weaker bonds will allow vibrations with less energetic radiation (from $\omega = (k/m)^{1/2}$).

Some molecules have a permanent dipole moment, that is, their natural shape has charges separated over the extent of the bond. Water is such a molecule. It is naturally bent and consequently has a permanent dipole moment (Fig 8.7). Other molecules can have unsymmetric vibrations that will induce a dipole as they vibrate leading to the same end – absorption of radiation at the characteristic frequencies. CO_2 is this type of molecule. It is linear, so it has two opposing dipoles, the C and the O on one side and the C and the O on the other. Their effects cancel out because they are oppositely directed. The CO_2 molecule can however, vibrate in two linear modes (see Figure 8.7): the two oxygen atoms go outward and inward while the carbon remains fixed in the center. This is a symmetric vibration, and so its two dipoles still cancel at all points on the cycle. It doesn't absorb the earth's radiation. The second linear mode is when one O goes outward from the C and the other O compresses toward the C; the C also moves. The process reverses to complete a cycle. This is an asymmetric vibration and as it occurs, a changing dipole moment is formed because the two O atoms are at different distances from the C atom at all times. In this picture, one of the two dipoles is smaller because its distance D between charges is smaller on one side than the other side of the carbon atom. There is a net dipole moment that remains. Finally linear molecules of more than two atoms can have a bending vibration which generally takes less energy and induces a dipole as you would intuitively expect from the example of H_2O. So we see that the carbon dioxide molecule has three vibrational modes but one is incapable of absorbing the earth's radiation. The other two are at wavelengths of stretching 4.26 microns and bending at 15 microns.

Another way molecules absorb radiation is by rotating; however, a dipole moment is still needed. Then, the incoming radiation will pull one end and push the other leading to a twisting force or a torque that will rotate the molecule. This interaction favors low energy radiation. Making a molecule rotate is easier and takes less energy compared with vibrations which require pulling the atoms apart. According to quantum theory, even the rotations occur at discrete frequencies but usually these are closely spaced together. If a vibration induces a dipole moment, it also makes it possible for the molecule to respond to rotations. This generates absorption at frequencies close to the vibrational mode that are a combination and represented as vibrational-rotational modes. This can add quite a few more frequencies, where a molecule can absorb. If the molecule has a permanent dipole moment, then it can also absorb energy in pure rotations which don't need vibrations to excite them. This leads to absorption by water at low frequencies where the radiation cannot excite vibrations.

There are two main mechanisms that allow molecules to absorb radiation at frequencies continuously both lower and higher than the natural frequencies of vibrations, rotations or their rotational-vibrational combinations (Figure 8.8). So, like the ball and spring, they will also take up energy at frequencies *near* the resonant modes, but on average, not as efficiently as *at* the resonant modes. One is the effect of air pressure. When it is high, as in the troposphere, it allows molecules to collide more frequently. A collision is when two molecules approach each other so closely that the electron clouds start to repel them and drive them away from each other. Since the bonds of the atoms are determined by the electrons

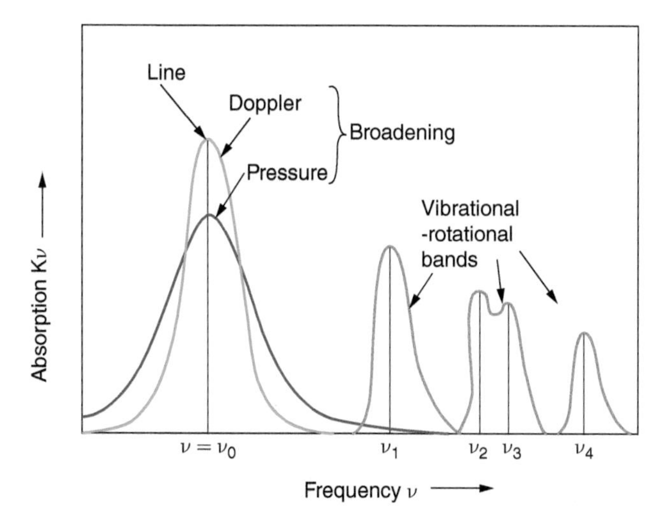

Figure 8.8 The absorption efficiency of the earth's radiation by greenhouse gases. The absorption efficiency is highest at one of the frequencies representing the normal modes of vibration. The vibrations can generate rotational modes and molecules can absorb at these rotational-vibrational frequencies. Finally, molecules absorb radiation at frequencies close to the lines due to their motion (Doppler effect) and because of collisions with other molecules (Pressure effect). These mechanisms lead to an absorption "band," rather than a "line" and increase the effectiveness of gases in absorbing the earth's infrared radiation.

around them, the "spring constant" is altered, shifting the normal modes lower or higher temporarily, but long enough that electromagnetic radiation of nearby frequencies is absorbed. The other is the Doppler effect. Suppose the molecule is stationary and it encounters electromagnetic waves of many frequencies. It will only absorb and vibrate with the frequency of the wave that is at its normal mode because it will experience a push every time a crest hits it at the resonant frequency. That happens at a rate equal to the frequency of the wave. When the molecule is moving toward the waves of radiation, it will encounter more crests per second than if it was stationary and hence will absorb a wave that is of a lower frequency because a lower frequency wave will compensate for the motion of the molecule and jiggle it at its resonant frequency. Similarly, if the molecule is moving away from the radiation, it will absorb waves that are of higher frequency. Since molecules are moving at various speeds, with a known distribution, they will collectively absorb radiation at all frequencies near their normal modes of vibration. This effect becomes more prevalent higher up in the atmosphere when the pressure dies down (Chapter 3). At in-between levels of the atmosphere both effects are important. In both cases, the efficiency of absorption decreases away from the peak at the normal modes (Figure 8.8). This is because only a few molecules move at speeds much lower or higher than the average and so the effect of both mechanisms falls off as we get further from the resonant frequencies (Endnote 8.4). These may seem like details, but they have an important role in the greenhouse effect and global warming. The absorption of radiation around the normal modes of vibration add considerably to the effectiveness of greenhouse gases than if the molecule only absorbed radiation at its one resonant frequency.

The outcome of these absorption mechanisms for the actual atmosphere and greenhouse gases is shown in Figure 8.9. The radiation from the sun and the earth is shown in the top

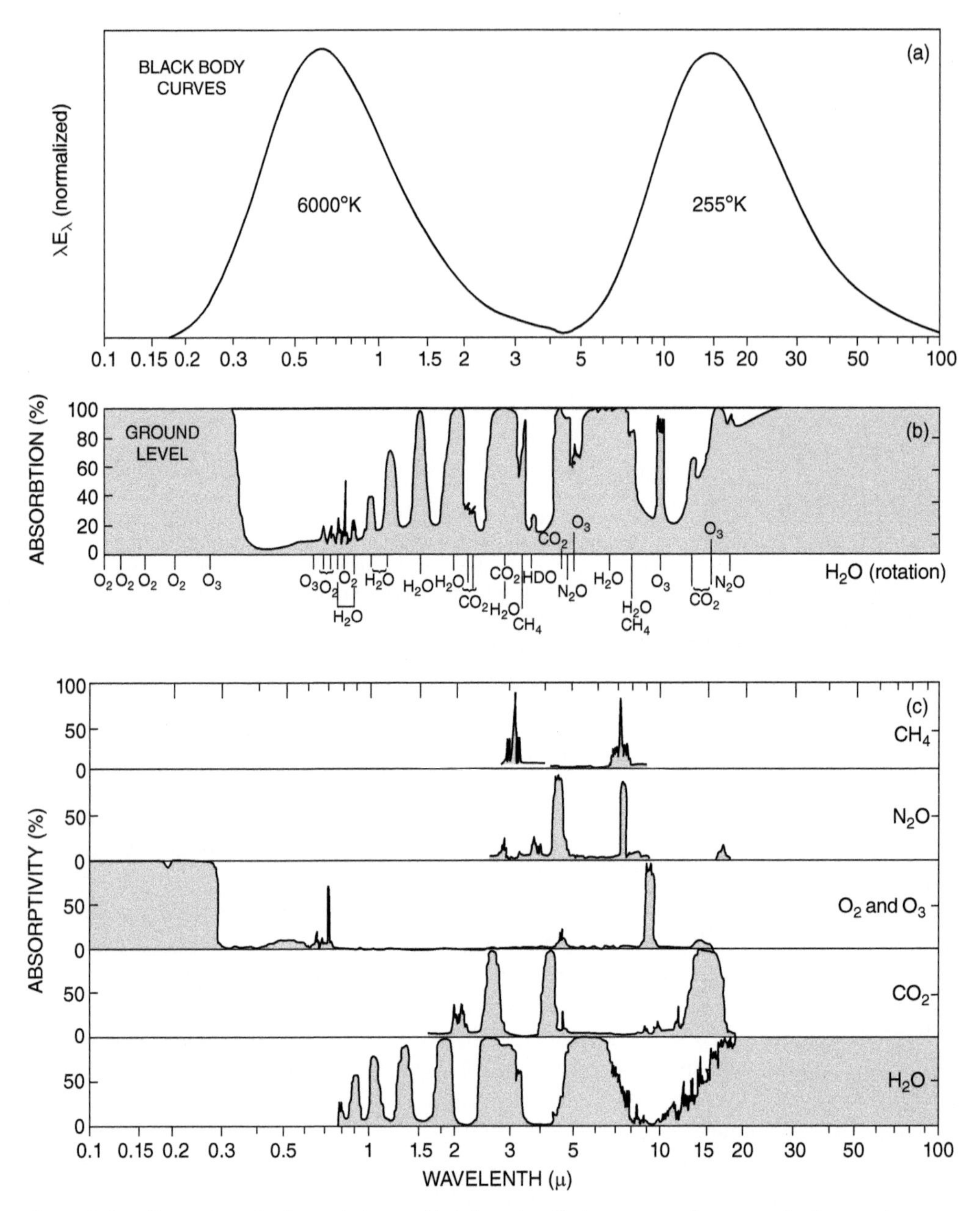

Figure 8.9 Absorption of solar and the earth's infrared radiation streams by atmospheric greenhouse gases. The emission and absorption are shown as a function of the wavelengths of radiation on a logarithmic scale. (a) Black body radiation curves. These are normalized for a clearer visual representation. (b) The absorption by all the important greenhouse gases. Not much of the solar radiation is absorbed, but most of the earth's infrared radiation is, justifying our estimate of 90% absorption. (c) The absorption by each greenhouse gas. The broadening of the normal modes is apparent. Absorption by the rotational bands of water appears toward the right (Endnote 8.6).

panel. These are based on Planck's black body radiation law and have been normalized meaning that the area under the curves is the same for both and equal to one. Moreover, the wavelengths are plotted on a logarithmic scale. If this were not done, the earth's radiation would have a much lower peak and would be much more spread out so that you would not be able to see the features of the peak emission or the range of wavelengths. As it is, we can see that the sun's radiation spans from about 0.17 to 5 microns, while the earth's radiation goes from 5 to 100 microns. There is virtually no overlap between the two radiation curves. Here are some salient points to observe. First note that the vibrations of CO_2, H_2O, CH_4, and other gases absorb quite strongly for frequencies around the normal modes, spreading out into absorption bands, rather than just at the single frequency of the mode as we discussed theoretically. Second, it is apparent that water vapor is the leading absorber of the earth's radiation, strongly absorbing virtually all radiation above 15 microns in its pure rotation bands and a lot of it between 5 and 7 microns in its 6.3-micron normal mode. Carbon dioxide has a major band at 15 microns that gives it its main importance in climate and global warming, supplemented by the 4.26 micron band, where there is not much radiation to absorb. Between 1 and 3 microns we get several significant rotation-vibration bands of CO_2 and water vapor. Between water vapor and CO_2 most of the radiation emitted by the earth is trapped, but neither is efficient between 7 and 10 microns. No other naturally occurring gas absorbs in this region either, except for a narrow band from ozone. This is called the "window region" from which some 10% of the emitted surface radiation escapes while the remaining 90% is absorbed as we discussed earlier.

8.3.3 Clouds

As we saw earlier, clouds are a major factor in the global albedo of the earth that results in a strong cooling influence. This is counter-acted by their significant greenhouse effect. Since some 60% of the earth is covered by clouds at any time, this effect is persistent. Clouds may absorb about 10% of the incoming solar radiation; only some of it is returned to the surface. The more significant effect is from the absorption of the earth's infrared radiation. Clouds have saturated water vapor and liquid droplets; both absorb the earth's infrared radiation as well as the incoming solar radiation. For atmospheric absorption by clouds, it means that they can take up the earth's radiation at all frequencies including the window region (8–12 micron wavelengths). So, in a circumstance when there are no clouds, radiation in the window region will escape to space unimpeded, and when there are clouds it will not. The composite cloudy atmosphere of the earth therefore absorbs the surface infrared radiation at all frequencies but still lets some 40 watts/m^2 get through the window region because here only clouds absorb, while at all other parts of the spectrum, there is virtually complete absorption by the combination of the greenhouse gases and clouds. The effect of clouds on the downward infrared radiation to the earth's climate is about 30 watts/m^2 which is about 10% of the total greenhouse effect. The dual role of the clouds in causing cooling by reflection (~60 W/m^2) and warming by absorption and re-radiation (~30–40 w/m^2) balance out to produce a net cooling effect in the present world and has done so for millennia, but it greatly complicates how they will behave in a warmer world.

8.3.4 Limitations to Warming

The environmental circumstances on every planet are different, but on the whole, there are significant limitations to how warm the surface can get. The surface temperature on any planet, and particularly the earth, depends on how the energy arriving at the top of the atmosphere plays out to determine the surface temperature. Let's consider a gas absorbing at one of the frequencies as shown in Figure 8.8. Since the absorption is most efficient at the center of the band, and there is a sufficient amount of the gas in the atmosphere, it may absorb all the radiation that the earth emits at and near this frequency, let's call it ν_0. This phenomenon, called "band saturation," limits or reduces the impact of adding more of the same gas to the atmosphere. A closely related phenomenon is that of "band overlap." It is most easily understood by considering the addition of a gas that has its fundamental absorption at frequencies that are also absorbed by other existing gases. In this situation, the added gas has to compete with the other gases to absorb the radiation in its characteristic band. If the other gases are already present in high enough concentrations to absorb most of the band, then the added gas will have a much diminished effect compared to the situation when the other gases are not present. Or more generally, gases are less efficient when they have to compete for the photons at the frequencies they can absorb. Does this mean that once the gas absorbs almost all the radiation in its characteristic band that adding additional amounts will have no further effect on absorption and hence no further global warming due to this gas? The answer is "no."

There are two mechanisms that allow a gas to continue to affect the atmospheric and surface temperatures. The first is the "absorption in the wings" which continues to occur as we add more gas to the atmosphere. This is the absorption of frequencies further from the center of the band (see Figure 8.8). At a given concentration C, a gas may absorb almost all the radiation at the center of the line, but further out the efficiency is low so it takes a lot more concentration to absorb all the radiation at such frequencies. Looking at it from the perspective of Eq. 8.5, which we can write as: $F_\nu = 1 - \exp(-\tau_\nu) = 1 - \exp(-\text{constant} \times k_\nu C)$. The constant $= \rho_0$ H (MW/MW_a). At and near ν_0, the mode of vibration of the gas, the

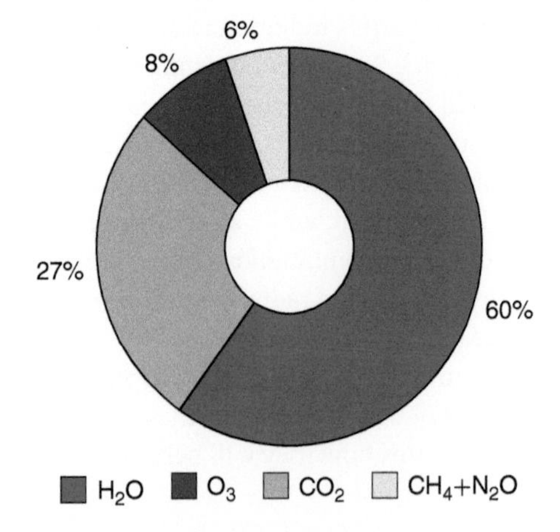

Figure 8.10 he calculated contribution to trapping of earth's infrared radiation. The relative importance of the gases would translate approximately in the same proportions to the greenhouse effect and its effect on the surface temperature. We see the large impact of water vapor, then CO_2, and relatively small effects of the other key gases. *Source:* Based on Keihl & Trenberth (1997).

value of $k(\nu = \nu_0)$ is large as shown in Figure 8.8, so τ (ν near ν_0) is large enough that exp($-\tau_\nu) \approx 0$ and the fraction absorbed F (ν near ν_0) ≈ 1. Adding more gas doesn't change this part – the absorption remains nearly 100% at these frequencies. For frequencies further away from ν_0, the value of k_ν is much smaller than it is nearer to ν_0 so exp $(-\tau_\nu)$ is not close to 0 at the given concentration C, and increasing the concentration will increase the absorption F_ν. This absorption is much less efficient than at the center of the band $\nu = \nu_0$, so increasing the gas concentrations has a diminishing effect on warming.

In the atmosphere band overlap is caused mostly by water vapor. For CO_2, the strong band at 15 μ has a partial overlap that fills in where water vapor leaves off extending the range of frequencies almost completely absorbed by the two gases. Ozone absorbs in the window region, so it has little competition with other absorbers and this fact makes it a contributor. Nitrous oxide has one band (17 μm) that interferes with water vapor and another around 6 μm that overlaps with methane. Figure 8.10 summarizes the discussion. It shows that about 60% of the atmospheric absorption is by water vapor and some 27% by CO_2. While methane, nitrous oxide, and ozone together round out the rest.

8.3.5 Atmospheric Radiation

We saw how the atmosphere absorbs most of the radiation from the earth. The last piece of the puzzle is the response of the atmosphere to this absorbed radiation. Like the earth's surface, it must radiate to balance the incoming energy and reach an equilibrium with the prevailing environmental conditions. We regard the atmosphere as a "gray" body, which has an absorptivity of F (Eq. 8.5), and will then have the same emissivity. If the earth radiates an amount of energy E, the atmosphere will absorb E(absorbed) $= F \times E$. It will acquire an average temperature T_a and radiate an energy of $2\ F\ \sigma\ T_a^4$ (Eq. 8.4) in watts/m^2 $=$ E(absorbed) from conservation of energy. It has two "surfaces," one points to outer space and the other toward the earth, hence the factor of 2. There is no implication here that the upward and downward energies will be the same, and figuring out the proportion is complicated and the gist of radiative transfer within the atmosphere. Conceptually we can see that it is inevitable that more radiation will be sent down than up since the effective lower surface of the atmosphere is warmer than the upper surface (Chapter 2). We will call it "more radiation down than up" phenomenon. It can be quantified as follows. In equilibrium the fraction of the E(absorbed) that is sent to outer space can be written as: $F\uparrow = \frac{1}{2}(1 - \delta)$ E(absorbed) and the fraction that goes to the surface is then: $F\downarrow = \frac{1}{2}(1 + \delta)$ E(absorbed), where δ is determined by the amount of greenhouse gases and their absorption characteristics, k and ρ_g as included in Beer's law applied to atmospheric radiation instead of the earth's or sun's radiation streams, as we did earlier. δ is a convenient way to express the disparity between upward and downward fluxes. From the definitions of up and down radiation it can be expressed thus:

$$\delta = \frac{F^{\downarrow} - F^{\uparrow}}{F^{\downarrow} + F^{\uparrow}} \tag{8.7}$$

If δ is 0, then half of the absorbed energy will go up and half will come down. The idea of $\delta = 0$ is the same as assuming that the atmosphere is at constant temperature so that the "top" and "bottom" radiate according to the same temperature in the Stefan-Boltzmann law in Eq. 8.1. But since the actual temperature of the atmosphere decreases with altitude, δ must be > 0 as mentioned before (Figure 8.11). As we add more greenhouse gases to the atmosphere, the value of δ increases further, causing the atmosphere to send increasingly

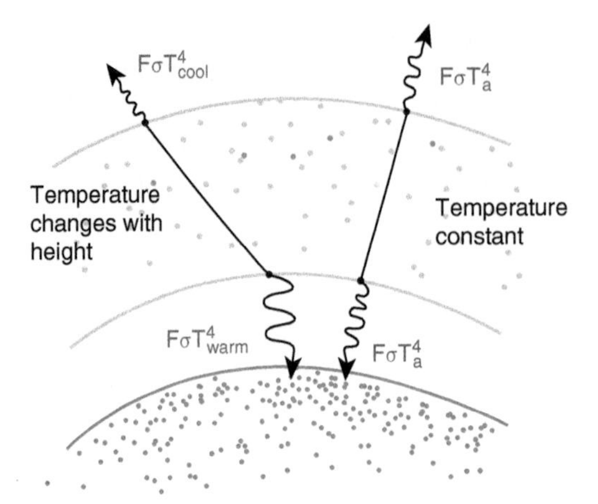

Figure 8.11 The atmosphere sends more radiation to the surface than to outer space.

more energy down than up, which in turn continues to increase the surface temperature. This statement is not proven yet and will be established in the next chapter. It is the second reason why the earth's surface temperature will continue to increase as we add more greenhouse gases, even though the atmosphere is already absorbing all the earth's radiation. Here, it is sufficient to note that this phenomenon by which more radiation is sent down is still subject to Beer's law, but complicated by the continual emission and re-absorption of photons within the atmosphere.

8.4 Heat Storage and Balance

One result of the continual absorption and emission of radiation, and radiative transfer is that the earth's system stores energy in the three components we isolated earlier – land, oceans, and the atmosphere – and there are continual exchanges of energy between these and with outer space. It is instructive to look at how much is stored in each, and the fluxes of energy that cause the prevailing climate.

The heat content of a reservoir is written as:

$$Q = mC_H T \tag{8.8}$$

where Q is the heat in the reservoir (j), m is the mass of the reservoir (g), C is the heat capacity (j/g-°K) and T is the absolute temperature. The mass of the reservoir can be written as: $m = \rho D A$, where ρ is the density, D an effective depth, and A the area. The effective depth is to isolate a part of the reservoir that is most effective in climate change over periods of decades to centuries. For the oceans, the mixed layer at the top 100–200 m absorbs much of the heat due to global warming and warms over the time scales of a decade or two, while transferring of heat to greater depths takes longer. Likewise for the soils, a depth of at most 50 m is where the active exchanges and biological feedbacks can occur; the deepest roots of trees do not go much beyond this depth. With these assumptions we calculate the heat content of these reservoirs as follows:

Oceans Mixed Layer: $Q = (\rho D A) C_H T = 1\ g/cm^3 \times 10^4\ cm \times (3.62 \times 10^{18}\ cm^2) \times (4.179\ j/g\text{-}°K) \times 288°K = 4.4 \times 10^{25}\ j$

All Ocean: $1\ g/cm^3 \times (3.5 \times 10^5\ cm) \times (3.62 \times 10^{18}\ cm^2) \times (4.179\ j/g\text{-}°K) \times 276°K = 150 \times 10^{25}\ j$

Land (Soils): $Q = \rho_s D_s A_{Land} C_S T_{Soil} = (1.5\ g/cm^3) \times (5 \times 10^3\ cm) \times (1.5 \times 10^{18}\ cm^2) \times (2\ j/g\text{-}°K) \times 278°K = 0.6 \times 10^{25}\ j$

Atmosphere: $Q = m_a C_a T_a = 4.8 \times 10^{21}\ g \times (1\ J/g\text{-}°K) \times 260°K = 0.13 \times 10^{25}\ j$

We see from this that most of the heat is stored in the oceans even if we consider only the mixed layer. This feature of the earth's environment plays an important role in the climate as we will see later.

The present energy balance is represented in Figure 8.12 showing how much energy moves from each reservoir and by each major process (Endnote 8.6). The units in the figure are fluxes of energy as w/m^2 which is the amount of energy moving from a surface, or across a surface, per unit area per unit time (Endnote 8.6).

Example According to the Figure 8.12 the earth's surface gets 161 w/m^2 of net radiation from the sun. How much total energy is that every year?

Solution: The total energy arriving per second is: Flux × Area of the earth's surface = 161 $w/m^2 \times 5.12 \times 10^{14}\ m^2 = 8.2 \times 10^{16}$ watts = 820 petawatts. The amount of energy that the

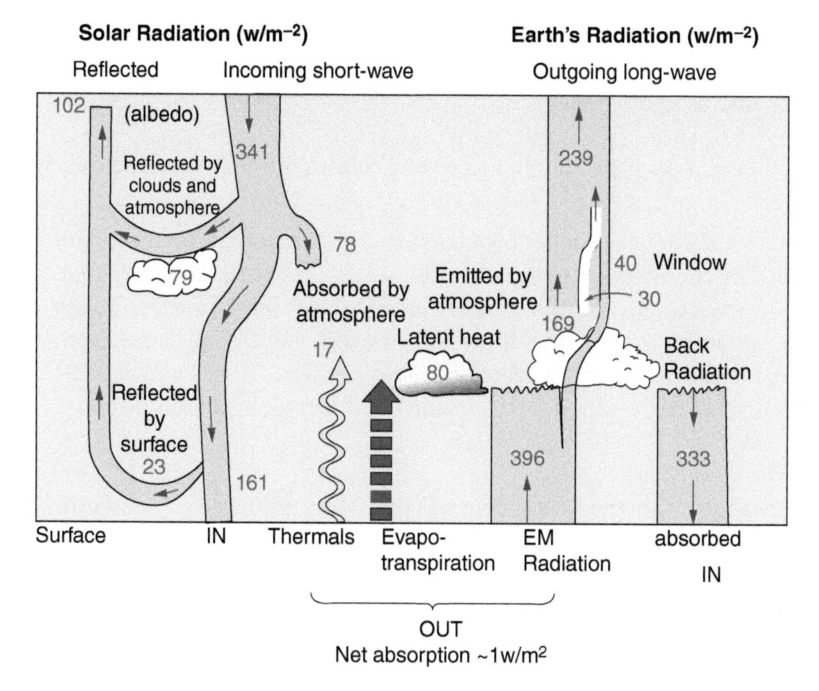

Figure 8.12 Energy balance of the earth. The flows of energy are shown between the atmosphere and the surface (land and ocean). Solar radiation and its movement is on the left-hand side. The greenhouse effect is reflected on the right-hand side whereby, the atmosphere absorbs 90% of the energy emitted by the surface. The fluxes of energy by latent heat and thermals are noteworthy at about 24% of the radiative flux. *Source:* Adapted from Trenberth et al., 2009.

earth's surface gets per year is 8.2 × 1016 j/s × 3.15 × 107 seconds/year = 2.7 × 1024 j/y – that's 2.7 gazillion joules!

Example What is the value of δ for the present energy balance in Figure 8.12?

Solution: $F\uparrow$ = Emitted by atmosphere (gases + clouds) = (169 + 30) w/m^2 = 199 w/m^2. $F\downarrow$ = Back Radiation = 333 w/m^2. Therefore, from Eq. 8.7, δ = (333 – 199)/(333 + 199) = 0.25. Is that a big effect?

Example: What are the average temperatures of the surface and the atmosphere according to the budget in Figure 8.12?

Solution: Surface: E (Radiation from Surface) = $\sigma\ T_s^{\ 4}$. Therefore, T_s = [E (Radiation from Surface)/σ]$^{1/4}$ = (396/5.67 × 10^{-8}°K^4)$^{1/4}$ = 289°K. Atmosphere: E (Radiation from Atmosphere) = $2\sigma\ FT_a^{\ 4} = (F\downarrow + F\downarrow)$ or T_a = (333 + 199 w/m^2)/[2 × (5.67 × 10^{-8}) w/m^2-°K^4 × 0.9]$^{1/4}$ = 269°K.

8.5 Precipitation

Precipitation is closely connected with the earth's energy balance and its temperature. The evaporation of water is driven by the sunlight that arrives at the surface. This causes a significant cooling shown in Figure 8.12 as the latent heat flux of 80 w/m^2. The other arm of evaporation is precipitation because all the water that is evaporated must be returned within a short time. From the water budget in Figure 7.1b we can see that the average precipitation is about 2.8 mm/day (≈ 1 m/year) which is consistent with the global precipitation shown in Figure 2.6.

Example Is the latent heat flux of 80 w/m^2 in Figure 8.12 consistent with the average precipitation of 1 m/year?

Solution: By definition of latent heat it takes L joules of energy to evaporate a kilogram of water, and likewise, this energy is returned when the water condenses back to liquid. Therefore, $Q = m\ L = (\rho_W\ D_{RainFall})\ L$. Here Q is the energy flux from the surface due to evaporation and transpiration (w/m^2), ρ_W is the density of water, and $D_{RainFall}$ is the depth of rainfall per square meter of the earth's surface or $D_{RainFall} = Q/(\rho_W\ L)$ = 80 w/m^2/(2.264 × 10^6 J/kg) × (3.15 × 10^{7}seconds/year) ≈ 1.1 meter/year, or an average of about 3 mm/day. Good to within the 10% rule.

The lifetime of water vapor (τ_W) in the atmosphere was previously estimated as about nine days. Over a year, most of the water that was evaporated in the net would have fallen as rain and a little as snow. So for annual time scales, the balance of water in the atmosphere can be represented by the steady-state equation (Eq. 4.4) or:

$$C_W = S_W \tau_W \tag{8.9}$$

Here C_W is the amount of water in the atmosphere, mostly in the lower part of the troposphere, in kg, S_W is the source in kg/y which is by evaporation and lesser amounts by plant transpiration. In the global atmosphere, as the warm air rises from the water surface, with a saturated concentration of vapor, it cools as we have discussed earlier. This cooling makes the water more condensable according to the Clausius-Clapeyron equation. In the aquarium example the water can condense on the cool lid and drip down, but where would the water condense in the atmosphere? Even at its highest concentrations of about 5%, there are so few molecules of water relative to the air that there is very little chance that two will come together, and the chances of forming a droplet are rarer still. If there was no surface available for the water vapor to attach to, the water molecules would stay apart in a "super-cooled" state and therefore would not meet the requirements of the Clausius-Clapeyron equation except near the ocean or other land surface. The atmosphere has a huge supply of tiny (< or ~0.2 μ) particles that come from many natural sources including gas-to-particle conversion processes, emissions from plants, fires, dust, oceans, and now some measurable amounts from human activities (Section 7.3). These are "cloud condensation nuclei"; however, any particle will do if it has an affinity for water. The smallest ones are not only abundant, because they have long residence times in the atmosphere, but also provide the most surface for their mass. If the air is warm, the forces of attachment will not be strong enough to capture as many water molecules which are moving fast and bounce off, as when the temperature is lower. So, the water vapor condenses on any surface that exists and droplets build in cooler conditions. This forms clouds and fogs, which are liquid and no longer vapor. Once the clouds are formed, a complex set of processes, including coagulation, and continued accumulation affected by the spherical nature of droplets, lead to larger drops that are heavy enough to fall as rain, or under cold conditions, as snow. In the end, we note that water, once put into the atmosphere, is mostly converted to rain or snow within a few days and that its atmospheric concentrations far from the ocean, lake, or soil surfaces is proportional to the saturation values, but not necessarily equal.

Let's look next at how precipitation may change with global warming, or cooling by writing the lifetime of water vapor as $\tau_W = \tau_{W0} + \beta \Delta T$, where $\Delta T = (T - T_0)$ is the temperature change by global warming and β is a constant that has to be determined by a deeper analysis of the mechanisms that change the lifetime or it may be found from observations. Then we can write the annual precipitation as P_W (precipitation) = S_W (evaporation) and from Eqs. 8.9 and 7.2 we get:

$$P_W = \frac{P_{W0} e^{0.07\Delta T}}{(1 + \beta \Delta T / \tau_{W0})} \tag{8.10}$$

This equation tells us how much the precipitation will increase with global warming because the evaporation, or source term for the water vapor balance must be equal to the precipitation. There are many complications that are ignored in deriving this simple expression, but it is instructive. If we can assume that the lifetime of water vapor won't change, then $\beta = 0$ and we would expect that the rainfall will increase at about 7% for every degree of global warming. It is noteworthy that this is due to the Clausius-Clapeyron equation which requires that the precipitation increase but it is not the only influence and what actually happens may be different. If the lifetime of water vapor increases ($\beta > 0$), then the effect of the Clausius-Clapeyron equation will be reduced and if it decreases ($\beta < 0$), then there will be even more rainfall or snow. The lifetime can be changed by the interaction of aerosols with clouds. If the changing budgets of aerosols cause cloud drops to become smaller, they will persist longer in the atmosphere and the lifetime will increase, but if climate change drives clouds to heavier and more frequent rain, then the lifetime may decrease. We expect however, that for small changes of the earth's temperature of a few degrees, these lifetime influences will also be small compared with the main driving force expressed by the Clausius-Clapeyron equation and global precipitation will increase close to the 7% per degree C.

Example Suppose the earth reaches a new steady state that is 5°C warmer in which there are more clouds because they rain out only half as fast as in the present. The effect of the cloud albedo change is already included in the 5°C warming because they are high clouds. Take the base lifetime of water vapor as 10 days. What is the effect of this global warming on precipitation?

Solution: We can apply Eq. 8.10. First we need to calculate β from $\tau_W = \tau_{W0} + \beta\,\Delta T$. We are given that $\tau_W = 2\,\tau_{W0}$ and $\tau_{W0} = 10$ days. Therefore $\beta = \tau_{W0}/\Delta T = 10$ days/5°C = + 2 days/°C. The residence time of clouds increases by 2 days for every 1°C warming. So, $P_W/P_{W0} = e^{0.07\,\Delta T}/[1 + \beta\,\Delta T/\tau_{W0}] = e^{(0.07 \times 5)}/[1 + 2\text{ days/°C} \times 5\text{°C}/10\text{ days}] = 0.7$, or rainfall will decrease by 30%. Note that in order for such a compensating result to represent the real world, mechanisms for such an extreme lengthening of the lifetime of water must be stated so that a β can be justified.

Review of the Main Points

1 The earth's natural climate is determined by solar radiation, albedo, and atmospheric constituents. Solar radiation is about 1360 watts/m^2 at the top of the atmosphere and it has a frequency distribution that is explained more or less by Planck's black body radiation law. Useful derived formulas are the Stefan-Boltzmann and Wein's laws.
2 The albedo of the earth is determined by reflection and scattering of sunlight. The reflection is mostly from clouds with a lesser role of the surface. Scattering is by aerosols and all atmospheric gases. The combined effect is that about 30% of the sun's energy is sent back to space without affecting climate. Atmospheric constituents, again including clouds, absorb another 20% of the sunlight. The often-expressed idea that the atmosphere doesn't absorb solar radiation is not correct. This leaves only 50% to be absorbed by the surface.
3 The earth radiates in the infrared part of the electromagnetic spectrum according to the laws mentioned in the text. Ninety percent of this radiation is absorbed mostly by the atmospheric greenhouse gases and lesser amounts by clouds. The gases are selective in the frequencies they absorb, but several mechanisms facilitate the absorption by broadening and adding to these fundamental frequencies. These mechanisms have a huge effect on the total absorption but also complicate the science. The absorption is represented by Beer's law – that radiation whether from the sun, earth, or the atmosphere, diminishes exponentially as it traverses the atmosphere containing absorbing material. As the atmosphere re-radiates the absorbed energy, it sends it reverberating throughout the atmosphere, still subject to Beer's law but with considerable complexity. The outcome is that the atmosphere sends more radiation down than up in the natural greenhouse effect.
4 Based on the science, water vapor emerges as the major greenhouse gas. Carbon dioxide is second. Methane, nitrous oxide, and ozone round out the natural greenhouse gases with small overall contributions. The energy involved in climate and climate change is stored mostly in the oceans with a lesser amount in the land. The atmosphere is virtually useless at storing energy relevant to the climate.
5 The effectiveness of greenhouse gases varies with the concentrations. When the concentrations are low, increasing levels lead to a linear increase in atmospheric absorption

and hence the greenhouse effect. As the levels rise, the gases are less and less effective in increasing the global temperatures.

6 The earth gains most of the energy by absorption of sunlight at the surface. Very little radiation from the surface is lost directly to space (through the window region). The loss of energy from the earth's system is mostly due to atmospheric radiation directed toward outer space.

7 Present rainfall is an average of about 1 m/year over the entire surface of the earth. It is part of the energy balance that includes the transfer of heat from the surface to the atmosphere by convective processes. It is likely to increase by roughly 6–7% per degree C for a few degrees global warming.

Exercises

1 Consider the temperatures of three stars – Spica, our sun, and Anteres. The temperatures are: 25,000°K, 5,800°K and 3,400°K respectively. The radii are 7.5 × R_{sun} for Spica and about 700 × R_{sun} for Anteres. In the following calculations, compare the results with those of our sun.
   (a) Plot the Planck black body radiation curves.
   (b) At what wavelength is the maximum amount of radiation?
   (c) How far do you have to put the earth so that the solar constant would be the same as it is with our sun?

2 When clouds are present they have a greater effect on the albedo than surface elements.
   (a) Show that $A_{Eff} = A_C + (1 - A_C) A_S - A_C A_S (1 - A_C)$. Here the subscript C is for clouds and S is for the surface. For simplicity, ignore cloud absorption. The last term represents the reflection from the underside of the clouds. The underside albedo is the same as at the top.
   (b) Calculate the effects of each process for $A_{Cloud} = 0.5$ and $A_{Surface} = 0.5$.

3 (a) Calculate how much net energy you radiate by black body radiation when the ambient temperature is 30°F and 100°F. The ambient environment provides a radiative energy at its temperature and assumes this is the only source of outside energy. Your body maintains a temperature of 98°F in both cases.
   (b) What happens if you fall in very cold water at 0°C? What happens then with radiation of heat from your body relative to other possible processes by which you can lose heat? Take your surface area to be 2 m^2.

4 A thought experiment was conducted and discussed in the text. In it 1000 molecules of N_2 and 10 molecules of CH_4 were confined to a box with which the molecules do not exchange energy. The box was irradiated with infrared light that caused a constant temperature after a while with all the molecules in the box contributing by thermalization. Suppose now, you take 1000 moles of N_2 and 10 moles of CH_4 and repeat the experiment.
   (a) How will that change the results of the experiment? Explain.
   (b) The CH_4 molecules are pulled out after equilibrium is reached. What will happen to the temperature? Explain.
   (c) With the CH_4 still removed, if we add conduction to the walls, so that energy is transferred as $dE/dt = -q(T - T_0)$ where T is the temperature of the gas in the box and T_0

is the temperature of the walls that does not change. Take $q = 0.1$ w/°K, C (heat capacity of N_2) =29 j/mole-°K, and T_I = the initial temperature of N_2 when CH_4 is removed = 273°K; T_0 = 0°K. How many days will it take for the temperature of the gas to reach 2°K?

Endnotes

Endnote 8.1 Planck's black body radiation law is: $E_\lambda = 2\pi c^2 h \lambda^{-5}/[e^{(hc/kT)\lambda} - 1]$ where h is Planck's constant, c is the speed of light, k is Boltzmann's constant, T is the absolute temperature, and λ is the wavelength of the light radiated. It should be noted that the radiation depends only on the temperature. As the temperature rises, the distribution shifts toward more ultraviolet emission. Many objects in everyday life and nature in general, follow this radiation law, including you and I.

Endnote 8.2 Kirchhoff's law relates to transfers of electromagnetic energy by absorption and emission by black bodies. It does not apply when the absorption of electromagnetic radiation fundamentally changes the nature of the absorbing medium, such as when it breaks a molecule apart, which happens often in the stratosphere. Additionally, there are other ways of gaining or losing energy for black bodies, such as conduction and convection, which are not subject to Kirchhoff's law.

Endnote 8.3 Scattering is an important atmospheric phenomenon that, for our interest, impedes solar radiation from reaching the surface, adding to the albedo. Air molecules scatter radiation preferentially as $1/\lambda^4$ where λ is the wavelength of light (Rayleigh scattering) which gives the sky its blue color. In this case light is scattered in all directions with about as much going backward toward the sun as forward toward the earth. The efficiency of the scattering is not large, so most of the sunlight gets past to reach the surface. Scattering from aerosols of interest in climate science comes mostly from fine particles which are subject to Mie scattering. This type of scattering sends smaller amounts of light back toward the sun and more of it forward to the earth. The total scattering of radiation is about 5% out of the total 30% albedo. There is more information in the textbooks cited in Endnote 8.6.

Endnote 8.4 Temperature is a measure of the average kinetic energy of molecules in a gas. Gas molecules move around at speeds that follow the Maxwell-Boltzmann distribution. The number of molecules moving at speeds near v is $n(v) = (m/2\pi kT)^{3/2} e^{-KE/kT}$, where KE is the kinetic energy of the molecules = $\frac{1}{2}$ m v^2, k is Boltzmann's constant, and T is the absolute temperature. All else being the same, at higher temperatures more molecules move fast than at lower temperatures. This causes more frequent and more energetic collisions. From this distribution it can be shown that the average speed of molecules of mass m, defined as $v(\mathrm{rms}) = (2RT/mN_A)^{1/2}$ where m is the mass of the gas molecule, N_A is Avogadro's number, and R is the gas constant. We see the remarkable result that the temperature, which we can feel and measure, is a reflection of the average kinetic energy or speeds of molecules in the gas.

Endnote 8.5 Formally, an effective optical depth can be stated for the absorption, by the atmosphere, of the earth's radiation or the solar radiation as: $F = 1 - \exp(-\tau)$

assuming that Beer's law applies. If we do this, then $\tau = 2.3$ for the absorption of earth's radiation found from inverting the equation above and taking $F = 0.9$ and it is $\tau(\text{solar}) \approx 0.2$ taking the absorption to be about 0.2. It may provide a sense of the magnitudes of the optical depth on the global scale.

Endnote 8.6 Sources: Figures 8.3 and 8.9 are from J.P. Peixoto and A.H. Oort, *The Physics of Climate*, American Institute of Physics, N.Y., 1992. Figures 8.4 and 8.7 are from D. Hartmann, *Global Physical Climatology*, Academic Press, San Diego, 1994. Figure 8.10 is from Table 3 in J. Kiehl and K. E. Trenberth, *Bulletin of the American Meteorological Society* (*BAMS*), vol. 78, p. 197, 1997. Figure 8.12 is from K.E. Trenberth et al., *BAMS*, p. 2, March 2009. Mathematically advanced readers may find more information on the topics of this chapter in R. Goody, *Principles of Atmospheric Physics and Chemistry*, Oxford University Press, 1995; D.L. Hartmann, *Global Physical Climatology*, Academic Press, 1994 and Wallace and Hobbs, 2006 in Endnote 7.4.

9

Instructive Climate Models

We have discussed key attributes of climate including the winds, precipitation, and humidity; now we come to the temperature, which is the iconic index of climate change. Our goal here is to delineate how the prevailing temperature of the earth is established and from it to understand how global warming will happen and what temperature change it can bring.

9.1 Base Temperature Model – Lessons, Flaws, and Resolution

Let's consider the simplest model that can tell us what the temperature of the earth should be considering what we have learned about the amount of energy arriving at the earth and radiative transfer in the atmosphere (Figure 9.1).

We will consider first the situation in steady state so that the temperature is constant. The land and oceans constitute the surface of an average temperature. We describe the atmosphere as a constant temperature layer, of unspecified thickness, that contains greenhouse gases and clouds capable of creating the known albedo (30%) and absorption of the surface radiation (90%). The atmosphere is expected to radiate half the energy it gets back to the surface and the other half to outer space because of the constant temperature assumption. Our goal is to calculate the temperatures of the surface (T_s), of the atmosphere (T_a), and by how much the greenhouse gases and clouds influence them. From Figure 9.1 we can write the energy balances per unit area for the surface and the atmosphere as:

$$\textit{Surface}: \qquad (1-A)S + F\sigma T_a^4 = \sigma T_s^4 \qquad (9.1a)$$

$$\textit{Atmosphere}: \qquad F\sigma T_s^4 = 2F\sigma T_a^4 \qquad (9.1b)$$

The left-hand sides are the energy received and the right-hand sides are the energy emitted. Note that we have used Kirchhoff's law liberally in Eq. 9.1b, whereby the emissivity of the entire atmosphere equals the absorptivity and S is the diurnally averaged solar radiation over the whole surface of the earth. These equations can be solved readily to get the temperatures:

Global Climate Change and Human Life, First Edition. M. A. K. Khalil.

Companion Website: www.wiley.com/go/khalil/Globalclimatechange

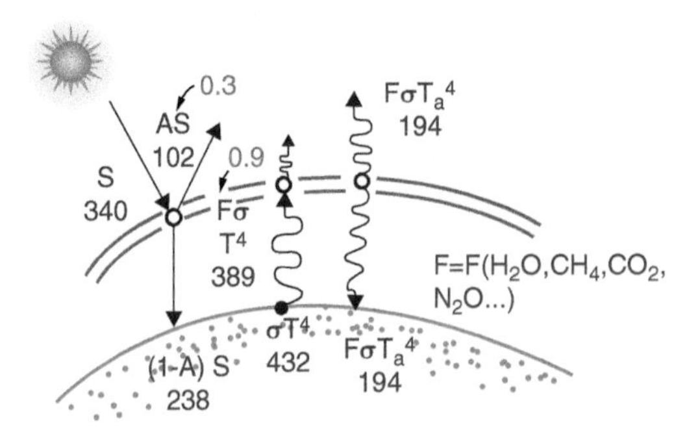

Figure 9.1 The simplest climate model. Radiative transfer is represented for sunlight on the left and for the earth's radiation on the right. The fluxes are given in w/m^2. The atmosphere radiates equal amounts up and down.

$$T_s = \left[\frac{(1-A)S/\sigma}{1-\frac{1}{2}F}\right]^{1/4} \tag{9.2a}$$

$$T_a = \left(\frac{1}{2}\right)^{1/4} T_s \tag{9.2b}$$

Putting in the numbers from Figure 8.12 (F = 0.9, A = 0.3, S = 341 w/m^2), we find that T_s = 296°K and T_a = 249°K. This result is not far from the actual temperatures of the surface and the atmosphere which were shown to be 289°K and 261°K in Section 8.5. It is noteworthy that our model is looking only at the vertical movement of energy. This is because solar energy comes in and the earth's energy leaves this way, making it the main dimension that controls the earth's average temperature. Energy is also transferred from one latitude to another and across land and the oceans, significantly affecting the environment and life, but not the net energy balance or the average temperature of the earth. We will come to these later.

We can learn more from this model. If there were no clouds or scattering, the albedo would be $A \approx 0.12$ similar to that of the moon, instead of 0.3. If all else were the same we will get T_s = 313°K or ~ 20°C warmer. If there were no greenhouse gases (F = 0) but the albedo was still 30%, the surface temperature would be T_s = 254°K or ~ 40°C cooler in round numbers. If there were neither clouds (A = 0.12) nor the greenhouse gases (F = 0) the temperature would be T_s = 270°K or a net effect of ~ 30°C cooler. This is more or less the state of the moon relative to the earth and so can be taken to represent the effect of the atmosphere. Readers can easily verify these results which are similar to what we get from more complete and thus more complex models. Without the atmosphere and its greenhouse effect, the earth's average temperature would be below the freezing point of water. There would be lots of ice and the world would probably be frozen as it might have been a long time ago. As a side, these findings show how mathematical models allow us to assess the effects of the different components controlling our system. It doesn't mean that it is possible to have a world with one and not the other of these elements. For instance, if we have clouds, we will have water vapor and consequently a greenhouse effect.

Between Eqs. 8.5, 8.6, and 9.2 the simplest model provides a picture of how the concentrations of gases determine the greenhouse effect. The fundamental variables are k's (absorptivities), C's (concentrations of absorbing gases), which determine F (the total absorptivity, including clouds); and the other two variables are A (albedo) and S (solar constant). While it gives us a lot of information for very little investment in complexity, examining its several flaws will teach us deeper aspects of climate theory.

The model gives a surface temperature that is too hot and an atmosphere that is too cold. From Figure 8.12, we note that this might be because we ignored two sources of heat into the atmosphere. One is the absorption of solar radiation, particularly by water vapor and clouds in the troposphere and the other is the convective transfer of energy from the surface to the atmosphere represented by thermals and latent heat (Section 5.1). Both these processes will cool the surface while warming the atmosphere. Alternatively we can realize that not all the energy that comes into the atmosphere is from the earth's radiation as assumed in the simplest model. If we add these two features in Eq. 9.1 and solve, we get $T_s = 270°K$ and $T_a = 253°K$. Now the surface and the atmosphere are both too cold. This is because we are now sending more energy to the atmosphere than before, cooling the surface, and the atmosphere is sending even more energy to space than before because it is a little warmer.

The previous calculation leads us to think that perhaps the most egregious assumption in Eqs. 9.2 is that the atmosphere radiates equal amounts of energy up as down. It is based on the assumption of a constant atmospheric temperature with altitude. While this simplifies the mathematics, it makes the model further from reality in more ways than one. We know from Section 8.5 that $\delta = 0.25$, for the current atmosphere that represents the disparity of upward versus downward radiation. It is uncomfortably far from zero required for equal streams. The steady-state equations including the missing pieces in Eq. 9.2, namely, convective fluxes, atmospheric absorption, and the "more down than up" phenomenon, are formulated and discussed in Endnote 9.1. When we put the numbers into these equations the temperatures of the surface and the atmosphere are 289°K and 268°K as represented in Figure 8.12 which is consistent with observations. It is noteworthy, however, that for thin layers of the atmosphere, the assumption of a constant temperature and hence equal radiation up as down is quite accurate. But then you have a more complex model with many layers to represent the whole atmosphere (Endnote 9.2).

Example Evaluate the percentage errors in the absorption of energy by the atmosphere, and the surface due to the assumptions of the simplest model. The assumptions are: (a) no absorption of solar radiation by the atmosphere, (b) no convection, and (c) same amount of energy sent up as down. Define percentage error as %Err = (Actual − Assumed)/Actual × 100%.

Solutions: Using the results in Figure 8.12: (a) No solar energy absorbed: % Err = $[1 - 356/(356 + 78)] \times 100\% = 22\%$. (b) No convection: $\%\text{Err} = [1 - 356/(356 + 97)] \times 100\% = 21\%$. (c) $\delta = 0$: $\%\text{Err} = [1 - 266/333] \times 100\% = 20\%$. All three assumptions violate our 10% rule, however the surface temperature from the simplest model is within 3% because of compensation by the factors and the influence of the 1/4th power in the temperature equation.

One further consequence of equal amounts of radiation up as down is that it leads to an artificial limit on greenhouse warming because it sends too much energy to space from the atmosphere. Here is why. Although the absorption is currently about 90%, the remaining 10% is escaping the atmosphere because the greenhouse gases cannot effectively absorb this radiation (window region). Even if we could increase the absorption to its maximum value of $F = 1$, it would give a surface temperature of 303°K. No matter how much more greenhouse gases we add, or which ones, the temperature in the model can never exceed this value. That seems unlikely to be the case on real planets. Indeed, as we fix the flaws of the simplest model, we will see next that it is impossible to avoid convection or a hotter lower atmosphere and therefore the δ will always be greater than zero.

Let's look at three scenarios (Figure 9.2). In the first, we have an atmosphere of uniform temperature as before; it does not absorb solar radiation and contains just enough mass of an absorber, to take up nearly all the infrared radiation from the earth ($F \approx 1$). According to Eqs. 9.2, the temperatures of the atmosphere and the surface will be:

$$\text{Surface:} \qquad T_s = 2^{1/4} T_e \tag{9.4a}$$

$$\text{Atmosphere:} \qquad T_a = T_e \tag{9.4b}$$

Here T_e is defined to be $[(1 - A)\, S/\sigma]^{1/4}$; it is the equilibrium radiating temperature (Endnote 9.3). T_S and T_a are 303°K and 255°K, respectively. The atmosphere would be sending the same amount of radiation, $\sigma\, T_e^4$ up as down. Since the temperature of the surface with no atmospheric absorber would be 255°K, the effect of total atmospheric absorption would be to increase the surface temperature by 48°C (303–255). In the second scenario, we double the amount of absorber in the atmosphere and assume that the lower half of the atmosphere has one temperature and the upper half another, but within the layers the temperature is constant. The lower half of the atmosphere by mass will have enough to absorb all the earth's radiation, and the upper half will be likewise.

In equilibrium, according to Figure 9.3, the equations comparable to Eq. 9.1 will be:

$$\text{Surface:} \qquad (1-A)S + \sigma T_1^4 = \sigma T_s^4 \tag{9.5a}$$

$$\text{Lower Atmosphere:} \qquad \sigma T_s^4 + \sigma T_2^4 = 2\sigma T_1^4 \tag{9.5b}$$

$$\text{Upper Atmosphere:} \qquad \sigma T_1^4 = 2\sigma T_2^4 \tag{9.5c}$$

And after some algebra the solutions comparable to Eqs. 9.4 will be:

$$\text{Surface:} \qquad T_s = 3^{1/4} T_e \tag{9.6a}$$

$$\text{Lower Atmosphere:} \qquad T_1 = 2^{1/4} T_e \tag{9.6b}$$

$$\text{Upper Atmosphere:} \qquad T_2 = T_e \tag{9.6c}$$

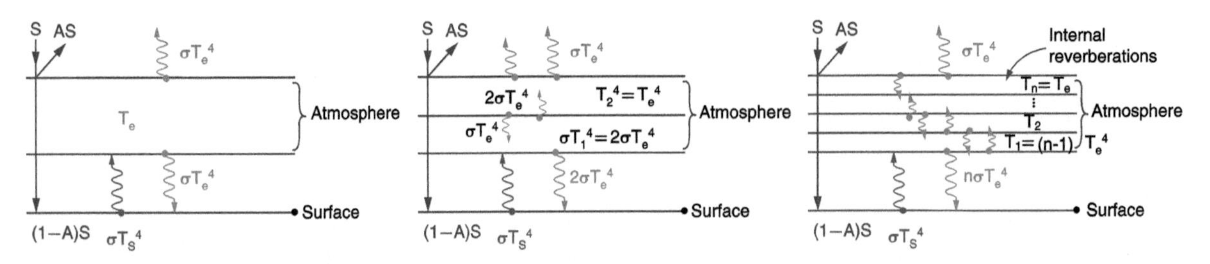

Figure 9.2 Energy absorption scenarios. Adding greenhouse gases and its effects on the surface temperature.

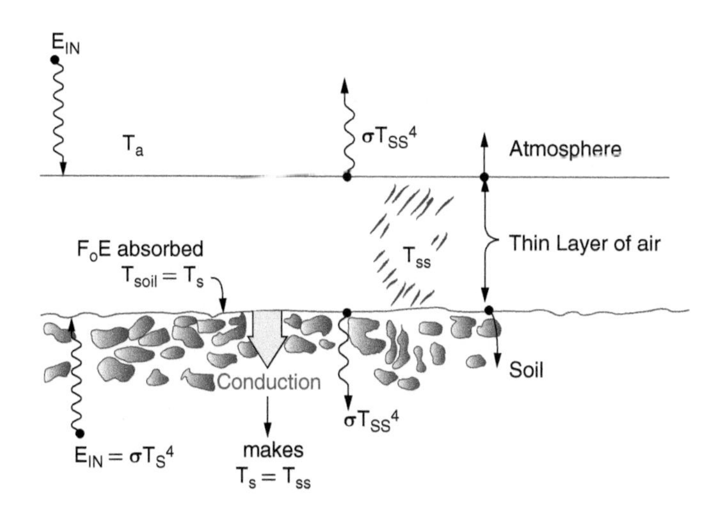

Figure 9.3 Conduction at the interface between the surface and the atmosphere. Radiation balance would result in an abrupt difference between the surface and air temperatures. These are eroded by conduction. Convection moves heat away from the surface, but first the heat has to get into the air.

or, $T_s = 336°K$, $T_1 = 303°K$, and $T_2 = 255°K$. Now the lower atmosphere is much warmer than the upper as an inevitable consequence of the fact that the sun's energy is absorbed by the surface, so the earth's emission is intercepted first by the layer closest. The upper layer is shielded from the direct emission of the earth and receives none of it because the lower layer has absorbed it all. It gets only half as much energy as the earth radiates because the lower layer splits it into two streams, half up and half down. Moreover, the upper layer loses all the energy it sends upward into space, but the lower layer gets energy both from the layer above and the surface below. The lower layer therefore, will always be warmer than the upper. If you now combine the two atmospheric layers in your mind and look at them as "the one atmosphere," you will conclude that the energy sent to outer space (Eq. 9.6c) $= \sigma\, T_e^4$ is only half as much as the energy sent back to the surface (Eq. 9.6b and Eq. 9.5a) $= \sigma\, T_1^4 = 2 \times \sigma\, T_e^4$. So, doubling the absorber causes more radiation to be sent to the surface than to space.

If we keep adding such layers, let's say you have "n" of them, readers can prove to themselves that the temperatures will follow the same pattern as in Eqs. 9.6 or: $T_s = (n+1)^{1/4}\, T_e$, the atmospheric layer nearest the surface will be at $T_1 = n^{1/4}\, T_e$ and the next one at $T_2 = (n - 1)^{1/4}\, T_e$ until we get to the top layer which will be at T_e. Each layer will cause a diminishing increase of the surface temperature, even though they have the same amount of absorber (greenhouse gas), but there would be no strict upper limit to the surface temperature as in our simplest model. Also instructive is to calculate the $\delta = (n-1)/(n+1)$ for the n-layer system. We see that adding more and more absorber increases the downward flux relative to the upward flux consistently.

We can divide the real atmosphere into many layers each thin enough that we can approximate it with a constant temperature and this process plays out the same way causing a warmer lower atmosphere and cooler above but the mathematics is more complicated because we have to take into account the fractions of the radiation stream that are absorbed and transmitted from each layer, and every layer influences every other layer. What remains the same qualitatively is that adding additional absorber increases the value of δ, sending increasingly more radiation down than up.

The phenomenon of sending more radiation down can be thought of in this alternative, and perhaps an interesting way. Photons representing the earth's heat are trapped and re-radiated multiple times. The more absorber we add, the more often the photon is likely to be trapped as it finds its way toward space. Each time it is trapped, it can be sent back to the earth or not, and so increasing greenhouse gases will always increase the chance that a photon is sent back to the surface thus increasing the greenhouse effect. The lowest layers of the atmosphere trap the photon first, so the higher the layer is in the atmosphere, the less chance there is that it receives one of the earth's "original" photons. Instead, it receives re-emitted photons from some layer below or above. And the number of the re-radiated ones is much less than the number being sent from the earth's surface because many have already been sent back and re-absorbed. In this way there is a continual decrease of temperature as we go higher up because there are fewer photons available for trapping, which represents the energy being absorbed to heat the air; the surface therefore receives much more of its own radiation back than goes out to space. While there may be a philosophical argument that when an atmospheric molecule absorbs a photon and later a different nearby molecule emits a photon of the same frequency, it is not the same photon. The effect is the same, however, and there is no way to tell one photon apart from another of the same frequency.

Within the two-layer model, we can calculate the lapse rate by assuming that the mixing ratio of the absorber is constant. In this case, recognizing that the density decreases exponentially according to Eqs. 3.1 (and Figure 3.2), we will find that half the atmosphere lies below 5.5 km. So, the two layers that contain equal amounts of the absorbing greenhouse gas are between 0 and 5.5 km, and 5.5 km to some very high altitude. If we assign the temperatures of these layers to an altitude that represents the middle of the absorber mass in each layer, then T_1 would represent the temperature at 2.3 km and T_2 at 11.1 km The lapse rate between the surface and the mid-lower layer would be about 14°C/km = (336 – 303) °K/(2.3 – 0) km and between the mid-lower and mid-upper layer it will be 5.5°C/km = (303 – 355)°K/(11.1 – 2.3) km. It is a general characteristic of pure radiative balances that the lapse rate is very high near the surface and decreases as we go upward. This always causes the surface temperature to be too warm and the atmosphere too cold compared with the real earth as we had noticed with the simplest model and the effect would only get worse if you made your model more refined with many layers. So, even the most complex radiative models will require the same fix as our simple model, namely, the inclusion of convective transport of heat, otherwise they will do worse than our one layer model! As we discussed in Section 5.1, under higher than adiabatic lapse rates, any small patch of air near the ground that gets warmer than the surrounding air will rise upward. This would be inevitable because the surface is not a uniform color or reflectivity. The rising parcel cools at the adiabatic lapse rate and the pure radiative lapse rate is so much greater near the surface, that the parcel will remain warmer than the surrounding air and rise rapidly. This would remove energy from the surface making it cooler than in purely radiative equilibrium and add it to the atmosphere. In the earth's system with a fluid atmosphere, and similarly on any planet, pure radiative equilibrium *cannot exist*. It will inevitably generate convection which will erode its very existence and lead to a more complex radiative-convective energy balance. Radiation forces convection, but convection does not require solar radiation and will occur whenever the surface is heated regardless of cause.

A related issue with pure radiation balances is that the temperature of a thin layer of air right above, or touching the earth's surface is abruptly lower than at the surface. The thin layer absorbs IR-radiation from the atmosphere above and from the earth's surface below. It radiates both up and down achieving the following balance and a temperature T_{ss} (Figure 9.3):

$$F_0 E_{IN}(From\ Atnosphere) + F_0 E_{IN}(From\ Surface) = 2\sigma F_0 T_{SS}^{4} \tag{9.7}$$

$$E_{OUT}(Surface) = \sigma T_s^4 \tag{9.8}$$

F_0 is the absorptivity or emissivity of the layer in Eq., 9.7, which cancels. Based on Eqs. 9.7 and 9.8 with E_{OUT} (Surface) = E_{IN} (From Surface), it is straightforward to prove that T_s will always be greater than T_{SS}, but we can also easily calculate these temperatures by using values of the energies on the left-hand side taken directly from Figure 8.12 as: E_{in} (From Atmosphere) = 333 w/m^2 and E_{in} (From Surface) = E_{out} (Surface) = 396 w/m^2 from which we calculate T_{SS} = 283°K and T_s = 289°K. This abrupt discrepancy of some 7°C, between the surface temperature and the air touching it, cannot exist in the real world except perhaps for short periods of time. When it does, the temperature gradient will cause a spontaneous flow of energy from warm to cold as required by the second law of thermodynamics and well known from human experience. Conduction will equalize the temperatures between the surface and the layer of air touching it, most likely followed by convection.

As the number of layers increase in Figure 9.2, we see that radiative transfer of heat energy becomes less efficient. The surface that is getting the heat of the sun retains more of it and gets very warm while the atmosphere is increasingly shielded and stays cool. In the atmosphere this process is broken by convection and also by the fact that the atmosphere doesn't have many completely absorbing layers. This is not so for soils, which completely absorb solar radiation effectively in thin layers, and there is no convection, so radiative transfer is not an effective means of transferring heat to lower levels of the soil. It leaves only conduction to move heat from the surface to lower levels retaining most of it within a few tens of meters. The oceans are in between, also absorbing within a fairly thin layer, but have strong mixing processes near the surface, driven by the winds above, that can move heat quickly at least within the mixed layer in the top 50–200 meters. Moving heat to the deeper oceans is much slower.

There are four lessons to note well from the multi-layer models. First, that increasing the amount of absorber, doubling in our case studies, cannot lead to more absorption of the earth's energy if it is already completely absorbed before the doubling, but more radiation will inevitably come back to the surface compared with what goes to space, causing continued global warming. Second, adding an amount of absorber causes a greenhouse effect, but adding as much more causes a lesser warming, and with each additional kilogram that we add, the warming effect is less and less. Third, we noticed that radiative processes will drive or compel convective activity causing the surface to be cooler than in pure radiative equilibrium. In the same context, conduction at the surface equalizes the temperature of the soils or water with that of the air right above it. This process can be a prelude to convection.

9.2 Radiative Forcing and Climate Sensitivity

Now we want to consider two widely used concepts: *radiative forcing* and *climate sensitivity*. We have two aims. One is to have a uniform agreed upon means of quantifying influences on the earth's temperature, and the other is to establish practical connections between increase of greenhouse gases and surface global warming.

The term "radiative forcing," though not evocative, is meant to describe the action of "forces" that determine the climate and cause climate change. In physics, forces are needed to move objects; here forces are needed to move the earth's temperature. Here is a way to think about it. At the top of the atmosphere the streams of energy that are coming and going are clearly understood. The streams are E_{IN} (w/m^2) = $(1 - A)\,S$ that contains the effect of the sun and the earth's reflectivity. For the incoming energy or radiation stream, the radiative forcing of the sun is just this $RF(\text{sun}) = (1 - A)\,S$. In our present world it is $(1 - 0.3) \times 341$ w/m^2 = 239 w/m^2. The radiative forcing of the albedo is $RF(\text{albedo}) = -AS = 102$ w/m^2. For the outgoing stream, it is a combination of the radiation that leaks through the atmosphere in the window region and the upward radiation of the atmosphere. In total it is $RF(\text{earth and atmosphere}) = (1 - F)\,\sigma\,T_s^4 + (1 - \delta)\,\sigma\,FT_a^4 = 40 + 199 = 239$ w/m^2 (based on Figure 8.12). The net radiative forcing, which is often taken as its definition, is $\Delta RF\,(\text{Net}) = E_{IN} - E_{OUT}$. Under equilibrium conditions, the $\Delta RF(\text{Net}) = 0$ indicating that the net force that can cause the earth's surface temperature to change is zero, and so the temperature will not change. Let's say the sun suddenly becomes warmer and the solar constant increases by 10 w/m^2 to 352 w/m^2. At that moment, before the various elements of the earth's system can react, the RF(sun) will become $(1 - 0.3) \times 352$ w/m^2 = 246 w/m^2, and therefore the $\Delta RF(\text{Net}) = (246 - 239)$ w/m^2 = 7 w/m^2. This will be called the radiative forcing caused by the change of the solar constant by 10 w/m^2. What happens next is that the earth's system responds by warming the surface and the atmosphere so the balance is restored again in a new equilibrium state. The question is by how much will that change the surface temperature if the solar condition persists. We can calculate the temperature change to expect from just this direct forcing when equilibrium is restored. It still requires an accounting of the energy balance, such as in the simple model we discussed earlier. It translates the additional in-coming energy to predict the change of temperature. From Eq. 9.2 it will be 298°K or a change $\Delta T_s = (298 - 296)$°K = 2°C. Refinements to the base concept of radiative forcing we have discussed are summarized in Endnote 9.4.

The *climate sensitivity* is defined as λ in the following equation:

$$\Delta T_s = \lambda\,\Delta RF \tag{9.9}$$

which in the case discussed above will be 2°C/7 w/m^2 ≈ 0.3°C/(w/m^2). A complex issue arises. If the solar energy does indeed increase by 10 w/m^2, how will the entire earth system react to that before it reaches equilibrium with the new reality? Will it lead to more water vapor because of the Clausius-Clapeyron law, adding more global warming? It is a virtual certainty. Will the ice melt, changing the albedo? Will it cause more convection, or more clouds, cooling the earth? These phenomena will occur to varying degrees and they will change the final temperature of the earth from the hotter sun. To what extent and by how much will depend on a much deeper analysis of the earth's response and on the forces we have not included in our model. We will take up this matter again in the next chapter as "feedbacks," but the radiative forcing will be well established from a known cause. So, in this scenario, we will agree on the radiative forcing, but will be skeptical about whether we have the right climate sensitivity without evaluating all the consequences. That doesn't detract from the definition of the climate sensitivity in Eq. 9.9 – just its value for a prescribed cause. Note that the ΔT_s in Eq. 9.9 is the final equilibrium temperature and the ΔRF is the initial disturbance. In the initial situation the other factors such as the albedo, and the absorptivity (F) are the same as now, but in the final temperature they may not be, making λ the focal point of taking the earth's response into account and a source of great uncertainty.

It is noteworthy that Eq. 9.9 is a general relationship that can be understood from theoretical arguments. In the earth's environmental system, the forces that affect the energy causing a net change of radiative forcing at the top of the atmosphere are solar radiation, albedo and the atmospheric absorption. If any of them change, the re-balancing of the energy will be mostly from the heating of the surface. This is because the surface receives the most energy in the incoming short wave radiation stream: $(1 - A - f)\, S$ and it is the biggest supplier of energy that heats the atmosphere. The outgoing energy that balances the incoming is made up of two sources – the leakage of the earth's radiation through the window region $(1 - F)\,\sigma T_s^4$, and the radiation from the atmosphere $(1 - \delta)F\,\sigma T_a^4$. But the atmospheric temperature depends heavily, though not entirely, on the surface temperature, so in the end, the re-balancing of any disturbance of the radiative forcing at the top of the atmosphere will end up coming mostly from a change of surface temperature. This leads us to say that the surface temperature T_s must be some function of the radiative forcing of the outgoing stream, especially when considering the effect of greenhouse gases, or $Ts = f(RF)$. We may not know what that functional relationship is, but we can say by using Taylor Series from mathematics, that any change of radiative forcing from an initial value of RF_0 will cause a surface temperature change that can be approximated as: $T_S \approx T_{S0} + \lambda\,(RF - RF_0)$ + more terms. This equation is the same as Eq. 9.9 if we can ignore the further terms. Here λ is the instantaneous rate of change of T relative to RF, dT/dRF, taken at the initial conditions before the change of radiative forcing took place. It is the total change of temperature (or total derivative) from the change of radiative forcing due to all variables that affect the temperature. We can note further that the λ should not depend on what caused the change of radiative forcing, whether it was albedo, the sun, or the addition of a greenhouse gas. This is because the surface completely absorbs both the short-wave solar radiation and the infrared atmospheric radiation and reacts by warming if these streams increase. All that matters is the amount and not the type of extra energy coming in. This line of reasoning justifies Eq. 9.9, although how a proper λ can be calculated is a different matter because it must include the contributions from all the factors that affect the temperature if the system is disturbed.

A discussion of how an imposed radiative forcing, let's say it is positive, affects the extra energy arriving at the surface may be informative. If re-balancing of the imposed ΔRF is entirely supplied by the surface warming, then the extra energy the surface is receiving from this disturbance will be just ΔRF and the radiative forcing can be interpreted as the extra energy arriving at the surface. That may be so just when the disturbance is initiated, but it cannot actually happen when the earth adjusts to a new equilibrium. Let's say the solar constant increases by 3 w/m^2 that will result in a $\Delta RF \approx 2$ w/m^2. Instantaneously the surface will be getting this much more energy than before the sun started acting up. But the surface warming it will cause will send more energy to the atmosphere which will absorb some of it since the absorption is $F\,\sigma\,T_s^4$, and the T_s has increased. The atmosphere then will radiate some of it up and the rest down, so the surface will now be getting more than the extra 2 w/m^2 that is coming from the sun. Based on our simple model (Eq. 9.1), the surface radiation will be

$\Delta RF/(1 - ½ F)$ or about 3.6 w/m^2 upon reaching equilibrium. So, in the new balance, the atmosphere will be getting this much more radiation, and not 2 w/m^2. The re-balancing of radiative forcings at the top of the atmosphere arise from the action of both the surface and the atmosphere. The effect only amplifies if the earth's system responds by changing its other components such as clouds, ice, and water vapor content which we will consider in the next chapter as feedback processes (Endnote 9.5). The following example illustrates the concepts.

Example Calculate the climate sensitivity of the simplest model in Eqs. 9.1 for small changes in (a) the solar constant S, (b) the albedo A, and (c) the absorptivity F. Are the sensitivities the same?

Solutions: We will look at "small changes" in F, A, and S. The exact amount of change does not affect the results as readers may verify. To start, we have (Eq. 9.1): $T_s = T_e/(1 - ½ F)^{1/4}$, and $T_e = [(1 - A)\, S/\sigma]^{1/4}$. Putting in numbers, $T_e = 254.5$°K and $T_s = 295.6$°K for $S = 340$ w/m^2, $F = 0.9$, and $A = 0.3$. (a) Change the solar constant by $\Delta S = 10$ w/m^2. This causes a change in radiative forcing of $\Delta RF = (1 - A)\,\Delta S = 7$ w/m^2 since we are not changing anything else. The new T_e will be: $[0.7 \times (340 + 10)\ \text{w/m}^2/(5.67 \times 10^{-8}\ \text{w/m}^2 - \text{°K}^4)]^{1/4} = 256.4$°K and $T_s = 256.4/(1 - 0.5 \times 0.9)^{1/4} = 297.7$°K. Therefore: $\Delta T_s = (297.7 - 295.6)$°K = 2.1°C, and $\lambda = \Delta T_s/\Delta\text{RF} = 2.1$°C/7 (w/m^2) = 0.3°C/(w/m^2). (b) Increase the albedo by 0.03. The change of radiative forcing is: $\Delta\text{RF} = \Delta\text{A S} = 0.05 \times 340$ w/m^2 = 17 w/m^2. The new $T_e = [(1 - 0.3 - 0.05) \times 340\ \text{w/m}^2/\sigma]^{1/4} = 249.9$°K, $T_s = 290.1$°K and $\Delta T_s = (295.6 - 290.1)$°K = 5.5 °C. Therefore, $\lambda = \Delta T_s/\Delta\text{RF} = 5.5$°C/17 w/m^2 = 0.32°C/(w/m^2). (c) Change absorptivity F by 0.02 to 0.92. At the top of the atmosphere, in this model, the outgoing energy is $(1 - F)\,\sigma\, T_s^4 + ½ F \sigma T_s^4 = (1 - ½ \text{F})\, \sigma\, T_s^4$. Here, the first term is the energy from the surface escaping through the window region, and the second term is the radiation of the atmosphere out to space, which is half of what it intercepts. Therefore, the change of radiative forcing due to a change in F is: $\Delta RF = -½\, \Delta F\, \sigma\, T_s^4 = -½\, (0.02)\, \sigma\, (295.6\text{°K})^4 = 4.3$ w/m^2. $T_s = T_e/[1 - ½ F(\text{new})]^{1/4} = 296.9$°K, $\Delta T_s = 1.3$°C and therefore, $\lambda = 1.3$°C/4.3 w/m^2 = 0.3°C/(w/m^2). We see that the climate sensitivity is the same regardless of which driver we change in the simple model.

9.3 Practical Relationships between Greenhouse Gases and Surface Warming

We have established much of the physics of how the greenhouse effect and climate change work, but not how to connect changes of greenhouse gas concentrations with actual or expected temperature change at the surface. That is the goal now. We could do so if we can connect F and δ with greenhouse gas concentrations or mass amounts in the atmosphere and took into account the flows in Figure 8.12. The relationships that evolve are not simple, however, and may not be possible to express in algebraic formulae such as the ones we have developed and applied to learn about the processes in this chapter (Eqs. 9.1–9.6). Climate models that take into account these processes to varying degrees of detail and dimensions, from vertical, to vertical and latitudinal, to three-dimensional, are all solved on high-speed computers using numerical techniques. These are very far from the goals of our discourse. To evaluate the nature of climate change and the human role in it, it is sufficient to connect the causes. Much can be learned from a semi-empirical approach that has both significant theoretical and practical benefits.

ΔRF can be calculated for all drivers of climate change from the science we have discussed, although with considerable difficulty. For greenhouse gases, these calculations are done numerically based on the known details of how each gas interacts with radiation, both solar and from the earth as discussed in Chapter 8. The results of such a calculation have been converted to approximate formulas given below (IPCC TAR, Endnote 9.6):

$$\Delta RF(CO_2) = \alpha \ln(C/C_0) \qquad (\alpha = 5.35\ w/m^2) \qquad (9.10a)$$

$$\Delta RF(CH_4) = \alpha\left(\sqrt{M} - \sqrt{M_0}\right) + \left[f(M_0,N_0) - f(M,N_0)\right] \qquad (\alpha = 0.036\ w/m^2 - ppb^{-1/2}) \qquad (9.10b)$$

$$\Delta RF(N_2O) = \alpha\left(\sqrt{N} - \sqrt{N_0}\right) + \left[f(M_0,N_0) - f(M_0,N)\right] \qquad (\alpha = 0.012\ w/m^2 - ppb^{-1/2}) \qquad (9.10c)$$

$$\text{Where} \quad f(M,N) = 0.47 \ln\left[1 + 2\times10^{-5}(MN)^{3/4} + 5.3\times10^{-15} M(MN)^{1.52}\right]$$

$$\Delta RF(CFC's) = \alpha(C - C_0) \qquad (\alpha = 0.22 \text{ and } 0.28\ \text{w/m}^2\text{-ppb for CFC-11 and CFC-12}) \qquad (9.10d)$$

C in Eq. 9.10a is the mixing ratio of carbon dioxide in ppm, M is methane, and N is nitrous oxide both in ppb and C for the CFCs in Eq. 9.10d is in ppt. The overlap of the methane and nitrous oxide absorption bands is represented by the function f(M,N). The square root effect for methane and nitrous oxide and the logarithmic increase in radiative forcing for CO_2 show the band saturation effects we discussed in Chapter 8.

Example What is the radiative forcing caused by the increase of CO_2 between pre-industrial and present times? Assume pre-industrial concentration = 280 ppm and present concentration = 400 ppm.

Solution: Radiative forcing is = 5.35 × ln (400/280) ≈ **2 w/m²**.

When the observed changes in greenhouse gas concentrations are used to evaluate the changes of radiative forcing during the last 100–200 years, the consensus result is as shown in Figure 9.4 (IPCC-AR5, Endnote 1.1).

Three points should be noted. First, that CO_2 has the largest effect on increasing radiative forcing, and is therefore the largest cause of global warming. Second, that other greenhouse gases, most notably methane, nitrous oxide and the halocarbons, have relatively small effects individually, but collectively they have been contributing about half as much to global warming as CO_2. Third, the effect of aerosols (sulfate and nitrate), direct and by interactions with clouds, is a significant cooling, but it is highly uncertain. The net result is an increase of about 2 w/m^2 with a disconcerting range of 1–3.4 w/m^2. It means that the earth's atmosphere and surface are retaining about 2 w/m^2 extra radiation now than it did a couple of hundred years ago just from the increase of the greenhouse gases and aerosols. It is transforming into the global warming we see as an unsteady increase of temperature by about 1.4°C on land and ~ 1°C over the oceans (Figure 1.1). From this, it may be concluded that the sensitivity $\lambda \approx$ 1.4°C/2 w/m^2 = 0.7°C-m^2/w. IPCC's AR5 also reports an

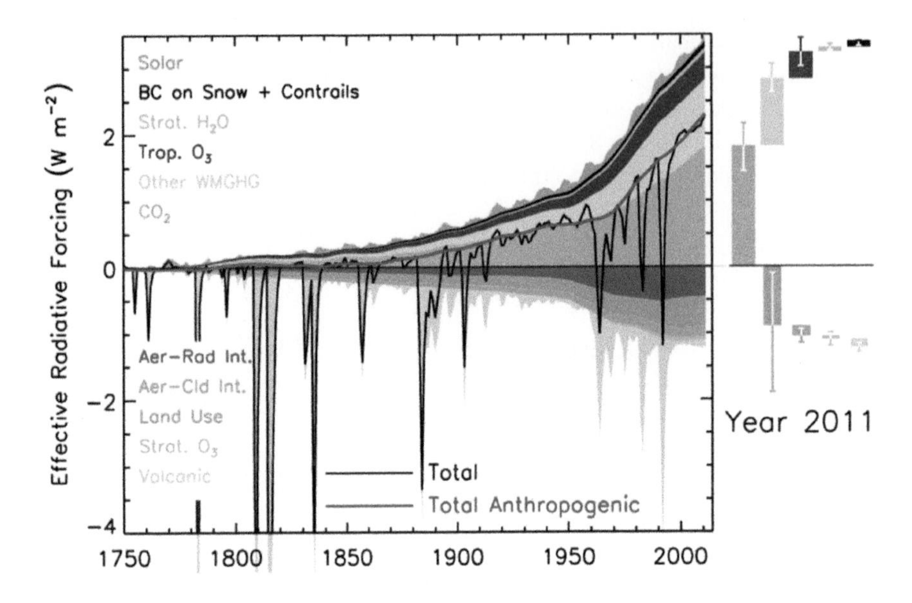

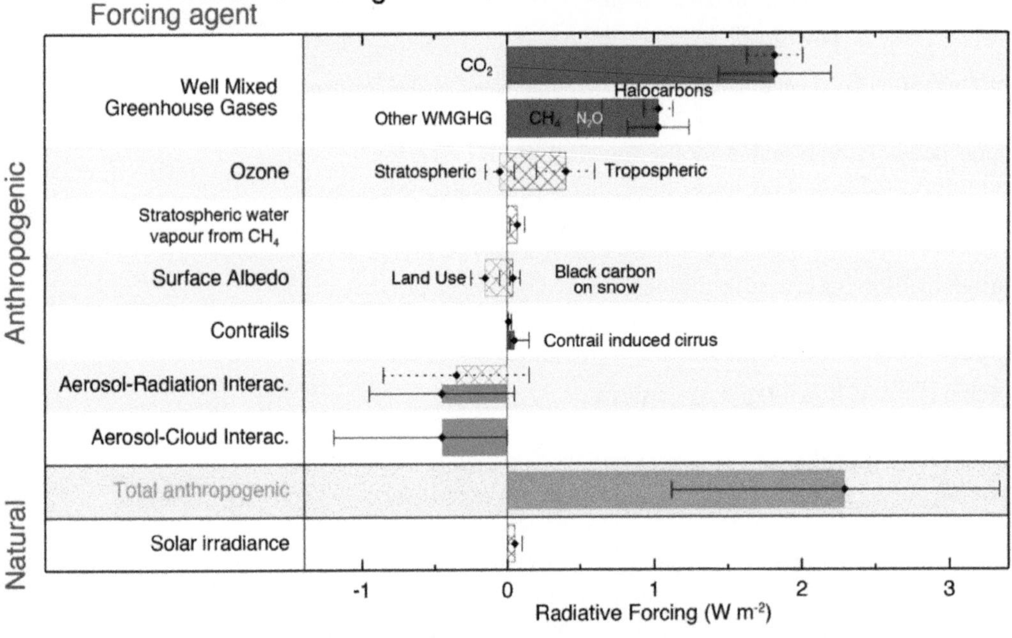

Figure 9.4 Radiative forcing changes over the last 200 years. The main contributors to radiative forcing during the last 200 years are shown based on model calculations. A major effect is from the buildup of carbon dioxide with additional contributions from non-CO_2 greenhouse gases. These increase the absorptivity causing "global warming." The effects of these gases are partially offset by "global cooling" from aerosols, aerosol–cloud interactions, and other smaller factors, which increase the effective albedo (IPCC AR5, Endnote 1.1).

average sensitivity of $\lambda = 0.7°C/(w/m^2)$ based on model calculations. The climate sensitivity will be a tool for us to calculate the expected global warming from the increases of greenhouse gases both past and future using Eqs. 9.9 and 9.10, being mindful of the assumptions we have made and the range of values consistent with the uncertainties.

Example In the worst-case scenario CO_2 is expected to reach 900 ppm from the pre-industrial levels of 280 ppm by 2100 (IPCC AR5). How much temperature change will this cause in equilibrium? What is the uncertainty?

Solution: $\Delta\ RF(CO_2) = 5.35\ w/m^2 \times \ln\ (900/280) = 6.2\ w/m^2$. $\Delta T_s = 0.7°C\text{-}m^2/w \times 6.2\ w/m^2 \approx$ **4°C**. For the uncertainty we consider only the range of radiative forcings between 0.3 and 1°C-m^2/w: ΔTs (Lower Limit) $= 0.3°C\text{-}m^2/w \times 6.2\ w/m^2 \approx$ **2°C**, ΔTs (Upper Limit) $= 1\ °C\text{-}m^2/w \times 6.2\ w/m^2 \approx$ **6°C**. This is a wide range. Two degrees would have much different impacts than six degrees. Perhaps we should revisit Figure 2.7.

There are several paths we must still explore to better understand the natural climate and its possible changes. The most significant large-scale impacts are from the oceans, clouds, and horizontal transport processes. These matters are next.

9.4 Role of the Oceans

Suppose we add a certain amount of a greenhouse gas to the atmosphere at one time; let's say it is inert, so the amount added will stay in the atmosphere forever. The gas will send some of the energy it absorbs back to the surface. This will be an ongoing process year after year and so heat will be constantly added to the surface causing it to get warmer. But we know that the surface will also radiate as $\sigma\ T_s^4$, and as it gets warmer it will radiate more. Additionally, the atmosphere will get warmer too and send more radiation back than before we added the greenhouse gas. Ultimately the system will reach a new balance in which both the surface and the atmosphere will be warmer and stable. How long will that take? The answer to this question is the delay in global warming that is caused by the earth's *thermal inertia* which is primarily determined by the oceans. This is because oceans are 71% of the earth's surface and have a high heat capacity causing them to absorb a lot of energy while warming only a small amount as we saw in Section 8.4. Moreover, the depth to which the heat mixes quickly is 50–150 m of the mixed layer, which is more than the effective mixing depths on land where the heat is moved by the slower process of conduction. Here is how we can understand and quantify this phenomenon:

We start with the climate in steady state with a surface temperature of T_{0s}. At the top of the atmosphere the energy will be balanced between $E_{abs} = (1 - A)S$ that is being absorbed and $F\uparrow$ that is leaving. Now we add a kg of an inert gas. An instant later the energy will be out of balance by the radiative forcing of that 1 kg and $\Delta RF = E_{abs} - F'\uparrow < 0$. That extra energy must heat the surface and atmosphere to restore the balance at some later time. When the balance is restored in a new steady state the surface temperature will be higher by $\Delta T = \lambda\ \Delta RF$ according to Eq. 9.9. During the transition between the steady states, the net energy being absorbed by the surface is: $dQ/dt = \Delta RF - T'/\lambda$. The temperature of the earth is $T_s = T_{0s} + T'$ so that the T' represents the heating of the surface during the transition starting at $T' = 0$ and ending with $T'(ss) = \Delta T = \lambda\ \Delta RF$. At the start the surface is heated at a rate about equal to ΔRF because the temperature hasn't risen yet to get rid of any of the additional heat and at the end the net heat being added is zero because the surface gets rid of as much as it receives. It is still getting more heat than before the addition of the gas, and will continue to do so because the gas is inert, but now it doesn't warm up any more. The surface, as stated earlier, is mostly ocean, so can be taken as the warming of the ocean mixed layer for now. We know from Section 8.4 that the heat being added raises the ocean temperature, per m^2, as: $dQ = (C_O\ m_O/A)\ d\ T_s$ where C_O is the heat capacity of ocean water and m_O is the mass of the ocean being heated and A = area of the earth. $m_O = \rho_W A D$, where ρ_W = density of water, and D = depth of heat penetration, so $dQ = (C_O \rho_W D)\, d\, Ts$. Therefore, from the discussion above, $dQ/dt = \Delta RF - T'/\lambda = (C_O \rho_W D)\, dT_s/dt$ and with $dT_s = d\, T'$, we can write the energy balance as:

$$\frac{dT'}{dt} = S_H - \frac{T'}{\tau_H} \tag{9.11a}$$

$$S_H = \frac{\Delta RF}{\rho_w D C_O} \tag{9.11b}$$

$$\tau_H = \lambda \rho_w D C_O \tag{9.11c}$$

Mathematically Eq. 9.11a is the same as Eq. 4.5 and will therefore have the same solutions. Moreover, we can interpret the first term on the right-hand side of Eq. 9.11a as the source of heat to the surface and the second term as the sink. Eq. 9.11a gives us a "residence time of heat" (τ_H) in the oceans and the final temperature $T'(\max) = \Delta T = S_H\, \tau_H = \lambda\, \Delta RF$ if a steady state exists. For the simple case of constant radiative forcing the solution of Eq. 9.11b is:

$$T' = \lambda \Delta RF \left(1 - e^{-t/\tau_H}\right) \tag{9.12}$$

Our goal in this venture is expressed in Eq. 9.11c, which tells us how long it will take to reach a new equilibrium. We can put numbers into this as follows: Taking D = 100 m (mixed layer), $\lambda = 0.7°\text{C}/(\text{w/m}^2)$, $\rho_W = 10^6$ g/m^3 and C_O = 4.2 j/g°C gives $\tau_H \approx 9$ years. It means that the oceans will delay a full global warming for about 20 years since it takes 2.3 τ_H to get to 90% of the final steady-state temperature. An important consequence of the results is that the global warming will continue for several decades even if we could hold the greenhouse gas concentrations constant at today's levels (see Sections 9.4 and 9.5). The delays will actually be longer because the heat will penetrate to deeper depths.

Example (a) How long does the land take to come to equilibrium with a disturbance to the radiative forcing? (b) How about the atmosphere? Assume heat penetrates rapidly to 10 m into the soils.

Solution: (a) For the soils: From Eq. 9.11c, $\tau_{Soil} = \lambda\, \rho_{Soil}\, D\, C_{Soil} = 0.7°\text{C}/(\text{w/m}^2) \times 1.5 \times 10^6$ g/m^3 × 10 m × 2 j/g-°C = 8 months. Despite uncertainties, we see that the land will equilibrate rapidly, perhaps 10 times faster, compared with the oceans. In deriving the λ in Section 9.2 from the observational data, the assumption was made that the land was approximately in equilibrium with the 2 w/m^2 radiative forcing from human causes. It seems so. (b) For the atmosphere: Similar arguments applied to the atmosphere lead to Eq. 9.11c with $D = H$ (scale height) and a factor of two in the denominator to account for loss of heat both up and down that causes a faster adjustment of temperature. Therefore, $\tau_{Atm} = \lambda\, \rho_{Atm}\, H\, C_{Air}/2$. Then, $\tau_{Atm} = 0.7°\text{C}/(\text{w/m}^2) \times 1280$ g/m^3 × 8000 m × 1 j/g-°C/2 ≈ 40 days.

One of the consequences of ocean thermal inertia is that by delaying global warming it can also buffer it. The meaning is illustrated in Figure 9.5 based on Eqs. 9.11. We take a scenario whereby a gas with a 10-year lifetime, like methane, is put into the atmosphere at one time

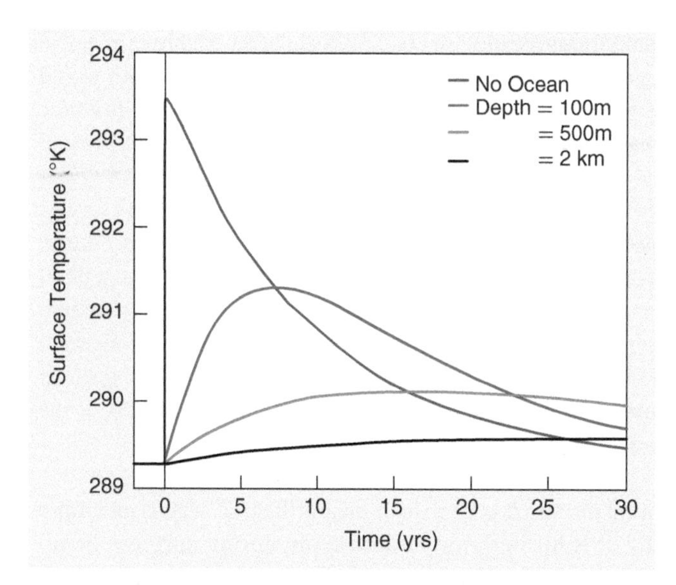

Figure 9.5 Ocean thermal inertia and buffering. For any disturbance to the earth's energy balance, as by increasing greenhouse gases, the oceans delay and reduce the peak global warming, but keep the earth warmer after the disturbance has subsided.

in a quantity sufficient to raise the temperature by 5°C if its effect were felt immediately, as would happen if there was no ocean. The effect of the ocean is shown if only the mixed layer took up the energy and the heat did not penetrate the deeper ocean, or if mixing extended to 500 m, or it mixed with the whole ocean to the bottom (3.5 km). Let's say that we consider a temperature rise over 2°C to be of concern. Then we see that without the ocean a high immediate rise in temperature will occur and will stay above our criteria value for some 10 years. Adding the mixed layer of the ocean makes the temperature rise slowly, with very little immediate effect and it reaches a peak in about 10 years but it is never as high as without the oceans. After the peak, however, the oceans will keep the earth warmer than the "no ocean" case, but still, our critical value of 2°C that we chose, will be exceeded for only 2 years. With a 500 m mixing, the temperature will never go over our critical value. Lots of warming, even over a short time can have more adverse effects than a small amount of warming spread over a long time. The oceans contribute to this benefit.

A few salient points should be noted. First, for simplicity, we discussed the situation in terms of an inert gas so that the radiative forcing would be constant, but this assumption is not required to derive Eqs. 9.11, but it is used to get Eqn. 9.12. The radiative forcing can change in time, but in that case the solution for the temperature T' will be more complicated than 9.12 and there may be no steady state until well after the radiative forcing stops changing. Second, although the discussion has been cast in terms of greenhouse gases, the same arguments say that the delay in warming should be expected from any cause of changing radiative forcing. Third, there is uncertainty in the estimate due to both the climate sensitivity and the depth D. The range from the former is ~ 10–30 years to reach 90% equilibration. The latter uncertainty may be even more significant. We have taken D to be the depth of the mixed layer of the ocean. By its very nature, the heat will be mixed in this layer quite readily so an accounting of the heat transport times is not needed because they are much shorter than the equilibration times. It is inevitable that heat will be transferred to the deeper ocean, but this will take transport time as well as the time to heat the water at the surface. If we could mix the additional heat of global warming in the entire ocean very quickly, instead of just in the mixed layer, the depth would be about 3.5 km and τ_H would become 350 years! We would not see a global warming from the increases of greenhouse gases that have occurred over the last century. More realistically, if heat is moved and stored deeper in the ocean due to ocean circulation cycles that may transfer heat more efficiently some years and less so in other years, the surface will warm more slowly, or perhaps not at all for some years despite the continued buildup of greenhouse gases. Fourth, it is noteworthy that the heat capacity tells us how long it will take to reach equilibrium, but it does not affect the final temperature that is reached. That depends only on the net radiation coming in and going out as seen in Eqs. 9.2 or their improved counterparts. This is consistent with Eq. 9.12 in which the result in equilibrium is $\Delta T = \lambda \Delta RF$, which does not involve the heat capacity. An example readers may find interesting, further illustrates these impacts of the oceans on the earth's temperature.

Example The earth's orbit around the sun is eccentric. It means that the orbit is an ellipse rather than a circle, so presently, it is further away from the sun during northern hemisphere summers and closer during winters. (a) Calculate the change of radiative forcing due to this annual effect. (b) Calculate the expected change of temperature due to this phenomenon for one orbit. (c) What would be the temperature change if there was no buffering capacity, that is, the heat capacity ≈ 0.

Solutions: The perihelion, distance closest to the sun is about $D_P = 1.496 \times 10^8$ km, and the aphelion, the distance furthest, is $D_A = 1.521 \times 108$ km. We can use the following relationship for the difference of the solar constant at these two times: $\Delta S = ½\, \sigma\, R_{Sun}{}^2\, T_{Sun}{}^4\, \delta D/D_P{}^3$ which is derived from Eq. 8.3 by Taylor Series, or we can just use Eq. 8.3 directly. The difference in the value of the solar constant is divided by 4 to account for the diurnal and lit side effects. Here $\delta D = D_A - D_P$. (a) The change of radiative forcing at the top of the atmosphere is $\Delta RF = (1 - A)\,\Delta S = (1 - 0.3) \times ½ \times (5.67 \times 10^{-8}\ \text{w/m}^2\text{-°K}^4) \times (6.95 \times 10^8)^2\ \text{m}^2 \times (5760)^{4}\text{°K}^4 \times (2.497 \times 10^9\ \text{m})/(1.496 \times 10^{11})^3\ \text{m}^3 \approx 8\ \text{w/m}^2$! That is 2.3% of the solar constant. (b) To calculate the effect on the temperature, we have to take into account the fact that the earth warms slowly, and this extra heat only comes for a short time while the earth is closer than usual to the sun. Let's take it liberally to be 3 months. Then, according to Eqs. 9.11c and 9.12, the change of temperature will be: $T' = \lambda\, \Delta RF\, (1 - e^{-\Delta t/\tau H}) = (0.7\ \text{°C/(w/m}^2) - \times 8\ \text{w/m}^2 \times (1 - e^{-0.25y/9\,y}) \approx 0.15\text{°C}$. This is not much. (c) With no buffering, $\tau_H \approx 0$ and so, $\Delta T = \lambda\, \Delta RF = (0.7\text{°C/(w/m}^2)) \times 8\ \text{w/m}^2 = 5.5\text{°C}$, which is a large fluctuation that would affect the climate. The difference illustrates the buffering effect of the oceans.

9.5 Role of Clouds

Perhaps the most unequivocal statement about the earth's climate is that water is the main force that determines it. Not only does water vapor have a major role in the greenhouse effect, so do clouds.

In the natural environment, we can consider clouds in three domains distinguished by their altitudes: low rain clouds, at 2 km or less, such as cumulus or stratus; high clouds, above 5 km, such as cirrus; and middle clouds or deep convective clouds that can take up much of the troposphere, such as cumulonimbus. Let's start with no clouds and look at the radiation balance at the top of the atmosphere. It will be $RF = (1 - A_{NO\text{-}C})\, S - F\uparrow_{NO\text{-}C} = 0$, where the $(1 - A_{NO\text{-}C})\, S$ is the absorbed solar radiation and $F\uparrow_{NO\text{-}C}$ is the radiation sent back to space. If we now put clouds into the atmosphere, it may change both factors: energy absorbed and energy radiated to space. This is seen as a change of radiative forcing: $\Delta RF = RF$ (with clouds) – RF (clear sky) or:

$$\Delta RF = \Delta F^{\uparrow} - \Delta AS \tag{9.13}$$

where the change of albedo, $\Delta A = A_{CLOUD} - A_{NO\text{-}C}$ and the radiation to space, $\Delta F\uparrow = F\uparrow_{NO\text{-}C} - F\uparrow_{CLOUD}$. The absorption by clouds increases the radiative forcing and the reflectivity decreases it, with commensurate effects on heating and cooling of the surface. With such counter-acting influences, a cloud will warm the earth if the absorption effect exceeds reflection and cool it if the opposite is true, or it may be neutral if the two effects cancel out.

At night all clouds cause warming because the albedo effect is gone. Experience shows that cloudy days are generally cooler and nights warmer than clear days and nights under otherwise similar conditions. Whether a cloud causes net cooling or warming depends on whether the ΔRF is positive or negative as a day–night average. From Eq. 9.13, it is expected that low clouds will cool the surface; the high thin cirrus clouds will warm the surface and the deep convective clouds or middle clouds may have a more or less neutral effect. The reasoning is that a cloud near the surface, particularly a thick one, will cause a big change in the albedo compared with the clear sky conditions, but it will not affect the total outgoing radiation very much because its temperature is close to that of the surface; therefore its black body radiation will be similar to the surface that is present whether the cloud is there or not. Since Eq. 9.9 is the energy balance at the top of the atmosphere, the low cloud will look like the surface with a high albedo. So, $\Delta A\, S > \Delta F\uparrow \approx 0$, leading to a cooling effect. The high cirrus cloud is thin so it does not affect the albedo much, but it absorbs the earth's radiation and sends some of it back to the surface day and night. This radiation would have gone out to space if the cirrus cloud was not there because at that altitude there is little water vapor and the impact of greenhouse gases is small since the concentrations (g/m^3) have dropped. So $\Delta A\, S < \Delta\, F\uparrow$ and consequently we will get a warming effect, especially since the warming goes on day and night. The deep convective type of cloud has a warm lower end that radiates heat toward the surface, and a cold top that radiates less to space than the clear sky condition. Both factors cause a warming for the surface by increasing ΔRF. But such a cloud has a very high albedo that significantly reduces the sunlight reaching the surface during the day causing a cooling effect. The two counter acting effects may be comparable in magnitude and therefore create a potentially neutral situation.

When the types of clouds present in the earth's environment are considered, a net cooling effect of some 20–30 w/m^2 is seen (Endnote 9.6). There is no implication in this theory about how the clouds will behave if the climate changes. If the albedo effect wins, they will reduce global warming; if the absorption effect wins they will exacerbate it, or they may do nothing. It will depend on which clouds change and by how much.

9.6 Horizontal Transport of Heat

One of the most significant aspects of climate is the large-scale transport of heat. This takes two forms – horizontal and vertical as we discussed earlier in Chapter 5. The vertical transport results in the removal of some 100 w/m^2 from the surface by convective processes that, as previously discussed, cause the surface to be cooler than it would be otherwise (Figure 8.12). The large-scale horizontal transport of heat in the atmosphere is driven by the Hadley circulation and related mechanisms. Additionally, the oceans transfer about the same amount of heat as the atmosphere from the warmer equatorial regions to the higher latitudes. In the end, equatorial regions are warmer by some 20°C compared with the higher latitudes. Without horizontal transport of heat the discrepancy would be perhaps twice as much; it makes a huge difference to all living things including human habitability. Here we will estimate this net transfer of energy from the tropical regions to the higher latitudes (see Figure 9.6).

To start with, our step one, we divide each hemisphere into two equal halves by area. This would be 0–30°, which is essentially the tropics and the subtropics, and from 30 to 90° which is middle and cold regions. Then, in step two, we use the model that is consistent with the energy balances in Figure 8.12 as stated in the Endnote 9.1 where T_s and T_a with convective fluxes and "more radiation down than up" are included. From this the temperatures of the two regions can be calculated without any transfer of energy out of the regions. We will find that the tropics are too hot and the higher latitudes too cold compared with the real world. Finally in step three, we calculate how much heat needs to be exchanged between the regions to so as to match the prevailing average temperatures of these regions.

We will use Eq. (1) in Endnote 9.1, which is the more complete form of Eq. 9.2 and consistent with the energy balances in Figure 8.12. Either from calculations or observations we find the annual average numbers required to calculate the temperatures, namely (S, A, Q) = (Solar constant, Albedo, Convective Fluxes), which for the equatorial box, = (405 w/m^2, 0.2, 150 w/m^2) and (S, A, Q) for the higher latitude box = (279 w/m^2, 0.42, 50 w/m^2). A constraint is that once we fix the values of these variables in one box, the values in the other box are also fixed because the average has to match the results shown in the energy balance Figure 8.12, that is (S, A, Q) for global = (342 w/m^2, 0.3, 100 w/m^2). The absorbed

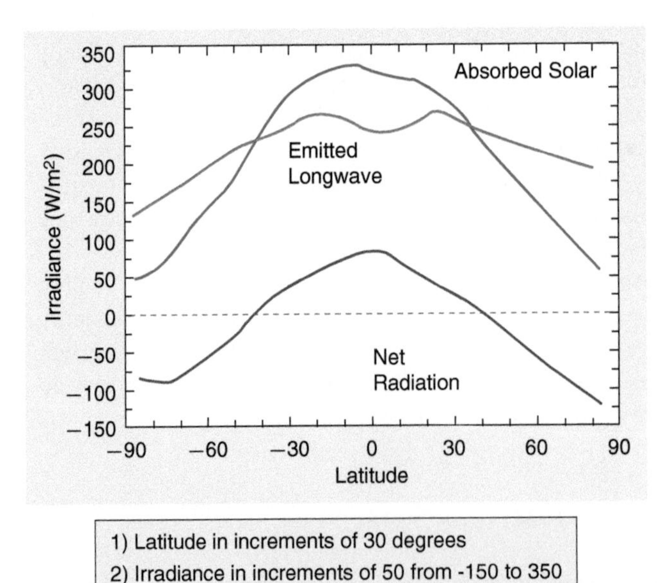

1) Latitude in increments of 30 degrees
2) Irradiance in increments of 50 from -150 to 350

Figure 9.6 Horizontal transport of energy. Energy is transported from the equatorial regions to higher latitudes by winds and the ocean as part of the general circulation and ocean drivers. This phenomenon has a major effect on global habitability. *Source:* Hartmann, 1994 / with permission of Elsevier.

energy in each box is $(1 - A)\,S$, which from these numbers is, 324 w/m^2 for the tropical box and 162 w/m^2 for the higher latitude box. Readers will recall that there is more solar radiation per square meter near the equator because the land is facing the sun more vertically than at higher latitudes where the land is tilted away (see Section 8.1); this results in the stated disparity of average solar radiation at the top of each box. The albedo of the tropical box is much lower because it is mostly ocean and that has a very low reflectivity, especially since the light is arriving nearly perpendicular to the surface. For the higher latitude box, the albedo is much higher, not only because of land reflection from drier desert areas, but also from the polar regions that include the ice caps and seasonal ice that extends further down. The convective fluxes (Q) are greater for the tropical box because it contains the upward arm of the Hadley cell which has a large effect on the transfer of heat from the surface. The other variables in Eq. (1) of Endnote 9.1, namely (f, F, δ) = (0.23, 0.9, 0.25), are assumed to be the same since these reflect the impacts of greenhouse gases and clouds which are not much different in the two boxes. With these values, Eq. (1) in Endnote 9.1 gives T_S(tropics – calculated) = 312°K and T_S (higher latitudes – calculated) = 264°K. If we look at the actual observed temperatures of these surfaces we find that T_S(tropics – observed) = 298°K and T_S (higher latitudes – observed) = 277°K (Figure 2.6). We note that without the horizontal transport the equatorial regions would be warmer by some 14°C and the higher latitudes would be cooler by about the same amount. This would make the climate significantly different than it is.

Finally, we can calculate how much energy has to be transported from the equatorial box to the higher latitude box for our model temperatures to match the observed. As before, each box gets energy from the sun which is the absorbed energy $(1 - A)\,S$ and loses it from the top to space, but now we add a horizontal transport to the other box to get: $E_{IN} = E_{ABS} = E_{UP}$ (at the top) + $E_{HORIZONTAL}$ (across latitudes). Therefore, $E_{HORIZONTAL} = E_{ABS} - E_{UP}$. Taking the actual surface temperatures, with the average temperature of the global atmosphere deduced from Figure 8.12, the energy loss at the top of the tropical box to be 281 w/m^2 so that $E_{HORIZONTAL}$ = 324 w/m^2 – 281 w/m^2 = 43 w/m^2. It means that the absorbed energy is more than what is leaving just at the top of the tropical box, so to balance it, the tropical box must transfer 43 w/m^2 to the higher latitude box (the m^2 refers to the surface of the earth in the box, not the interface). If we do the same for the higher latitude box we will find that it is losing more energy at the top than it is getting from the sun, so it must be importing some from the adjacent box in about the same amount. These transfers of energy are usually stated in petawatts (10^{15} w) to stress the fact that these are movements of energy per unit time. Since the surface area of each of our boxes is a quarter of the earth's surface, we can multiply these fluxes by 1.28×10^{14} m^2 which would give the horizontal energy transfer as about 5.5 petawatts that leaves the tropical box and enters the higher latitude region. That is a substantial amount of energy that cools the tropics and warms the higher latitudes.

The land maintains a higher average temperature than the oceans resulting in a similar transfer of heat. The net radiation reaching the land and the oceans, say, at similar latitudes, is the same, made up of the solar and atmospheric radiation streams. Convective processes, however, remove heat from the oceans more effectively than from land due to the evaporation of water and its latent heat, and perhaps by a greater convective activity due to the large scale circulation. We can write the energy balances for the land and oceans as: $E_{IN} - Q_L = \sigma T_L^4$ and $E_{IN} - Q_O = \sigma T_O^4$, where the Q's are the convective heat transfers. It is apparent that if $Q_O > Q_L$ then, $T_O < T_L$. The temperature differences this mechanism generates are eroded by atmospheric transport processes, but cannot be equalized. The land then, remains warmer than the oceans in steady state.

Review of the Main Points

1 The key variables that define the earth's climate: the albedo, solar radiation, and the earth's infrared radiation, combine to create specific temperatures of the surface and the atmosphere. In this chapter we calculated these temperatures based on the simplest model that any reader can reproduce. We found that while it comes close, there are some flaws.
2 Fixing the flaws was instructive because it showed us additional critical mechanisms that make up the climate. The important roles of convective transport, shielding of the upper atmosphere from the earth's radiation and the absorption of the solar radiation by the atmosphere were delineated. Two conclusions were that a pure radiative balance will inevitably lead to convective activity that will reduce surface temperatures and that adding more greenhouse gases will inevitably cause the atmosphere to send more radiation down than up, increasing the greenhouse effect and global warming.
3 The important concepts of *radiative forcing* and *climate sensitivity* were defined. Radiative forcing is an imbalance in w/m^2 that is created at the top of the atmosphere if one of the forces that determine the earth's temperature is altered. More complex definitions are currently in use (Endnote 9.4). It is calculated by radiative theory and is quite certain for greenhouse gases and less so for the effect of clouds and aerosols. Climate sensitivity is the ratio of the change in surface temperature in a new equilibrium that we expect from a change in radiative forcing. Climate sensitivity can be determined empirically or calculated by climate models – it is not very certain. Every climate model, from the simplest to the most sophisticated, can be used to arrive at a value of the sensitivity parameter. Once we accept a value, we can calculate the predicted global warming from the increase of any greenhouse gas. Our lack of certainty in such a calculation comes mostly from the uncertainty in our estimate of the earth's climate sensitivity.
4 We looked at three "big picture" features of the climate – the roles of clouds, oceans, and horizontal transport. (a) In addition to absorbing some solar radiation, clouds absorb the earth's radiation at almost all frequencies, including the window region, where the greenhouse gases cannot absorb. They have a dual role: cooling by reflecting sunlight and warming by absorption. The net effect in the current state of the atmosphere is cooling, but it complicates what they will do in a warmer world. (b) Oceans buffer temperature change from global warming. They slow down warming trends delaying the peaks by some 20 years and even longer as the heat penetrates to the deeper oceans. They also reduce the peak surface temperature change from any disturbance to the greenhouse effect. The present levels of greenhouse gases have unrealized potential for causing global warming due to ocean thermal inertia. (c) The winds, which drive surface ocean currents, combine with the net ocean movement of heat to significantly reduce tropical temperatures and warm the higher latitudes. This is currently a major boon to life on earth which proliferates in both tropical and middle latitudes from this cause. (d) The higher temperatures on land compared with the oceans under similar conditions were explained by differences in the convective transfer of heat by evaporation.

Exercises

1 Radiating Temperatures of Planets.

(a) Calculate the equilibrium radiating temperature for the following planets:

Use the following data: (Albdo, Surface Temperature in °K) or (A, T_S): Venus (0.77, 740), Earth (0.3, 288), Mars (0.15, 243), Jupiter (0.58, 163).

Note: Here T_s is the surface temperature of the planet. Calculate the greenhouse effect on each of these planets. State your assumptions.

(b) Suppose the sun's energy output drops to 50% its current value. Re-calculate the temperatures and compare to the real situation at present. By what percentage do the temperatures drop? Comment on the results.

2 Consider the simple radiative model of the earth's temperature. The absorptivity $F = 0.8$ and albedo A=0.3 in the natural atmosphere (base case). Due to increases in greenhouse gases the absorptivity goes up to 0.95.

(a) By how much will the earth warm under these conditions?

(b) You have a means to increase the earth's albedo. By how much should you change the albedo so that the temperature returns to the same value as the base case?

3 A star puts out 10 times the energy per second as our sun (or it is 10 times brighter). A planet orbits this star at a distance three times as far as the earth is from our sun. The albedo of this planet and the fraction of the long wave radiation absorbed by the atmosphere are the same as the earth (A and F). What would be the surface temperature of this planet? Do you think there could be life on such a planet similar to the earth?

4 A planet has no atmosphere but there is a thin shell surrounding it. If electromagnetic radiation comes to the shell from one side (reflective side) 90% is reflected and 10% is transmitted. If radiation comes from the other side (transparent side), 90% is transmitted, and 10% is reflected. There is no absorption by the shell. The albedo of the planet surface is 0 and S, the amount of solar energy per square meter per second is the same as at the earth. What is the temperature of the surface if:

(a) the reflective side faces the sun?

(b) the transparent side faces the sun?

5 For the earth and other planets, show that the atmosphere will always be colder than the surface. The environmental conditions are that the atmosphere is not very thick, does not absorb most of the sunlight, but absorbs the planet's radiation. For simplicity, assume that the atmosphere radiates the same amount up as down. Construct a hypothetical planet with a surface and atmosphere where this would not happen.

6 Calculate the effects on the surface temperature from, clouds, surface albedo, and atmospheric absorption of solar radiation and earth radiation (separately). Use Figure 8.12 and the advanced model in Endnote 9.1.

7 By how much would you have to change the solar constant to compensate for the global warming by a doubling of CO_2? State in w/m^2 and percentage.

8 Based on the eccentricity of the earth's orbit discussed in Section 9.4, estimate the effect on the temperature of the moon by calculating the difference of solar radiation between perihelion and aphelion. State your assumptions before calculating the answer.

9 It is claimed that if we turn off all power plants that generate energy from fossil fuels, the temperature of the earth will go up rather than down, at least for a time after the event. The idea is that the sulfate particles that come with fossil fuel burning will fall

out, but the CO_2 will remain, so the cooling effect of the sulfate will be removed. For simplicity, assume the CO_2 concentration declines at 2% per year for 10 years, and the sulfate has a lifetime of a month. Evaluate the merits of this claim by the following calculations:

(a) estimate the concentrations of CO_2 and the aerosols after 1 year.

(b) Estimate the effect on radiative forcing and global warming in 10 years, assuming nothing else changes. Use the *RF* estimates for various processes in Figure 9.4.

Endnotes

Endnote 9.1 We can modify the energy balance and the temperature Eqs. 9.1 and 9.2 by adding the missing pieces – convection, evapo-transpiration, atmospheric absorption of solar radiation, and more energy sent to the surface than space as in Figure 8.12. This makes the resulting equations more complicated but useful for some of the discussions in the main text. Writing the balance for the surface and the atmosphere we get:

$$\text{Surface:} \quad (1-f-A)S+(1+\delta)F\sigma T_a^4=\sigma T_s^4+Q$$
$$\text{Atmosphere:} \quad fS+\sigma F T_s^4+Q=2\sigma F T_a^4$$

These equations can be solved for T_s and T_a:

$$T_s=\left\{\frac{\left[(1-A)-\tfrac{1}{2}f(1-\delta)\right](S/\sigma)-\frac{1}{2}(1-\delta)(Q/\sigma)}{\left[1-\tfrac{1}{2}(1+\delta)F\right]}\right\}^{1/4} \quad (1)$$

$$T_a=\left\{\frac{\left[f+(1-f-A)F\right](S/\sigma)+\left[1-F\right](Q/\sigma)}{2F\left[1-\tfrac{1}{2}(1+\delta)F\right]}\right\}^{\frac{1}{4}} \quad (2)$$

Here f is the fraction of the solar radiation absorbed by the atmosphere including the ultraviolet absorption by the ozone layer and the spectrum of absorption by clouds ($f \approx 0.2$, Figure 8.12); A is the albedo, as previously described (A = 0.31); δ accounts for the greater downward flux from the atmosphere than the upward flux as described in Eq. 8.7 ($\delta \approx 0.25$); F is the fraction of the earth's infrared radiation absorbed by the atmosphere ($F \approx 0.9$); Q is the non-radiative flux of energy from the surface to the atmosphere representing the latent heat from the equatorial arm of the Hadley cell and other water reservoirs, transpiration by plants, and sensible heat that rises as thermals all over the world ($Q \approx 100$ w/m^2); and the remaining symbols are as in the simplest energy balance model. Intuitive insights can be gained by looking at how much each process affects the earth's surface temperature. Atmospheric absorption = −10°C; convective fluxes = −15°C, "more down than up" = +15°C.

Endnote 9.2 Defining and evaluating the average temperatures of the surface and the atmosphere have some subtleties that require discussion. We can assume that the temperature is more or less uniform with changing longitudes (going around a latitude circle), so only latitudinal variations need to be taken into account, which are shown in Figure 2.6b. For the surface, the temperature should be calculated by taking the average of temperatures in "bins" constructed over the various latitudes so that the area of the surface each of them represents is the same. Mathematically, it means that the horizontal temperature can be represented as a function of the sine of latitude and then this function is integrated between -1 (South Pole) and +1 (North Pole). Half the earth's surface lies between 30°N and 30°S latitudes which is the tropics and sub-tropics of both hemispheres; the other half is the higher latitudes beyond 30° to the poles, in each hemisphere. From Figure 2.6b we see that the temperature distribution is more or less symmetric in the hemispheres and ~25°C in the equatorial, and 5°C for the higher latitudes. Since these regions are of equal areas, the average surface temperature is ~15°C or 288°K.In the vertical, the average temperature can be defined in different ways. Lower in the atmosphere, the air is warmer ($T_0 =$ 288°K) compared to say near the tropopause ($T_T =$ 210°K). The straightforward average then would be the temperature half way to the tropopause or 288°K −6.5 °C/km × 6 km ~250°K. If our interest is to find an effective average temperature that represents the heat content of the atmosphere ($Q = m\ C\ T_{\text{Eff Avg}}$), it can be shown that this would be about 260°K, which is the temperature at 4.3 km. Our interest here is the "radiating temperature of the atmosphere" that corresponds to the greenhouse effect. That is the temperature T_a which represents the net radiation from the atmosphere (Figure 8.12), that is, Net Radiation = $2\sigma F\ T_a^4 =$ 532 w/m^2 so that $T_a =$ 268°K (Figure 8.12). It is also the T_a in Endnote 9.1 above. This is the temperature of the atmosphere at about 2.8 km.

Endnote 9.3 The concept of *radiating temperature* of the earth, and of other planets by extension, is often useful. If you look at the earth from outside the atmosphere, you will see that the infrared radiation coming to you from the earth is exactly $(1 - A)\ S$ under equilibrium conditions, because of the law of energy conservation. It balances the amount of energy the earth absorbs from the sun and converts it to infrared frequencies. It will not follow Planck's black body radiation law because it is a combination of emissions from the surface and various parts of the atmosphere which are at different temperatures and have various emissivities. We can, however, calculate an effective temperature (T_e) that would occur if the earth's outgoing infrared radiation could be represented by an equivalent black body:

$$(1 - A)\ S = \sigma\ T_e^4$$

T_e is called the *equilibrium radiating temperature* which, from previous calculations, we know to be 255°K for the earth. It does not depend on the greenhouse gases, surface temperature or radiative transfer. For the earth, the departure of the surface temperature from this value is the net impact of the atmosphere which includes the greenhouse effect – a total of about 33°C. In general though, under some circumstances, the energy is not balanced and the planet may be radiating more or less than what it is receiving. The earth for instance is out of balance at this time due to accumulating greenhouse gases. The radiating temperature therefore is less than the equilibrium radiating temperature by small amounts.

Endnote 9.4 The original simple concept of radiative forcing is stated as the instantaneous imbalance of energy at the top of the atmosphere as discussed in the text. What is calculated is the ΔRF with all temperatures in the atmosphere fixed. One refinement is to keep the temperature of the troposphere fixed and let the stratosphere adjust to the change of *RF* that has occurred. This doesn't take long. The imbalance of energy after this adjustment is taken at the tropopause rather than the top of the atmosphere. The rationale is that this gives a more accurate value for λ and hence a better prediction for the change of surface temperature because it focuses on the changes in the troposphere and the surface processes which drive the surface temperature change. This definition has been used by the IPCC and climate scientists for some time and it is important because their consensus estimates of which agents are causing how much radiative forcing over the last 250 years are used to state the man-made climate change so far. For some forces that act on the climate, this is seen as "not good enough" and an "effective radiative forcing" is calculated. Further discussion of these ideas can be found in: IPCC (2001) TAR and (2015) AR5, Chapter 8 (Endnote 1.1) and "Radiative Forcing of Climate Change" (National Academy Press, 2005, Washington DC).

Endnote 9.5 For the simplest model in Eq. 9.1 we can get a formula for the λ_0. The outgoing energy at the top of the atmosphere is $RF = (1 - ½F)\,\sigma T_s^4$ and $\Delta T_s/\Delta RF = \lambda_0 = 1/[4\,\sigma T_{s0}^3\,(1 - ½F_0)]$, for only the direct forcing. For this model, $\lambda_0 \approx 0.3°C/(w/m^2)$, as shown in the text. This relationship is the same regardless of how we change the radiative forcing at the top of the atmosphere, whether by the solar radiation, albedo or atmospheric absorption. For any more complex model, such a formula may not be possible to derive.

Endnote 9.6 The complex role of clouds is discussed further in Hartmann 1994, p. 75; Keihl and Trenberth 1997, and Trenberth et al., 2009 (Endnote 8.6). The radiative forcing formulas are from IPCC (2001): TAR. These and radiative formulas for most other gases of interest in climate science can be found in Joos et al. *Global Biogeochemical Cycles*, v. 15, p. 981–1000, 2001. The ocean response in Section 9.3 and Figure 9.6 is from Hartmann, 1994, p. 37 and pp. 337–338 and is also in Wallace and Hobbs, 2006, p. 446 (see Endnote 8.6).

10

Climate Feedbacks

By now we are familiar with the comings and goings of energy in the earth's system. Starting from this state, if we add a greenhouse gas to the atmosphere and keep its concentration constant for a while, the energy balance will shift to a new equilibrium after some time. Radiative theory tells us only how much more back radiation will be caused due to the addition of the greenhouse gas at the top of the atmosphere. That it will increase the surface temperature is certain, but by how much is not. In the new steady state of the climate, factors, such as the albedo, convection, and absorptivity, that determine the earth's surface temperature may be different from the state we started with and not just the back radiation that we initially changed by adding a greenhouse gas. Consequently, the change of temperature between these steady states will not be determined by just the change of direct radiative forcing, but also by the environmental changes it may cause that affect temperature. These are feedback processes and they are the subject of this chapter. The idea of the climate sensitivity λ is designed to provide a single measure of how the actual temperature will change due to the addition of a greenhouse gas when a new equilibrium is reached. It is therefore built on the effect that would arise directly from the change of the concentration of the gas on the surface temperature, and an additional effect that comes from the feedbacks it generates. It should be noted that any initial disturbance to the climate system – not just the addition of a greenhouse gas, such as changes in the solar constant or the albedo, will likewise drive feedbacks that will increase or decrease the impact of the original driving disturbance.

10.1 How They Work

To see how feedbacks work we will use our simplest climate model once again for illustrative purposes. Although the model has limitations, the numbers are within the bounds of reality and calculated between steady states.

Case 0: Base Case: Let's start with Eq. 9.2a: $T_s = T_e/[1 - ½ F]^{1/4}$, $T_e = [(1 - A) S/\sigma]^{1/4}$. $F = 0.9$, $A = 0.3$, $S = 340$ w/m$^2 \Rightarrow T_{s0} = 295.6$°K, and $T_{e0} = 254.5$°K

Case 1: Now let's double CO_2 and figure out what the new temperature will be in equilibrium if all else remains the same. In this model, the outgoing long-wave radiation at the top of the atmosphere is $(1 - ½ F)\, \sigma\, T_{s0}{}^4$, where T_{s0} is the initial surface temperature before we doubled CO_2. Therefore, the change of radiative forcing is $\Delta RF = \Delta F\, (½\, \sigma\, T_{s0}{}^4)$

Global Climate Change and Human Life, First Edition. M. A. K. Khalil.

Companion Website: www.wiley.com/go/khalil/Globalclimatechange

because it is defined immediately following the disturbance before the surface temperature can adjust. In this model, F is the only variable that describes the absorption by greenhouse gases. For doubled CO_2, $\Delta RF = 5.35 \times \ln 2 = 3.7\ \text{w/m}^2$ from Eq. 9.10a. Therefore, $\Delta F = 0.017$. Now:

$2 \times CO_2 RF$ Only: $F = 0.917$ (original 0.9 + 0.017 from above), $A = 0.3$, $S = 340\ \text{w/m}^2$ result in $T_s = 296.7°\text{K}$ (from Eq. 9.2a as stated above). Therefore the surface temperature change and the climate sensitivity are:

$\Delta T_s = 1.1°\text{C}$

$\lambda_0 = \Delta T_s/\Delta RF = 1.1°\text{C}/3.7\ \text{w/m}^2 = 0.3°\text{C}/(\text{w/m}^2)$

Case 2: We know that any temperature change will increase water vapor based on the Clausius-Clapeyron equation discussed in Section 7.2. In equilibrium, whatever change of water vapor occurs, it increases F for this model by another 0.015 bringing it to 0.932. We haven't proven it is this much more, but let's say it is. Now:

$2 \times CO_2$ with Water-Vapor Feedback: $F = 0.932$, $A = 0.3$, $S = 340\ \text{w/m}^2$ result in $T_s = 297.8°\text{K}$

$\Delta T_s = 2.2°\text{C}$

$\lambda = \Delta T_s/\Delta RF = 2.2°\text{C}/3.7\ \text{w/m}^2 = 0.6°\text{C}/(\text{w/m}^2)$

Case 3: Further, this will melt some of the polar ice and lower the global albedo. This is the well-known ice-albedo feedback. In the new equilibrium state, compared with Case 1, $\Delta A = -0.01$ and the $\Delta F = 0.033$ due to water vapor, or another 0.018 above that in Case 2. The water vapor has to increase more than Case 2 because the ice-albedo feedback will also increase the temperature in addition to just the doubling of CO_2, which in turn will add more water vapor.

$2 \times CO_2$ with Water-Vapor and Ice-Albedo feedbacks: $F = 0.95$, $A = 0.29$, $S = 340\ \text{w/m}^2$ result in $T_s = 300.0°\text{K}$ and the new $T_e = 255.4°\text{K}$

$\Delta T_s = 4.4°\text{C}$

$\lambda = \Delta T_s/\Delta RF = 4.4°\text{C}/3.7\ \text{w/m}^2 = 1.2°\text{C}/(\text{w/m}^2)$

From these cases a big picture of the feedbacks emerges. Here are the salient points: First, all our calculations are for a single initial change to the environment, namely doubling CO_2 and all use the same radiative forcing from Eq. 9.10a, but the outcome for the earth's temperature is quite different depending on what else happens to cause feedbacks. The climate sensitivity is $0.3°\text{C}/(\text{w/m}^2)$ for no feedbacks but increases substantially depending on which feedbacks we invoke. The range can be similarly wide even if we use the same feedbacks but estimate different strengths for them. It is noteworthy that any energy balance, such as the one in Eqs. 9.1 or another more complex model, does not have any built-in mechanism for invoking feedbacks or estimating their impacts. To do so requires a more comprehensive set of equations that deal with the global environmental system in addition to its energy balance, such as including the Clausius-Clapeyron equation to take the water vapor effect into account. Second, readers will note that the radiative transfer theory provides us with an unequivocal change of radiative forcing calculated precisely and expressed in equations such as 9.10. From that we know how much heat the added gases can trap, but that's not enough to figure out how much the surface temperature will increase. To get that we need to understand the response of perhaps a large number of variables that affect climate and find how each will respond to a temperature change. This is disconcerting because we know well why human activities are changing the greenhouse gases and by how much, but we cannot translate it with certainty into a global warming without understanding the response of the entire natural system. Third, feedbacks have a complex inter-connection with each other. One can drive the other, so you can't add them easily. While doubling CO_2 has a prescribed impact on radiative forcing, each feedback does not, because its effect depends on the other feedbacks in the system. This matter will be discussed in more detail later in this chapter. And fourth, the discussion casts feedbacks as one of the main attributes of "climate change" and sets it apart from the understanding of "the climate." To understand how climate changes affect human life or the environment adds a further dimension that defines the new science of global change that is much more complex than climate theory can address and involves the whole earth system, and living things too.

The feedbacks we have looked at in the case studies are positive. It is an important characteristic of positive climate feedbacks that they move the temperature further from the stable conditions at which the original disturbance was initiated. The idea is most clearly illustrated by the ice-albedo feedback. If we increase the temperature, by whatever means, ice on the surface will melt causing a decrease of the albedo that will add to the original cause of the warming. But, starting from the same initial state, if we decrease the temperature, as perhaps by reducing greenhouse gases, more ice will be formed and cause an increasing albedo that will drive the temperature further down than without the feedback. Positive feedbacks make a warming world warmer and a cooling world cooler. Negative feedbacks work in the opposite way. They will bring the temperature toward the stable point from which the climate was disturbed. Suppose, for the sake of illustration, that warming the earth increases low clouds and cooling reduces them. If we suddenly cool the earth, low clouds will start disappearing and when an equilibrium is reached the earth will be warmer than if the clouds were unaffected by temperature. Similarly, if a warming

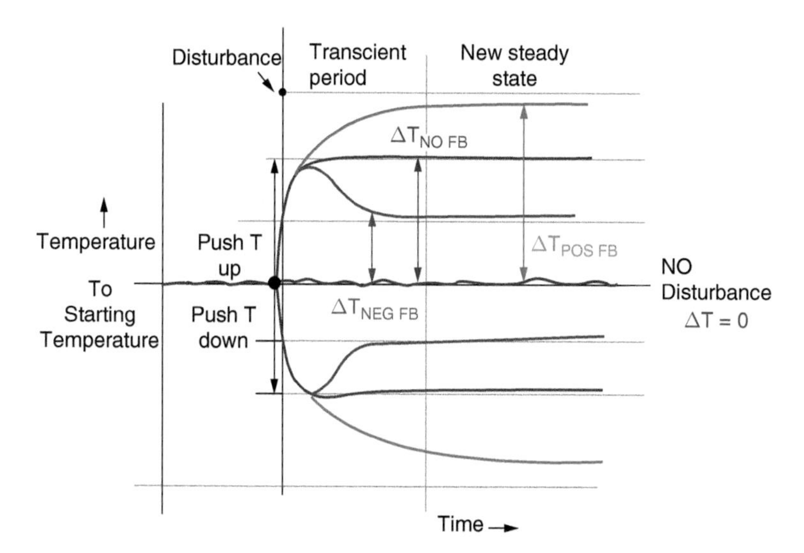

Figure 10.1 The course of positive and negative feedbacks. Positive feedbacks push the system further from the starting point, whether toward warming or cooling. Negative feedbacks bring it toward the starting point.

occurs it will be offset because more clouds will form and the earth will return closer to the starting point (Figure 10.1).

Although feedbacks are pervasive in the environment, our interest is mainly in those that affect the global temperature. For our inquiry into climate, we will say that a climate feedback is any process that affects the Earth's temperature beyond the direct effect caused by an initial change of the absorptivity, albedo or solar radiation. When we examine the known feedbacks, we will see that most of them are positive. Yet the climate of the earth has been quite stable for hundreds of thousands of years, punctuated by ice ages with quantifiable and credible causes in the changes of solar radiation due to shifts in the earth's orbit or orientation. The earth would be a cold planet without the greenhouse effect as we saw in Section 9.1. Positive feedbacks are a way to make it warmer with less effort from living things that control most of the greenhouse gases. In this way positive feedbacks are favored and have created a more habitable planet out of the earth. But when it comes to maintaining the earth at some stable climate, negative feedbacks are necessary. Can we push the earth's climate system enough to allow the positive feedbacks to run away as on Venus, or will negative feedbacks, identified or yet to be discovered, keep the climate stable? No one knows for sure.

Let's review the "no feedbacks" climate sensitivity so that we can distinguish it from the sensitivity when feedbacks occur. We saw in Case 1 above (and earlier in Section 9.2) that if we consider the impact of just the change of an imposed radiative forcing on the surface temperature and include no other responses of the earth's environmental system, then $\lambda \approx 0.3°C/(w/m^2)$. We will take this as the "no feedbacks" sensitivity which can also be deduced from general considerations (Endnote 10.1).

In the previous chapter we saw that the sensitivity of the earth to changes in radiative forcing is $\lambda \approx 0.7°C - m^2/w$ based on observations of temperature changes and model results. This response includes all the processes and feedbacks as they have played out in the observed changes of temperature. If we calculate how much the change of temperature should be without feedbacks, we saw just above that the climate sensitivity is only $\lambda_0 = 0.3°C - m^2/w$, which means that during the recent century, the temperature should have warmed by around 0.6°C (Eq. 9.9), based on the ~ 2 w/m^2 radiative forcing due to greenhouse gas increases and other causes shown in Figure 9.4. The actual temperature change is 1.4°C warming on land. Feedbacks more than double the direct effect of increasing greenhouse gases or other drivers of climate change. We will see that the major known feedback is from water vapor which may be responsible for a substantial part of this overall response. It illustrates the important role water vapor plays in global warming because it is affected by the surface temperature itself, as explained by the Clausius-Clapeyron equation.

Eq. 9.9 can be extended to make the role of feedbacks explicit:

$$\Delta T_s = G\lambda_0 \Delta RF \tag{10.1a}$$

$$G = \lambda / \lambda_0 \tag{10.1b}$$

As before, the λ_0 represents the "no feedbacks" case. The G is for "gain" in the climate sensitivity that is caused by feedbacks. For a single feedback, $G > 1$ represents a positive feedback that amplifies global warming and $G < 1$ represents a negative feedback that ameliorates it.

When more than one feedback is active at the same time, which is the case in the earth's environment, the effect on the temperature doesn't simply add. In other words, if we add a positive feedback, it will increase the sensitivity from λ_0 to a higher value leading to an expected change of temperature as in Eq. 9.9. But if we have two positive feedback processes, the change of temperature due to one, will drive the other further than if the first feedback wasn't there. Similarly, the second will drive the first, until some equilibrium is reached and the final result is greater than adding the λ's from each feedback. Another way of describing feedbacks is by dimensionless factors, *f*'s, which do add and provide a method to calculate the effect of many feedbacks operating simultaneously. Feedback factors work as follows:

Step 1: Let's say a doubling of CO_2 and holding the concentration produced an initial temperature change which led to the water-vapor feedback. We know from before that this initial change is $\Delta T_0 = \lambda_0\, \Delta \mathrm{RF}$ which we have calculated to be about 1°C.

Step 2: The water vapor concentration will increase according to the Clausius-Clapeyron equation, but we don't know how much. This is because the increase of water vapor due to the initial 1°C temperature increase will drive the temperature up by some amount, but that additional warming will cause more water to evaporate and cause more warming and so on. Eventually a stable steady state will arise in which there will be some additional amount of water vapor in the atmosphere $\Delta\, C_{\mathrm{WV}}$ and the temperature will be higher by ΔT. The additional water vapor can be written as: $\Delta\, C_{\mathrm{WV}} = (\Delta\, C_{\mathrm{WV}}/\Delta T)\ \Delta T$. The term $(\Delta\, C_{\mathrm{WV}}/\Delta T)$ is the rate at which water vapor changes with a change of temperature.

Step 3: The feedback effect of this amount of additional water vapor is: $\Delta T_{\mathrm{FB}} = (\delta\, T/\delta\, C_{\mathrm{WV}})\ \Delta\, C_{\mathrm{WV}}$, where $(\delta\, T/\delta\, C_{\mathrm{WV}})$ is the rate of change of temperature due to the change of water vapor. The small δ is used distinguish it from the big Δ in the previous step. There are two reasons. First, we want to emphasize in δ that we are calculating the change of temperature due to changes of water vapor while holding all other causes of temperature change constant. The second reason is that this calculation $(\delta\, T/\delta\, C_{\mathrm{WV}})$ is fundamentally different from the calculation of $\Delta\, C_{\mathrm{WV}}/\Delta T$ and not merely an inverse. The former answers the question: how much does the water vapor increase if we increase the temperature by a small amount? We can get the answer by using the Clausius-Clapeyron equation. The latter answers the question: how much does the temperature change if we increase the water vapor by a small amount? This question is answered by radiative theory with the absorption bands of water vapor in the infra-red emission of the earth. The two pieces are quite different, aren't they? They require different theories to evaluate.

Step 4: Now we can write the temperature change we expect as: $\Delta T = \Delta T_0 + \Delta T_{\mathrm{FB}}$. The $\Delta T_{\mathrm{FB}} = (\delta\, T/\delta\, C_{\mathrm{WV}})\ \Delta\, C_{\mathrm{WV}}$ (from Step 3) $= (\delta\, T/\delta\, C_{\mathrm{WV}})\ (\Delta\, C_{\mathrm{WV}}/\Delta T)\ \Delta T$ (same as the previous step, except we have multiplied and divided by ΔT, which we can always do – most of the time with no benefit). Therefore: $\Delta T = \Delta T_0 + f\,\Delta T$ where $f = (\delta\, T/\delta\, C_{\mathrm{WV}})\ (\Delta\, C_{\mathrm{WV}}/\Delta T)$ is the *dimensionless feedback factor*, for water vapor in this case. We can simplify this further by solving for ΔT as: $\Delta T = \Delta T_0/(1 - f)$. The concept can be extended to many feedbacks:

$$f_I = \left(\frac{\Delta X_I}{\Delta T}\right)\left(\frac{\delta T}{\delta X_I}\right) \tag{10.2}$$

Here $I = 1, 2,.., N$ representing N feedbacks and the X_I represent a feedback cause, such as water vapor. We can deduce that the gain from all feedbacks combined is (Endnote 10.2):

$$G = \frac{1}{(1-f)} \tag{10.3}$$

$$f = f_1 + f_2 + \ldots + f_N \tag{10.4}$$

From Eq. 10.3 we can estimate the effect of all feedbacks combined, but to isolate the effect of each one is generally ambiguous and not helpful because of synergistic influences. For positive feedbacks f_I are positive > 0 and for negative ones f_I are negative < 0. The value of the factor expresses the strength of the feedback which comes from the $(\Delta X_I/\Delta T_s)$ term. As the sum of the feedback factors approaches 1, a runaway greenhouse effect will occur causing potentially very high surface temperatures.

When more than one feedback is present, the synergistic effect is illustrated by the following example:

Example Suppose there was one strong climate feedback with $f_1 = 0.7$. What would be the expected temperature change for a change of CO_2 from 280 ppm during pre-industrial times to 500 ppm at a future time? Now suppose that a new, but weak feedback was activated with an $f_2 = 0.2$. What would be its effect on the temperature if it was the only feedback? What is the combined effect of both feedbacks?

Solution: The change of CO_2 would cause $\Delta RF = 5.35 \times \ln(500/280) = 3.1$ w/m^2 and a temperature change without feedbacks of $\Delta T_0 = \lambda_0\, \Delta RF \approx (0.3°C\,/(w/m^2) \times 3.1\, w/m^2) =$ **0.9°C.** For the first feedback using Eq. 10.3: $G_1 = 1/(1-0.7) = 3.3$ and $\Delta T_1 = 3.3 \times 0.9°C =$ **3°C** (Eq. 10.1a). Similarly, the second feedback acting by itself would give: $G_2 = 1/(1-0.2) = 1.3$ and $\Delta T_2 = 1.3 \times 0.9°C =$ **1.2°C**; but if it is added on top of the first one, or however it happened that both feedbacks were operating at the same time: $f = f_1 + f_2 = 0.9$, $G_{1+2} = 1/(1-0.9) = 10$ giving a temperature change of $\Delta T_{1+2} = 10 \times 0.9°C =$ **9°C** !

Note three points from the example. First, that the effect of the separate feedbacks does not add: $\Delta T_1 + \Delta T_2 \approx 4°C$ is not $= \Delta T_{1+2}$. Second, the radiative forcing is fixed by the change of greenhouse gas concentrations (here it is 3.1 w/m^2). The effect of any feedback is not fixed but depends on which other feedbacks are present. And third, the feedback factors are expected to be less than one for every feedback. But the sum of these factors can exceed one, and that would generate a runaway greenhouse effect.

10.2 Feedbacks Classified and Delineated

We see that feedbacks cannot be ignored in the understanding of climate change. There are many climate feedbacks: known, yet to be discovered, and negligibly small. The task for us is to isolate those that have the most influence for the climate and perhaps some that may become important in the future. Feedbacks occur over different time scales. If we confine ourselves to those that operate quickly, certainly within decades to a century, it narrows the list. This doesn't mean that the long-term feedbacks are inoperative at this time, just that they cause a change in temperature over a long time, which amounts to very little for our time scales of interest. Moreover, the drivers of climate change such as increasing greenhouse gases may not persist for long enough to allow these feedbacks to play out. We will therefore focus on the fast feedbacks. Within this category, feedbacks can be stimulated by physical, chemical, or biological factors and they start with a direct change to one of the driving elements of the climate system, namely, the absorptivity, albedo, or solar radiation. Based on this context, the physics-based feedbacks are: water vapor, lapse rate,

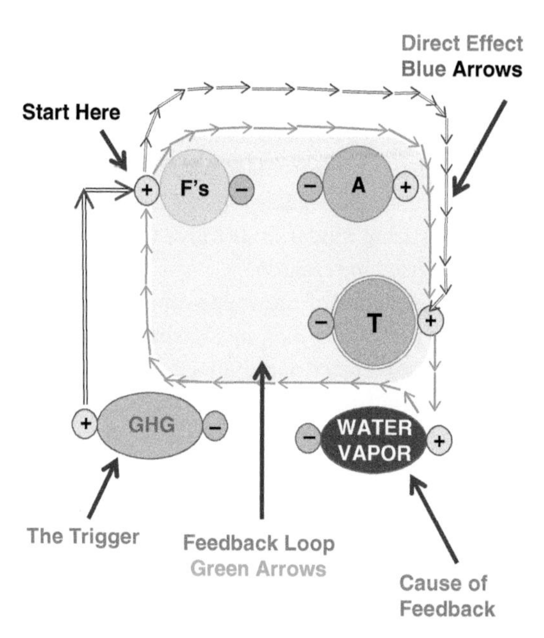

Figure 10.2 Climate feedbacks affecting the earth's temperature. Feedbacks are often described as a loop. The diagram emphasizes that climate feedbacks are started by a disturbance, such as increasing greenhouse gases. This increases the absorptivity of the earth's radiation causing a feedback that pushes the temperature higher than it would go without the feedback, here represented by water vapor.

ice-albedo, clouds, and perhaps thermo-haline; and the biological feedbacks are from terrestrial plants, soils, and arctic permafrost. Chemical feedbacks that affect climate may arise from changes in hydroxyl radicals, or ozone precursors or may be the mechanisms for biological feedbacks. Most of the fast-acting feedbacks of interest are triggered by global warming initially caused by the addition of greenhouse gases which increases absorptivity; or the addition of sulfate aerosols that affect climate by changing the albedo. Figure 10.2 is a representation of how climate feedbacks of our interest work.

10.3 Physical Feedbacks

10.3.1 Water Vapor Feedback

The largest feedback is from more water vapor in the atmosphere that accompanies any increase of surface temperature regardless of the cause. If we add any gas that has the ability to trap the earth's radiation, such as CO_2, CH_4, or N_2O, the increase in the surface temperature will compel a greater evaporation of water from the surface, mostly from the oceans, which can be described by the Clausius-Clapeyron equation if we make some assumptions. The first is that we expect or assume that the average relative humidity will remain the same even if we make changes to the global temperature. The water vapor concentration in the atmosphere (C_W) is:

$$C_w = \alpha q_s \tag{10.5}$$

Here q_S is the saturation water concentration (Eq. 7.1) and α is the average relative humidity. We saw that each°C of temperature change raises the saturated water content by about 7% so it will raise the actual water content by the same percentage which can then be put into a climate model of the temperature to evaluate the feedback (see Case 2 above). Estimates of the strength of the water vapor feedback from model studies give the feedback factor $f \approx 0.56$. It is included in every climate model used in research.

We can look at some numbers to validate the importance of the water-vapor feedback. Taking the $f_{WV} \approx 0.56$ tells us that the gain is ≈ 2.3 (Eq. 10.4) which means that it doubles the impact of the direct change of temperature. For an increase of radiative forcing of 2 w/m^2 as from pre-industrial to present times, the temperature change without feedbacks is: $\Delta T_s = 0.3°C/(w/m^2) \times 2\ w/m^2 = 0.6°C$ but with the water vapor feedback it is about 1.4°C in equilibrium.

10.3.2 Lapse Rate Feedback

As the surface warms, the reduction of the lapse rate may be the most significant negative feedback in the present earth's climate system. The global average lapse rate can become less than the present average of 6.5°C/km. Changes of lapse rates can be considered as convective feedbacks if they are driven by surface energy transfers. If more water is evaporated it takes more heat into the atmosphere causing the temperature at altitude to increase and the surface to change less, therefore reducing the lapse rate. The feedback is taken to be related to the water-vapor feedback in the sense that both come from the increased evaporation of water in a warmer world. Because of their common cause the two feedbacks are sometimes combined. Although uncertainties remain, it is found that the lapse rate factor is −0.26 making a combined positive feedback factor of about + 0.3 from the IPCC assessment (Endnote 10.3)

10.3.3 Cloud Feedbacks

Clouds under current environmental conditions have a net cooling effect (Section 8.4), but it does not imply that the net cloud feedback will be negative. The lifetime of water in the atmosphere is a combination of the lifetime in the vapor phase and as clouds. If both lifetimes stay the same in a warmer world, then the Clausius-Clapeyron equation implies that clouds will increase. Clouds reflect sunlight and they also absorb both sunlight and the earth's radiation. The consensus based on model experiments is that the net feedback is positive, but the uncertainties are very large. From the previous discussion about clouds we can see that the magnitude of these feedbacks depends on many details of how the clouds will change and not just on the idea that there will be more of them. If, for instance, the bigger change of clouds is lower in the atmosphere, the negative feedback of albedo will win because, as we saw in Section. 9.2, low clouds lead to a net cooling. If the higher clouds increase, then the positive feedback from the radiation absorption wins. It is the leading cause of the large uncertainties in the total climate sensitivity in model predictions. The estimate of the net feedback factor is 0.18 with a range of 0 (no net feedback) to 0.37 among fourteen models reviewed by the IPCC (Endnote 10.3, Figure 10.3).

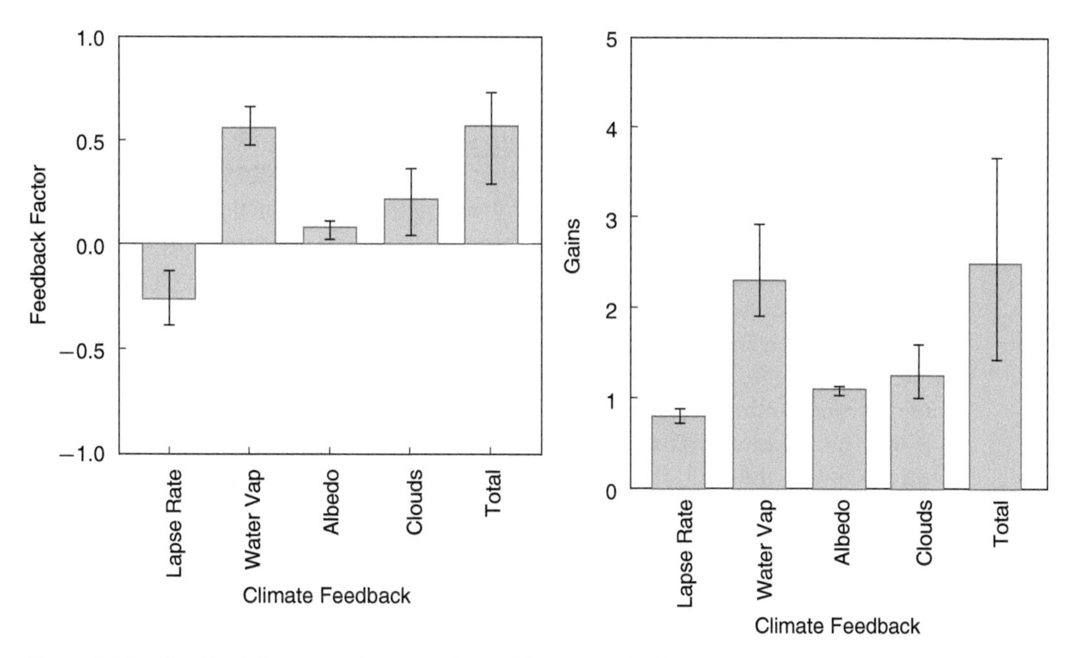

Figure 10.3 Feedback factors and gains estimated from models. The main feedbacks discussed in the text are shown along with their strengths in the climate system. The error bars are the range of estimates (Endnote 10.3).

10.3.4 Ice-Albedo and Thermohaline Feedbacks

We encountered the ice-albedo feedback in Case 3. It is perhaps the most easily understandable feedback of all; however, it has its greatest impact during the ice ages, if they are forced by changes in the solar radiation as the earth's orbit or tilt shifts over the millennia, as in the Milankovich theory of ice ages. When the polar ice caps and snow-covered ground extends deeper into the hemispheres, more light is reflected per square meter. As ice recedes to higher latitudes the solar radiation is much less per square meter, so the effect of the feedback is progressively reduced. The ice-albedo feedback is clearly limited in the extent of global warming it can cause. Once the ice is melted, the feedback stops. It is less limited in how much cooling it can cause if the ice grows. Clouds too interfere with the effectiveness of the ice-albedo feedback. Since clouds reflect light before it gets to the surface, the net effect of reflection from ice is reduced. Current estimates of the feedback factor are $f = 0.08$ which gives a gain of less than 10%.

The ice-albedo feedback can combine with the thermo-haline feedback to cause climate cooling especially in parts of the northern hemisphere. This combined feedback has an unknown magnitude, but a fascinating, though farfetched speculation that the earth can be pushed into an ice-age by global warming which was dramatized in the popular movie "The Day After Tomorrow." The mechanism of the feedback is as follows. It is well known that a deep slow ocean current continuously transfers heat from the equatorial regions to the North Atlantic and parts of the southern oceans. This current is a global cycle that is driven by salinity and temperature of the waters that give it its name. The current earth's circumstances favor the combination of cold and saline water to form near the surface in the north Atlantic and parts of the Arctic. It happens because as ice forms, it leaves the salt behind making the remaining water very saline. Cold, saline water is heavier than the water below causing it to sink. The sinking water spreads out near the ocean floor replacing the water there. Warmer water currents near the surface replace the sinking water bringing heat to the north Atlantic. A slow supply of heat from the surface keeps the cells flowing. The time scales for the completion of the circuit are quite long, but enough heat is brought to the north Atlantic and parts of the Arctic seas to make these regions warmer by 5 to 10°C. If this current is diverted to lower latitudes, or shut off, it could lead to a cooling of the high northern latitudes. Since the regions involved are at the margins of polar ice, additional ice could form leading to the ice albedo-feedback, further cooling the climate until a new balance is reached. If the ice grows too far south, it could reduce the natural cycles of greenhouse gases and water vapor thus reducing the greenhouse effect until we end up in an ice age. This is due to the fact that we are invoking positive feedbacks so if we start heading toward cooling, they will take us further in that direction. Any circumstance that would cause the sinking water to become less saline, warm or both, would cause the circulation to weaken or stop, at least for decades to centuries. Global warming may provide one such circumstance of reducing salinity and increasing the temperature of the source waters. It reduces the salinity by reducing ice formation and adds fresh water from melting polar ice caps and glaciers. Increased precipitation arising from lower latitudes may also add to the effect. A runaway ice albedo feedback is not likely, but it is possible that the circulation would weaken and ameliorate global warming for some regions. The climate event called the Younger Dryas is cited as a possible case study of this phenomenon. After the last ice age, ~ 20,000 years before present, as the earth was warming, a sudden cooling took place, ~ 12,000 years before present, in the northern high latitudes when the temperatures fell by some 5°C lasting about a thousand years. Although many causes are possible, the one that would invoke the thermo-haline feedback is a sudden intrusion of fresh water from destabilized land ice as the earth climbed out of the ice age. Some evidence for such a freshwater intrusion event has been found, but in the end, the causes remain speculative.

From Figure 10.3 the total feedback factor from the known feedbacks is 0.56 with a range from 0.29 to 0.73. From these the climate sensitivities range from 0.4 to 1°C/w/m^2, which would lead to an estimated average temperature increase of 2.3°C with a range of 1.5 to 3.7°C for a doubling of CO_2. It is a curious fact that in these model results, the lapse rate, albedo and cloud feedbacks cancel to nearly zero, leaving the total feedback factor nearly the same as that for water vapor alone.

10.4 Role of the Living World

10.4.1 *Biological Feedbacks*

Biological feedbacks are certain to occur, but their strengths are not known and a great scientific challenge to estimate. For the influence of temperature on living things, including plants and microbes, we can write the dependence of the population on temperature most simply in a parabolic form (Figure 10.4):

$$P = P_{Max}\left[1 - \frac{(T - T_0)^2}{\Delta^2}\right] \tag{10.6}$$

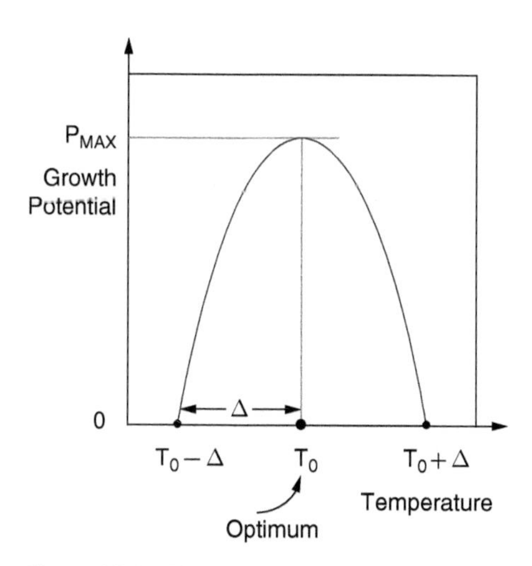

Figure 10.4 Temperature and the idealized population potential of biological species. The simplest model of the steady-state population is a parabolic function of temperature as shown here. At T_0 the population is maximum, and it goes to zero at $T_0 + \Delta$ and $T_0 - \Delta$.

$$\lambda = \lambda_0 + \lambda_{FB} = \frac{\Delta T_0}{\Delta RF} + \sum_I \left(\frac{\Delta X_I}{\Delta RF}\right)\left(\frac{\delta T_s}{\delta X_I}\right)$$

Here P is the population in steady state, T_0 is the temperature at which the population can reach its heyday or the optimal temperature from its perspective, and 2 Δ is a temperature range over which the species can exist on earth, or in a specific environment. If the temperature is either below $T_0 - \Delta$ or above $T_0 + \Delta$ the species dies. The values of T_0 and Δ are different for different species incorporating the idea that some species do better in primarily warm conditions and others are better off in colder climates. If for any species, the current temperature is lower than T_0, global warming will lead to more favorable conditions and a higher population, and if it has already past the optimum, global warming may lead to decline and toward potential extinction. Since the temperature of the earth is not uniform, we expect that the populations of many species could shift toward higher latitudes with global warming. We will return to a further discussion in Chapter 12.

Real species follow more complex growth curves that are moderated by crucial factors other than the temperature. As such, Eq. 10.6 is an idealization of the real world in which a decline of population occurs as we get further from the optimum environmental conditions. Nonetheless, it makes sense that there are limits of cold and hot that any species can tolerate, including us. Biological climate feedbacks arise if temperature increases lead to higher or lower sustainable populations or densities of organisms, such as bacteria and plants, that produce greenhouse gases. The net emissions of methane, nitrous oxide, and carbon dioxide are affected in this way by temperature changes. The reduction of emissions has been observed based on ice core measurements during ice age conditions. Methane concentrations fall to about 350 ppb during ice ages. Reductions of both methane and nitrous oxide have been observed during the little ice age some 200–500 years ago. And CO_2 concentrations go down to 200 ppm during ice ages and are ~ 300 ppm during warm periods (Endnote 10.3).

10.4.2 Permafrost Feedback

One of the pending biological feedbacks from global warming is the release of methane and carbon dioxide from frozen below-ground deposits of organic material in permafrost at the high latitudes and especially the polar regions. When the soils are heated from above, the heat moves deeper over time; however, this is a slow and inefficient process because it occurs by conduction. Nonetheless, two stores of carbon exist that can be affected. One is in hydrates and the other in permafrost. The hydrates are deep deposits in which methane is trapped within frozen ice structures made of water. They are 0.5–10 km below the surface and occur under continental shelves and in the polar regions; it is unlikely that the heat from global warming can penetrate to those depths within century time scales especially if they are at the sea floor and under water. The deposits are estimated to be very large, ~ 3000–10,000 PgC, and are considered possible sources of future fossil fuel energy if the global warming impacts can be managed.

The more likely feedback is from the shallower permafrost. It is frozen ice that contains a large amount of organic material from eons ago. It exists from near the surface to a few hundred meters below. It is likely that this layer will melt slowly putting the frozen organic material into a liquid suspension. There is methane buried in the permafrost; it can move quickly to the atmosphere, but it is not a large amount. More significant is the estimated ~ 1700 PgC in organic material buried in the permafrost (Figure 7.2). When it melts, it may be decomposed by the master mechanism for methane production discussed in Section 7.2. Methanogenic bacteria will emit methane, which will encounter oxidation by methanotrophic bacteria as it reaches the surface ending up as CO_2. How these bacteria come together in this process also depends on whether plants grow in the region with warming trends and what type of plants they might be. This leaves two aspects about the feedback that are virtually unknown. One is a matter of the split between the methane and carbon dioxide in the flux. The carbon dioxide is favored because of the slow movement of material with few plants and the efficiency of the methanotrophic bacteria which also have a greater response to increased temperatures than the methanogens and will exist in the warmer upper parts of the polar soils. If most of the flux is carbon dioxide, it will have a lesser greenhouse effect than if most of it is methane. Readers are reminded that, per molecule, methane is not only more efficient at trapping the earth's radiation than carbon dioxide, but that it will react with hydroxyl radicals and convert to carbon dioxide within a few years continuing the warming impact for a much longer time. The releases of carbon from permafrost, whether directly as carbon dioxide, or after the oxidation of methane, would be fossil carbon dioxide that will increase the atmospheric concentrations similar to the burning of fossil fuels. The other matter is that the methanogens will be in very cold environments below the surface and their ability to grow would therefore be much less than what we observe in the wetland and rice ecosystems in the warm parts of the world. And without them, the methanotrophs cannot function. This may slow down the decomposition rates so that the emissions may not amount to much over the time scales of decades or even centuries. Global warming, at least from human activities may not persist that long.

Example: Suppose the top 50% of the permafrost melts and is converted to methane before being released over 100 years after some critical temperature is reached in a warmer world. It will convert to fossil-derived CO_2 in the atmosphere. Once formed, about 50% of the CO_2 will remain in the atmosphere over the 100-year period. Assume the store is 1700 PgC (Figure 7.3).

(a) What will be the annual flux of methane? (b) What will be the concentration at the end of the 100 years? (c) How much global warming will that cause assuming that at the time of this melting, methane concentration was already 2000 ppb? (d) How much will that change the CO_2 concentration if it is 400 ppm to start with? (e) And how much global warming will the additional CO_2 cause?

Solutions: That's a lot of questions. Let's see if we can answer them in round numbers.

(a) Flux of $CH_4 \approx 0.5 \times 1700$ PgC × (16 g CH_4 per mole/12 g C per mole) × (1000 Tg/Pg)/100 years ≈ 11,000 Tg/y compared with about 500 Tg/y from all sources at this time.

(b) This part of the emissions will come into equilibrium in about 20 years, so by the end of 100 years, C(From Permafrost) $= S\tau =$ (11000 Tg/y) × 0.4 ppb/Tg × 8 y ≈ 35,000 ppb.

(c) From Eq. 9.10b, the change of radiative forcing will be: $\Delta RF = 0.036\ \text{w/m}^2 \times (37000^{1/2} - 2000^{1/2}) \approx 5.3\ \text{w/m}^2$ and $-0.7\ \text{w/m}^2$ (overlap effect) $= 4.6\ \text{w/m}^2$. Therefore, $\Delta T \approx 0.7°\text{C/w/m}^2 \times 4.6\ \text{w/m}^2 \approx 3.2°\text{C}$.

(d) If all methane converts to CO_2, and 50% remains, there will be about 400 PgC as $CO_2 = 400$ PgC × 0.5 ppm/PgC ≈ 200 ppm CO_2.

(e) From Eqs. 9.10, $\Delta RF = 5.35 \times \ln(600/400) = 2.2\ \text{w/m}^2$ and $\Delta T \approx 0.7°\text{C/w/m}^2 \times 2.2\ \text{w/m}^2 = 1.5°\text{C}$. The total temperature change will be ≈ 5°C. We have stretched the climate formulas quite far, so it may not be that bad. Or if it all comes out as CO_2, the temperature change will be the 1.5°C calculated here instead of 5°C. We see that release as methane is much more potent for global warming.

10.4.3 Biological Feedbacks and Daisy World

Next, we want to consider the possibility that living things, not only change their environment, but do so in directions that make their survival easier or more probable. We, humans, have done that to varying degrees and as we have seen, living things in general are responsible for controlling the greenhouse gases on the earth that raise the temperature so that the world is not frozen. J. Lovelock's Gaia hypothesis is an articulation of this idea. To explore it further, we will consider the case of "Daisy World," an allegorical model constructed by A. Watson and J. Lovelock to illustrate these ideas. It adds to our understanding of biological feedbacks in the real world.

In daisy world only daisies grow and they control the temperature to generate and sustain a livable climate through negative biological feedback processes. There are two types of daisies in Daisy World – black and white. The temperature of the planet is $T_e = [(1-A)S/\sigma]^{1/4}$. The daisies reflect or absorb light according to their colors thereby controlling the albedo. The climate is therefore entirely controlled by two factors. One external, the solar radiation and one internal, the albedo. The sun's luminosity, and hence the radiation reaching the world is constantly increasing linearly. This is the source of climate change and global warming on this planet. The populations of the daisies are determined by a growth rate that is represented by a parabolic function of temperature. It would lead to steady state populations that can also be represented in the parabolic form of Eq.10.6. The albedo is calculated according to how many daisies there are of each type and stated as the area of the planet they cover so the albedo is: $A = \alpha_W A_W + \alpha_B A_B + (1 - \alpha_W - \alpha_B) A_{BL}$ where the α's are the fractions of the land occupied by the white (subscript W) or black (B) daisies and the *A*s are the albedos, including the bare land (BL). Moreover, heat is transferred by conduction. So, if one patch is hotter than another, heat will be exchanged and there will be a net flow from the warm to cold creating a more uniform average temperature for the planet and creating a synergy between the two types of daisies. This will affect the growth of the daisies depending on whether they need heat or are too hot (see Endnote 10.4). Can the daisies survive on such a planet and for how long? Does it help the daisies if they are diverse – black and white, or perhaps even shades of grey, or is just one color better?

To answer these questions, the calculated daisy world model temperatures are shown in a series of Figures 10.5a–e (Endnote 10.4). We start by looking at what happens if there is only one type of daisy. Figure 10.5a shows the temperature of the planet with only black daisies. At a temperature $T_0 - \Delta$ the black daisies start to grow and quickly cause a stable temperature that persists for about 200,000 years, but then they all die because the temperature has gone above $T_0 + \Delta$. For comparison the planetary temperature without any living things is also shown and it constantly rises as would be expected with the increasing solar radiation. Similarly in Figure 10.5b we see what happens if we allow only white daisies to grow. They do better because they are able to reflect more radiation as their populations grow. In this manner they are able to keep the temperature of the planet habitable for 400,000 years, but not very stable. A global warming occurs during most of their existence. If we put both color daisies together the result is strikingly different as shown in Figure 10.5c. Now the combined effects of the two colors of daisies regulate the planetary temperature for some 500,000 years and also keep it nearly constant that long, despite the ever-increasing solar radiation. If daisies of many shades of grey are added, the result is an even longer time of stable climate as shown in Figure 10.5d.

We can draw some lessons from daisy world. The environment and survival of the daisies is defined by the planetary temperature. Daisies control it to extend their survival. In our real world the connection is more complex, but living things regulate the environment including the temperature, as the daisies do in their world. Life on earth does not merely adapt to environmental circumstances, but can exploit modifications that improve survival and global habitability. It is hard to imagine an earth at this time if there was no life on it – it would be an alien planet, as it once was, and will become one day – but that is beyond our time scales of interest which are decades to centuries, however, it reflects the fate of daisy world. Life has been evolving and modifying the environment for a long time. The existence of plants provides oxygen in the atmosphere allowing animals to co-exist. Even the main constituent of the atmosphere, nitrogen, is cycled by bacteria in the soils, releasing nitrous oxide. Living things, mostly plants and marine organisms, have made the world warmer by some 20°C by contributing greenhouse gases to the atmosphere. Life has generated the ozone layer from the oxygen it controls. This may have helped evolve complex species by protecting them from ultraviolet radiation at vulnerable stages. The important role of life in the earth's climate continues as human activities push toward a warmer world, not only by increasing the biogenic gases, but also by putting nearly permanent technological ones into the atmosphere.

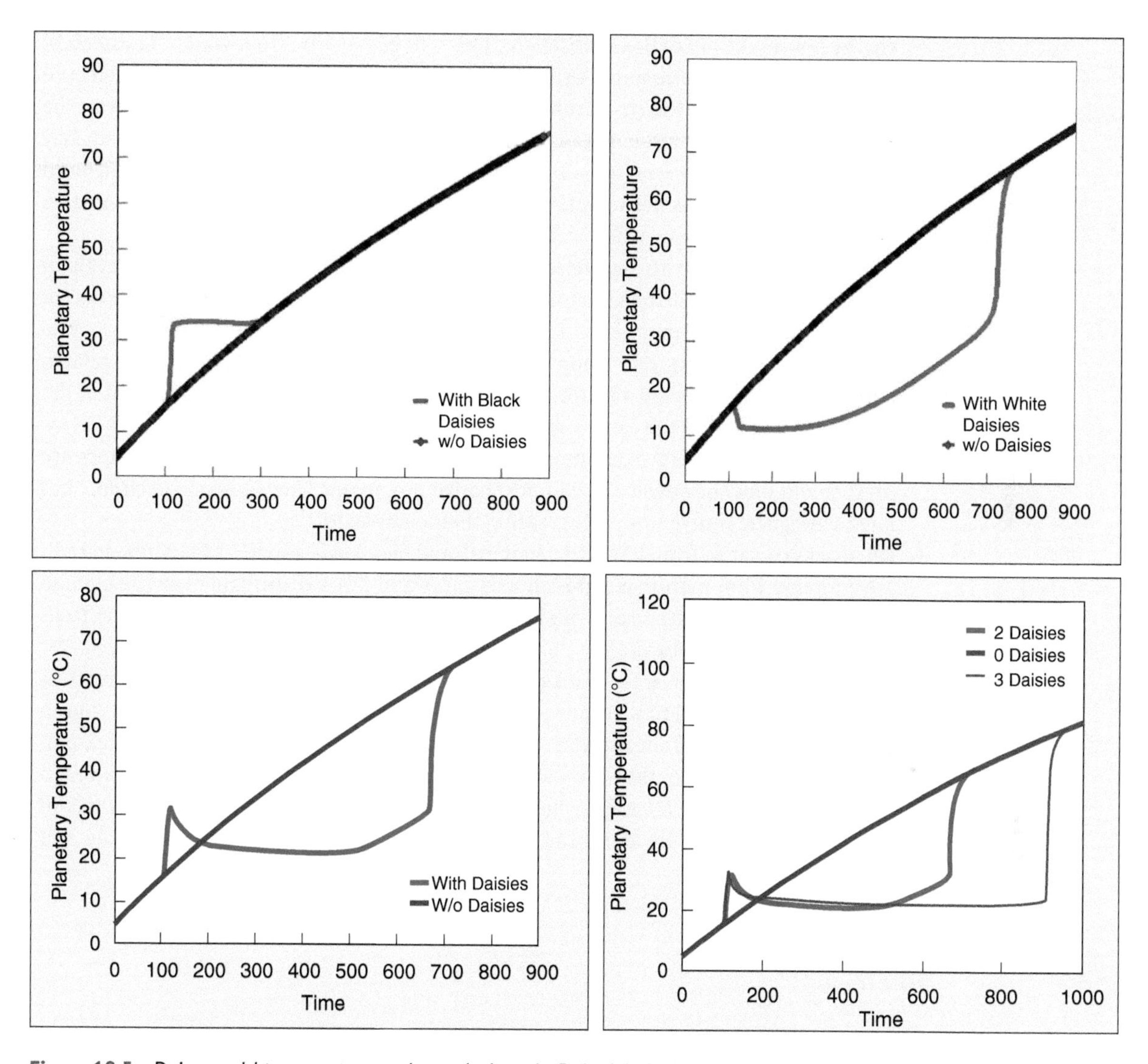

Figure 10.5 Daisy world temperature and populations. In Daisy World, black and white daisies counteract global warming and regulate climate change caused by constantly increasing solar radiation. They are able to maintain constant and livable temperature for long periods of time.

Review of the Main Points

1 In writings on climate science "feedbacks" are defined variously. We see it this way: If we alter one of the fundamental drivers of the earth's surface temperature, that is, the solar constant (S), albedo (A), or the atmospheric absorbers (mostly greenhouse gases and ozone, F,f) then that change will generate a radiative forcing, or an imbalance, at the top of the atmosphere. If nothing else is altered, in particular the remaining fundamental variables, a surface temperature change will occur that will re-balance the

energy flow at the top of the atmosphere. This will be our base from which the effects of feedbacks will arise. The initial driver, and its resulting temperature change will cause additional changes in the environment that may exacerbate or reduce the impact on the earth's temperature relative to the direct base effect. These are the feedbacks and they may include changes in the remaining variables (out of S, A, F, f), which may be caused by myriad connections between the climate and the physical and living environments of the earth.

2 Feedbacks are known to significantly increase global warming from rising levels of greenhouse gases and indeed from any initial cause. All climate models, however sophisticated, require theories of the earth's larger environmental system to incorporate the effects of feedbacks. Our limited knowledge of feedbacks particularly those related to aerosols and clouds are a source of significant uncertainties in the prediction of the near future climate of the earth.

3 Feedbacks do not have a fixed impact on climate but rather depend on which ones are present and how they evolve as climate change progresses. They do not add simply but have synergistic contributions, increasing their complexity.

4 Feedbacks can arise from physical, chemical, and biological responses to increased radiative forcing from man-made greenhouse gases and other disturbances to the climate system. The physical feedbacks are best understood, except for clouds. Biological feedbacks are least understood.

5 Aspects of how biological feedbacks may regulate the temperature of a planet are illustrated by "Daisy World" – an allegorical model. In it the rising temperatures due to increasing radiation from the sun are controlled by the interplay between black and white daisies. Together they are more effective than either by itself in maintaining a constant temperature for a long time. It illustrates many aspects and complexities of feedbacks and especially about the effects of living things on the earth's climate and environment.

Exercises

1 Consider a spherical drop of water of radius r. Find how the surface to volume ratio changes with radius to illustrate how a cloud with smaller drops favors a greater absorption of water vapor and other gases than one with larger drops with the same water content. Suppose there is one drop of radius 10 microns. Split it into 1000 drops of equal size. How much more surface area is made available?

2 For the expected global temperature change during the last century, calculate the effect of each of the feedbacks stated in Section 10.3 and Figure 10.3, if they acted alone and the combined effect for all acting together. Calculate the effect of the feedbacks for steady state conditions with a radiative forcing of 2 w/m^2 from greenhouse gases and aerosols, as currently seen.

3 In the case studies in the text, add a fourth case in which there is only ice-albedo feedback and no water-vapor effect. Use the feedback factor $f_{I\text{-}A} = 0.25$. How much will the temperature change? Show that the temperature changes by the water vapor and ice-albedo feedbacks do not add up to the temperature change from the combination.

4 The earth is ice covered due to the ice albedo feedback. Time has passed and an equilibrium has been established. What is its temperature? Use an albedo of 0.7 for ice. Make whatever assumptions are necessary to decide how the greenhouse effect will be affected. State your assumptions to justify your final calculations.
5 Calculate the climate sensitivity of the simplest model in Eqs. 9.1 for changes in
(a) the solar constant *S*,
(b) the albedo, and
(c) the absorptivity *F*. Are the sensitivities the same?

Endnotes

Endnote 10.1 The "no feedbacks" sensitivity is a general concept. One way to look at it is to note that the energy at the top of the atmosphere consists of the short-wave radiation from the sun, $E_{IN} = (1 - A)S$ going toward the earth's surface (A = albedo, S = solar constant, F = atmospheric absorptivity) and $E_{OUT} = f(T_s, T_a(T_s, F,..),..)$, the long-wave energy from the earth going to space. In equilibrium, $\Delta R = E_{IN} - E_{OUT} = 0$ and the radiating temperature is $T_e = [(1 - A)S/\sigma]^{1/4} = 254.9$ °K. If we alter any of the fundamental variables that determine the climate (A, S, f, or F), then the $\Delta R > 0$. Depending on what happens, it could be < 0, but that doesn't affect the reasoning, so for simplicity we will take it as positive, and let's say that after our action, $\Delta R = 10$ w/m^2. That is, 10 more w/m^2 are going into the system than coming out. This could happen if the sun gets brighter, the albedo decreases, or we add some more greenhouse gases into the atmosphere that can trap this much radiation. Then instantaneously the new radiating temperature will be $T_e' = \{[(1 - A)S - 10 \text{ w/m}^2]/\sigma\}^{1/4} = 252.2$°K, so the $\Delta T_e = 2.7$°C (cooler). When a new equilibrium is reached, 10 w/m^2 more radiation has to be sent back out at the top of the atmosphere. Much of this must come from the surface, and we have not allowed any other variables to change. So we may say that $\Delta T_s \approx \Delta T_e$ and therefore the $\lambda_0 = \Delta T_s/\Delta R \approx 2.7$°C/10 (w/m^2) = 0.27°C/(w/m^2), which is about 0.3°C/(w/m^2) that we saw in the text, but here we have not used any "models" or defined a cause of the initial disturbance. The only assumption, commonly made, is that the change of the radiating temperature between the time the disturbance was initiated and the state of equilibrium is about the same as the change of the surface temperature. This is not true, but in the earth's present climate, it is close. From the simple model we discussed: $\Delta T_s = \Delta T_e/(1 - ½ F)^{1/4}$ or $\Delta T_s = 1.16\, \Delta T_e$. Although the difference is outside the 10% rule, it has no practical consequence, but it affects the concepts we have studied. The discrepancy becomes worse the thicker the atmosphere gets (F, or $(1+\delta)F$ become large) and disappears if there is "no atmosphere" ($F = 0$). If there is no absorbing atmosphere, the re-balancing will have to come only from the surface temperature change, so then it is simply the blackbody sensitivity to radiative forcing.

Endnote 10.2 The derivation of Eq. 10.4 is as follows: By definition $\lambda = \Delta Ts/\Delta RF$, which we can write as the direct effect plus the effect of feedbacks:

$$\Delta T_s = G\lambda_0 \Delta RF$$

By the same arguments as for the factors, each feedback term is made up of two parts. It should be noted however, that these individual terms λ for each feedback, work only in the sum as a composite feedback sensitivity λ_{FB}. We want to convert the feedback term in Eq. 10.6 to its dependence on changes of equilibrium temperature (ΔTs) instead of changes of radiative forcing (ΔRF). To do this, we multiply the second term by $\Delta Ts / \Delta Ts$, which we can always do because it is 1. Then we re-write this as $(\Delta X_I/\Delta RF)(\Delta Ts / \Delta Ts) = (\Delta X_I/\Delta Ts)(\Delta Ts/\Delta RF) = (\Delta X_I/\Delta Ts)(\Delta Ts/\Delta RF)(\delta T_s/\delta X_I) = f_I (\Delta T_s/\Delta RF)$. Since the term multiplies each f_I it can be pulled out of the sum and the equation above becomes $\lambda = \lambda_0 + \lambda f$, and so, $\lambda = \lambda_0/(1 - f)$ leading to Eq. 10.4 when compared with Eq. 10.2.

Endnote 10.3 Sources: Analysis of feedbacks shown in Figure 10.3 are from IPCC, *Climate Change 2007: The Physical Science Basis*, Chapter 8.6 (see Endnote 1.1). These are converted to the feedback factors as used in this book. The feedback factors and sensitivities are discussed further by Wallace and Hobbs, 2006, Chapter 10 and Hartman 1994, Chapter 9 (full references in Endnotes 7.2 and 8.7). The feedback effect of temperature on greenhouse gases is seen in: J.-M. Barnola, D. Raynaud, Y.S. Korotkevich, and C. Lorius. Vostok ice core provides 160,000-year record of atmospheric CO_2, *Nature* 329, 408–14. 1987. M.A.K. Khalil and R.A. Rasmussen, Climate-Induced Feedbacks for the Global Cycles of Methane and Nitrous Oxide. *Tellus 41B*, 554–559, 1989.

Endnote 10.4 The Daisy World Model. The graphs in Figure 10.5 are based on the following mathematical representation of the model. The temperature of the planet is $T_p = [(1 - A) S/\sigma]^{1/4}$ since there is no greenhouse effect. $S = S_0 + \beta t$ where β is the rate of increase of the solar constant. The daisies affect the albedo through which they regulate the planetary temperature. The total albedo is $A = \alpha_b A_b + \alpha_w A_w + \alpha A_p$. The subscript "b" is for black daisies, "w" for white daisies and "p" is for planet. The α's are the areas covered by the daisies of the two colors. The α's are proportional to the populations of the daisies and evolve by the equation: $d\,\alpha_{b,w}/dt = \alpha_{b,w}(x\,\beta - \gamma)$. In this equation $\beta = 1 - a\,(T_0 - T_{b,w})^2$ is the parabolic growth rate. It can lead to populations that follow Eq. 10.6. $T_{b,w}$ are the temperatures of the daisies. $x = p - \alpha_b - \alpha_w$ is the remaining available area where p is the habitable area of the planet without daisies. γ is a constant death rate. $T_{b,w} = q\,(A - A_{b,w}) + T_p$ where q is a constant representing conduction. By this equation, the daisies exchange energy and homogenize the temperature of the planet. Their growth depends on these temperatures with the optimum at T_0. The daisies grow when T_p reaches a value that makes the temperature of the daisies warm enough to grow, and from there the populations increase by the population equation driving the albedo change that regulates the temperature for some length of time. There are quite a few variables that you can choose and adjust to create an outcome. The equations can be extended to add daisies with shades of gray. The model comes from: A.J. Watson and J.E. Lovelock. "Biological homeostasis of the global environment: the parable of Daisyworld." *Tellus* B. 35B 286–9, 1983. It illustrates aspects of the Gaia Hypothesis from: J.E. Lovelock, *Gaia: A New Look at Life on Earth*, Oxford University Press, 1979 and 2000.

11

Match of Climate Change Observed and Modeled

We should look now whether, all things considered, the present early twenty-first-century state of global warming and climate change is explained by the theory, models, and arguments we have discussed. The answer will be "yes," but perhaps the more important matter is that, because of uncertainties, it still leaves flexibility for significantly different futures. Nonetheless, societal interest in climate change is *driven* by the results we will discuss here.

The manifestations of climate change can be direct or more subtle. Direct influences are increases of global temperature, rainfall, and shifts of the earth's major heat and atmospheric transport processes. Any practical and valid theory of climate change must explain these aspects. The most important matter is to know whether the temperature change that is observed can be fully explained by our theory. After that everything is an effect, including precipitation and wind changes. Some physical effects may not translate into impacts on global habitability, but they can serve as tests to validate our understanding of the earth's climate system or reveal deficiencies in the theory. It is noteworthy that because the earth's system is complex with many interconnected parts, and models are approximations, it is generally easier to explain an observed aspect of climate change, than it is to predict it. Our goal in this chapter is to examine clearly observed physical aspects of climate change to see whether they are consistent with our understanding of the science.

11.1 What Is Global Warming?

To start with, let's decide what global warming is and what it isn't. One hundred and fifty years of global monthly average temperatures archived at CDIAC are analyzed to look at some key features (Figure 11.1, Endnote 11.1).

The data visually show that the temperature has increased unsteadily by about 1°C over the last century. The increase has been more on land and less over the oceans (Figure 1.1). The unsteadiness can be translated into rates of change over contiguous periods lasting a decade or more and expressed as dT/dt in °C per decade. For the six periods selected, three show global cooling and three show warming for every month of the year (Figure 11.2a). Yet the net change is positive because the time over which the temperature increased (~ 100 yrs) is longer than that over which it decreased (~ 50 yrs).

Let's look at it another way. We can ask: How often is one year warmer than the previous and how often is it colder? Using the data, we find that it is 50–50, or compared to the

Global Climate Change and Human Life, First Edition. M. A. K. Khalil.

Companion Website: www.wiley.com/go/khalil/Globalclimatechange

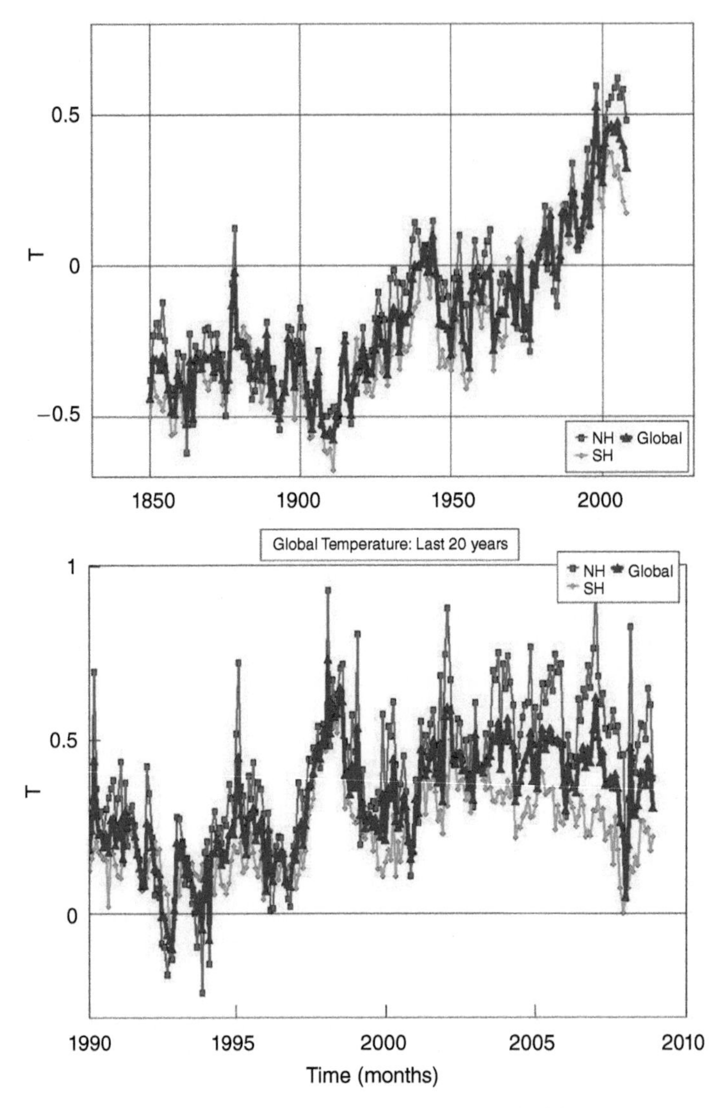

Figure 11.1 Annual average temperature of the earth. (a) Annual average temperatures in the two hemispheres and the global average. (b) A warming hiatus between the late 1990s and 2010 (monthly). Temperature increases have picked up since.

previous year, each year is as likely to be warmer as it is to be colder! We can extend this question to ask, how often is the trend of temperature positive over some N number of years. That is, if we calculate the rate of change, using a linear regression, over all contiguous N year periods starting with years 1,..,N, then years 2,..,$N + 1$, then years 3,..,$N + 2$, and so on, how many times will those trends be positive and how many times will they be negative? The results are shown in Figure 11.2b.

The remarkable result is that for one- to five-year periods, there is more or less as much chance of warming as there is of cooling. On the other hand, if we look at all 100-year periods, and there are 50 of them, the trend is always positive (100%). The percentage at which we are willing to accept that the earth is warming is somewhat arbitrary, but it

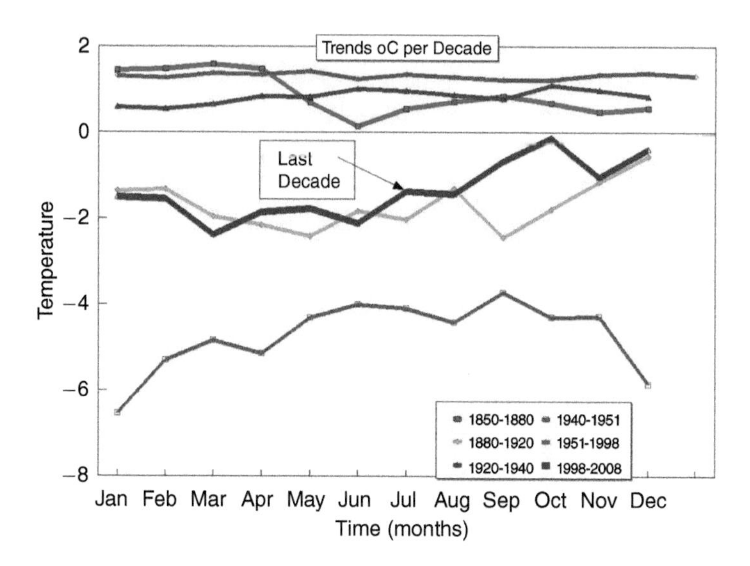

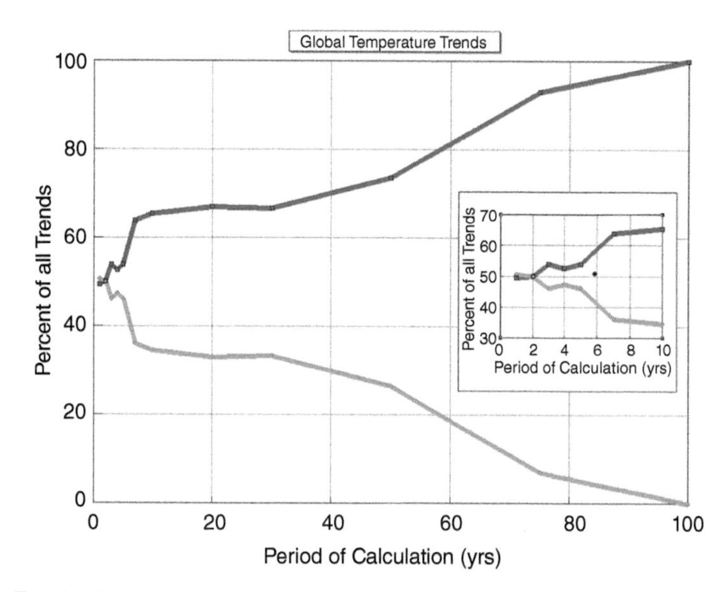

Figure 11.2 Trends of the global temperature. (a) The rates of change are calculated over selected times for each month of the year. The warming or cooling is seen in all the months during each period. (b) A look at the percentage of positive and negative trends over different periods of time. The first datum on the left is for the percentage of time the one year is cooler than the next. That happens about 50% of the time. The next point is how many times the rate of change over 2 years is a warming and then, how many times all contiguous 3 years show a warming and so on. A warming trend emerges for periods 10 year or longer.

probably shouldn't be less than 2/3 to 1/3 probabilities, which means that the earth is warming if the positive trends are twice as probable as the negative ones. This happens around $N = 10$ years suggesting that what constitutes global warming should be a positive trend in temperature over at least a decade. The year-to-year variability in the observations is caused by the lack of complete global coverage, measurement imprecision or even inaccuracies and, perhaps most importantly, the natural variability of the climate for which all the reasons are not known or may not be knowable. We can improve on the first two, especially with satellite observations, but the significant inherent climate variability will still remain.

If there are trends that last a few years or longer, or are dramatic, then some explanation may exist. Of these, several are seen with some regularity – volcanic eruptions causing cooling; solar cycles and El-Nino events causing variations in the global temperature. None of these are large influences, but they can be quantified. Volcanoes and El-Nino events can have dramatic localized influences on climate that can be felt by people. Global warming, on the other hand, is not perceptible on a year-to-year basis, but can only be felt after decades of change as we have seen in the analysis of temperature records. Global warming has been about 1°C in a century, leading to an average increase of just 0.01°C/year, or during the most rapidly warming decades, it has been at most about 0.02°C /year. Such small changes can neither be measured nor be felt from one year to the next. And people who have lived a hundred years to see the 1°C change may have faulty memories by now.

The increase of greenhouse gases is quite steady (Figure 7.1). You might think that this should cause a similarly steady increase in the temperature, but it doesn't (Figures 11.2 and 1.1). When we see continued increases in greenhouse gases and no temperature change, as in times between the mid-1990s and 2015, it causes people to question whether global warming is occurring and whether it is connected to the man-made greenhouse gases. But again, the longer-term observations restore the consistency between global warming and human influences. Because of the unsteadiness of the temperature increase despite the steady rise of greenhouse gases, it is evident that there is more going on, that determines the earth's temperature, than just the radiative forcing from increasing CO_2 and other gases.

These results establish that any dramatic changes we see from one year to the next should not be attributed to global warming, or the lack of it. It is a curious fact that the temperature of the earth may be constant on average for a decade, but you may still see the warmest years on record during this time. This happens when the temperature has risen to higher than past levels and then becomes constant for a while such as the period shown in Figure 11.1b. During this time, there are several years that are the hottest on record up to the year they happen even though the temperature is not warming at the decade time scale. Moreover, these excursions to the high temperatures are not caused by the increase of greenhouse gases during that year since no large concentrations are observed. In such a situation, that has occurred in recent years, the cause of the hottest year is the interaction of natural variability of the climate with the long-term temperature increase and thus adds complexity to an attribution of cause.

11.2 Causes of Observed Warming

Let's look now at the consensus model results compared with observations summarized by the IPCC (Figures 11.3 and 11.4, Endnote 1.1). These are a benchmark of our understanding of global warming. In these calculations the long-term observed changes of the greenhouse gases (Figure 7.1), volcanoes, aerosols, and the solar radiation are used as input and combined with their effect on the energy balance of the earth's land and oceans resulting

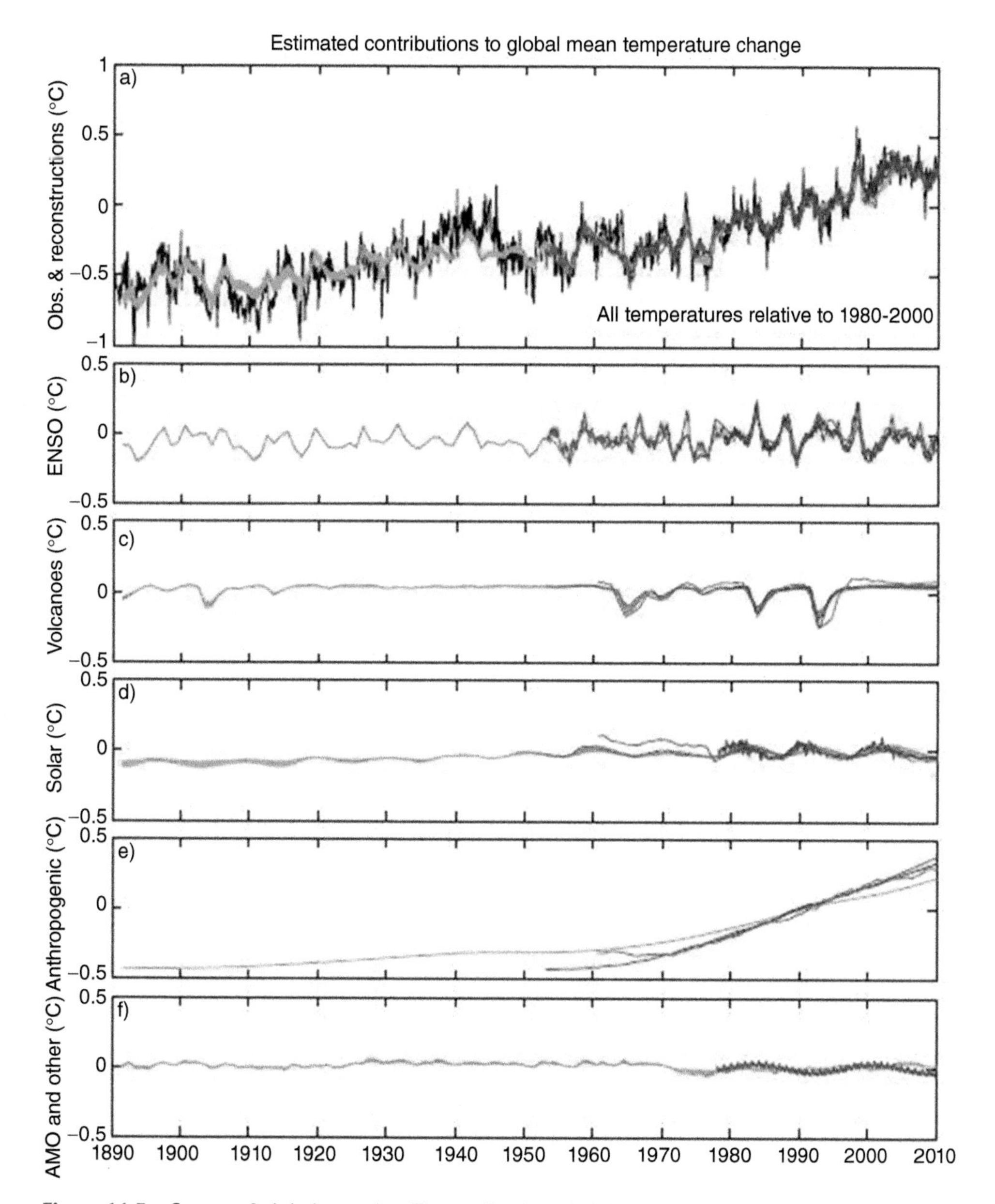

Figure 11.3 Causes of global warming. The attribution of global warming to various causes. The most significant long-term impact is from the increase of greenhouse gases and aerosols caused by human activities. Natural causes result in cycles as from the Sun or aperiodic changes from volcanoes and El-Nino events (from IPCC AR5, Endnote 1.1).

in a calculated surface temperature change. The models are based on the principles we have discussed but incorporate many details feedbacks and connections. They cannot be written like the formulas we have developed in this book; rather, they are solved using numerical methods on high-speed computers. Although this lack of transparency is disturbing, we should note that the results are consistent in magnitude and timing with what we expect from the theories, simpler models, and observations we have delineated and discussed in this book so far (Endnote 11.2).

The salient points are the following: that the measured concentrations of the greenhouse gases, CO_2, CH_4, N_2O, and halocarbons, along with aerosol and cloud effects, correctly predict both the timing and the magnitude of the observed temperature changes over the last 150 years when direct temperature measurements have been taken as seen most clearly in Fig 11.4 where the measurements are plotted along with model calculations. In these calculations both greenhouse gases and aerosol influences are included. Using only the greenhouse gas increases over-predicts the expected temperature increases, and thus cooling influences cannot be ignored. It complicates the calculations and increases the uncertainties. The magnitude of the man-made increase is about 1–1.4°C and is dominated by the increases in long-lived greenhouse gases (see Figure 9.4). The natural contributions to temperature change are solar cycles, volcanos, and el-Nino events. The solar influences amount to fluctuations and small trends that are generally around 0.05°C and hardly measurable due to the internal variability of the models, which is $\approx \pm 0.15$°C (not shown on the graphs). Major volcanic eruptions, however, reduce the average temperature by 0.1–0.2°C for several years and can be seen in the temperature record above the natural variability. It is noteworthy that the global warming hiatus, as it is called, is not explained by the model calculations as is evident in Figure 11.4 because during the time, greenhouse gases were increasing unabated and there were no counteracting cooling influences such as volcanoes.

Although this consensus picture of global warming is tightly woven, there is an extreme uncertainty in our predictions of future climate change. It comes from the effects of aerosols and cloud interactions, and the internal variability of the earth's climate reflected in its sensitivity to environmental changes. At the least, the observed change of the earth's temperature and the societal concern it has generated has taught us a lot about the nature of

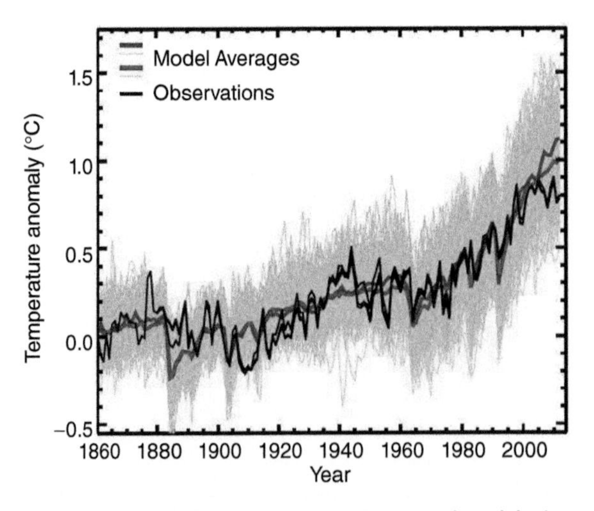

Figure 11.4 Match between observations and models. An ensemble of models were used to match the expected global warming from all the known causes with the observed temperature record. The timing and magnitude of the observed warming are both explained by the combination of the human and natural drivers of climate (from IPCC AR5, Endnote 1.1).

our world and the inherent and persistent uncertainty in our knowledge, motivating us to seek more satisfying answers (Endnote 11.3).

11.3 Differential Effects of Climate Change

As stated earlier, the changing temperature of the earth generates many physical effects. Some may not affect our lives, but test whether the climate theory is right or not. There is a growing list of such phenomena. Some are detectable now and others will become so, as the warming trends progress. We look at them next.

11.3.1 Stratospheric Cooling

One of the most significant tests of global warming by greenhouse gases is that it predicts a cooling of the stratosphere! This phenomenon is made more important by the fact that if solar radiation, the decrease of albedo, or heat from the core of the earth are the causes of surface warming, then the stratosphere will not cool. As it happens, observations show that the stratosphere is cooling, thus supporting both the climate theory and greenhouse gases as the cause. This would be much clearer were it not for the simultaneous depletion of the ozone layer caused by the man-made chlorofluorocarbons, which also causes cooling because, as we discussed, the ozone reactions and absorption of the solar and the earth's radiation streams, heat the stratosphere. Less ozone means less heating. So let's see how this works. The energy balance of the stratosphere is given in Figure 11.5.

Right away, readers will note that there is a little less energy being sent back to the troposphere than out to space. Unlike the troposphere which is warmest at the bottom, the top half of the stratosphere by mass is warmer than the bottom half. The stratosphere is so rarefied that it does not have a lot of re-radiation that was a leading cause of more radiation being sent

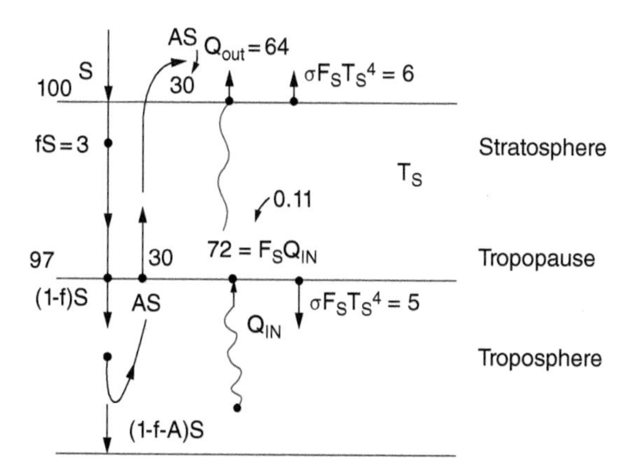

Figure 11.5 Energy balance in the stratosphere. The energy balance of the stratosphere shows that global warming is cooling the stratosphere. The symbolic version of the energy balance of the stratosphere is to isolate the effects of changes in greenhouse gases, ozone layer, and albedo on the stratospheric temperature. The numbers shown are radiation fluxes in % of the solar constant at 342 w/m^2.

down in the troposphere than up. The absorbed energy is due almost entirely to the O_3 in the ozone layer and CO_2, which is well mixed in the stratosphere. Of the Earth's infrared emissions, the stratosphere captures some from the window region, because of ozone, and some from the upward radiation sent by the troposphere; we can combine these as $F_s\,Q_{IN}$. It is the fraction of the infrared radiation arriving from below that is absorbed by the greenhouse gases in the stratosphere. From Figure 11.5, we can write the energy balance as:

$$fS + F_S Q_{IN} = 2\sigma F_S T_{Strat}^{\ 4} \tag{11.1a}$$

$$Q_{IN} = (1 - f - A)S + \sigma F_S T_{Strat}^{\ 4} \tag{11.1b}$$

Here f is the fraction of the solar radiation absorbed by the ozone layer, F_s is the fraction of the infrared energy coming from below (Q_{IN}) that is absorbed by the stratosphere, A is the albedo, and T_{Strat} then is the average stratospheric temperature. We can substitute Eq. 11.1b into Eq. 11.1a and solve for T_{Strat}. The stratospheric temperature in equilibrium, after simplifications is (Endnote 11.4):

$$T_{Strat} \approx \left(\frac{S}{2\sigma}\right)^{\frac{1}{4}} \left[\frac{f}{F_S} + (1 - A)\right]^{\frac{1}{4}} \tag{11.2}$$

If S and A are constant, as observations indicate, Eq. 11.2 is $T_{strat} = 234°C\ [f/F_s + 0.7]^{1/4}$. We see that the increase of greenhouse gases will cause the stratospheric temperature to decline because F_s will increase. Similarly, a depletion of the ozone layer will reduce "f" and that too will cause stratospheric cooling. If, on the other hand, global warming is caused by an increase of the solar constant or the net decrease of albedo, then in both cases the stratospheric temperature will increase according to Eq. 11.2 (Endnote 11.4).

In non-mathematical terms, global warming caused by increased greenhouse gases leads to surface warming, but the stratosphere does not get much more heat because it is shielded by a large mass of air to begin with and now more is being trapped. Yet if the greenhouse gases, including CO_2, increase in the troposphere, they will increase in the stratosphere too. That means they emit more radiation into space (by Kirchhoff's law). So the stratosphere doesn't get any more energy, but it radiates more into space causing it to cool.

We can put some numbers into Eq. 11.2 to see the effects of ozone depletion and greenhouse gases on stratospheric temperatures. To start with, using the data in Figure 11.5, we can calculate the base case with $f = 0.03$ and $F_s = 0.11$ as T_s(base) $= 234°C\ [(0.03/0.11) + 0.7]^{1/4} = 232.3°K$. For the maximum ozone depletion of 4% from observations, reducing the solar energy absorption by 4% we get, T_s (ozone depletion) $= 234°C\ [(0.03 \times 0.96)/0.11) + 0.7]^{1/4} = 231.7°K$. So ΔT_S (ozone depletion) $= -0.6°C$. Next let's look at the effect from CO_2 increases, which are a major cause of global warming. From 1960 to around 2000 the increase is about 17%. We assume that the stratospheric absorption of infrared radiation is mostly due to CO_2. Therefore increasing F_s by 17% gives T_s (CO_2 increase) $= 234°C\ [0.03/(0.11 \times 1.17) + 0.7]^{1/4} = 229.9°K$ and ΔT_S (CO_2) $= 229.9°K - 232.3°K = -2.4°C$. The total change is about $-3°C$ with the majority of the influence due to CO_2. Indeed, the observations of the stratospheric temperature are consistent with roughly this amount of stratospheric cooling over the period (Endnote 11.5, Figure 11.6).

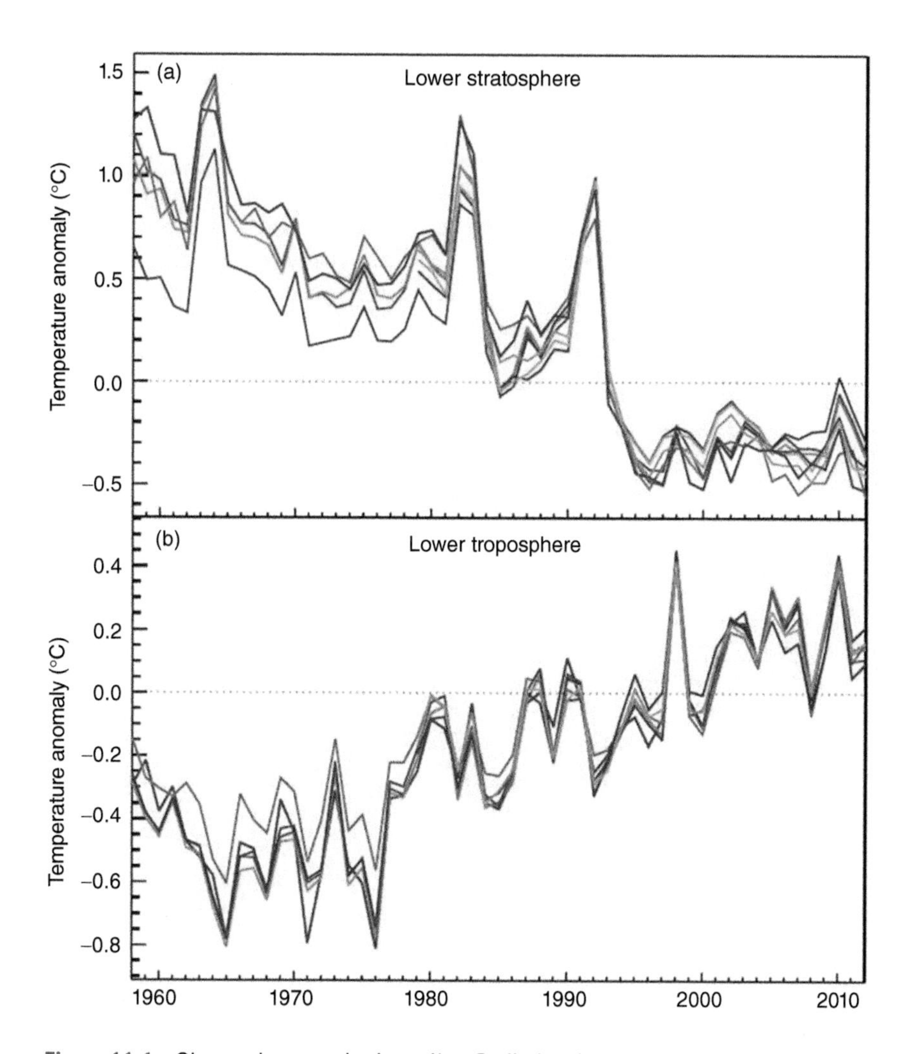

Figure 11.6 Observed stratospheric cooling. Radiative theory predicts that if the surface warming is due to the increase of greenhouse gases, the stratosphere will cool. Other causes of surface warming do not produce this result (From IPCC AR5, Endnote 1.1).

11.3.2 More Tests and Differential Expectations

The greater increase of temperature in the polar regions has been predicted for some time based on the ice-albedo feedback which has its largest effect there. It is expressed as the polar amplification factor $f(\text{Polar}) = \Delta T(\text{Polar})/\Delta\, T(\text{Global})$. For the Arctic, this factor is estimated to be 1.5–4.5 for various models under doubled CO_2 conditions and observations support the results with a measured factor of about 3 during the last 50–100 years. The main driver is the diminishing sea ice, even though the permanent polar ice is more stable (Figure 1.1a). Changes in circulation, clouds, and oceans are also thought to be significant contributing factors. The effect has a spillover to warmer winter temperatures over a larger part of the higher northern latitudes. The phenomenon may also hasten the loss of ice sheets, particularly in the northern hemisphere.

With global warming the tropopause height is expected to increase. As the lower atmosphere warms it would expand somewhat like a hot air balloon and since the stratosphere cools at the same time, it would contract. Both factors would result in a higher tropopause height. Increases in the tropopause height are not uniform with latitudes, but have been observed amounting to some 300 m over the 1°C warming of the earth.

Increases of rainfall and water content of the atmosphere are still relatively small as might be expected from the Clausius-Clapeyron equation from a 1°C warming, but observations are consistent in showing increasing trends. Unlike the temperature record, there is more variability in the observations and the records are not as long, so clear relationships such as that for the temperature are not seen.

Review of the Main Points

1 Global warming at its current rate is not defined by whether one year is warmer than the previous one, or even that it is the warmest year on record. It is defined instead by whether one decade is warmer than the previous decade. This is because the expected annual global warming from greenhouse gases is so small that it cannot drive changes large enough to be observable from one year to the next.
2 When all the known effects, natural and man-made, and feedbacks are taken into account, models representing the climate theory we have discussed are able to reproduce both the timing and the magnitude of the observed temperature changes. The major anthropogenic driver is the increase of carbon dioxide from burning fossil fuels. There are some discrepancies such as prolonged periods of slow or little warming even when the greenhouse gases are increasing unabated. Such periods point to the variable nature of the earth system as heat is transferred between various parts of the system, including the oceans.
3 The troposphere must warm if greenhouse gases increase that will in turn cause a surface global warming. Yet, radiative theory predicts that such a circumstance will cause the stratosphere to cool! And other potential causes of global warming do not produce this result. A stratospheric cooling is observed and the magnitude is consistent with expectations lending considerable support to the validity of the overall understanding of climate change caused by human activities.

Endnotes

Endnote 11.1 Sources: The temperature data are from "Carbon Dioxide Information Analysis Center, Global and Hemispherical Temperature Anomalies," P.D. Jones et al. 2009 (DOI 10.3334/CDIAC/cli.002).

Endnote 11.2 The match between temperature change caused by man-made greenhouse gases and the observed global warming has been known for a long time. Both the timing and magnitude have matched in reproducible climate science models once the ice-core data for concentrations became available. See: J. Hansen, J.A. Lacis and M. Prather (1989), "Greenhouse Effect of Chlorofluorocarbons and Other Trace Gases," *J. Geophysical Research*, 94, pp. 16417–16421 and R.M. MacKay and M.A.K. Khalil (1991), "Theory and Development of a One Dimensional Time Dependent Radiative Convective Climate Model," *Chemosphere* 22(3–4), pp. 383–417. We can see the consistency of the observed temperature changes by looking at the greenhouse gas increases in Fig. 7.1, the radiative forcing of these gases as in Eqns. 9.10 and the warming effect represented by Eqns. 9.11.

Endnote 11.3 The first attempts to connect carbon dioxide and climate is attributed to Svante Arrhenius in 1896. The original paper is: S. Arrhenius, "On the Influence of Carbonic Acid in the Air upon the Temperature of the Ground," *Philosophical Magazine* and *the Journal of Science*, 41, pp. 237–276, 1896. At the time, the greater concern was with ice ages. A fascinating story follows to this day. Interested readers are referred to Spencer Weart's "The Discovery of Global Warming," Harvard University Press, Boston, 2008.

Endnote 11.4 Equation 11.2 is obtained by considering an energy balance at the tropopause. The energy leaving the tropopause going downward is $(1 - f)\, S + \sigma\, F_s\, T_{Strat}{}^4$. That is, the solar energy not absorbed by the ozone layer (first term) and the infrared radiation of the stratosphere downward. What's coming back must be nearly the same, but $A \times S$ is the reflected part of the solar radiation that just goes through the stratosphere without interaction, so the infrared part, that can be absorbed by carbon dioxide and ozone in the stratosphere is the Q_{IN} of Eq. 11.1b. In going from Eq. 11.1 to Eq. 11.2, two approximations are made. Since F_s and f are small, $1 - ½\, F_s \approx 1$, and $1 - f - A \approx (1 - A)$. Without these assumptions the equations are more complicated, but the result of the calculated total temperature change is smaller at ≈ 2.6°C and closer to observations. The stratospheric cooling approach here is similar to D. Hartmann's, *Global Physical Climatology*, 1994, p. 332. Figure 11.5 is adapted from this source and the values of "f" and F_s are taken from there.

Endnote 11.5 The albedo is expected to have increased somewhat in response to the aerosols and their interactions with clouds (Figure 9.4). The change of radiative forcing is $\Delta RF = -S\, \Delta A$, therefore, $\Delta A \approx 1\ (\text{w/m}^2)/342\ \text{w/m}^2 = 0.0029$ which is too small to measure. The expected temperature of the stratosphere would be: $234°C\ [(0.03\ /0.11) + 0.6971]^{1/4} = 232.2°K$ giving a cooling of ΔT_s (Albedo) = −0.1°C. It is neglected in view of the larger influences of ozone and CO_2 discussed in the text. The decrease of ozone also reduces F_s thus offsetting some of effect from less absorption of solar radiation represented by f, but this effect too is expected to be small compared with the significant increase of CO_2 during the time we are considering and the major greenhouse effect of CO_2 in the stratosphere.

12

Population, Affluence, and Global Change

We are now ready to make a crucial connection between climate change and human life. It is our last component in articulating global change science as laid out in Chapter 1. Now we will focus on the social forces and behaviors that cause climate change including the role of population and its intersection with global warming. In the subsequent chapters we will examine how this affects a societal response and what that response might accomplish.

12.1 Basic Relationships

The emissions of greenhouse gases come from the basic needs of human life – the consumption of food, energy, and material goods. We saw earlier that emissions of CO_2 are fundamentally related to burning fossil fuels for energy; some of it goes into food production. Methane and nitrous oxide from human activities are mostly from agricultural processes – cattle, rice agriculture, and fertilizers. As such, population becomes an underlying variable that determines the rates and accumulation of these gases in the atmosphere that are at the heart of climate change. The relationship between population and emissions, however, is not simple. Affluence and technology are major confluential factors. For our purposes, affluence will be defined for countries and groups of countries and measured by one or more related economic indices of wealth such as the per capita gross domestic product (GDP/c), or the per capita energy consumption.

The emissions of greenhouse gases and aerosol precursors can be written as a product of several factors. This separation allows us to examine each variable and its role in climate change.

$$E_I = K_I \times C_I \times P_I \tag{12.1a}$$

$$\Delta E(Global,t) = K(t)C(t)P(t) - K_0 C_0 P_0 \tag{12.1b}$$

In Eq. 12.1a, "I" is an index; it represents the type of population. It may be from any classification of convenience such as a specific country or a category of affluence. For the sake of discussion let's say the "I" represents a country, which relies on fossil oil for its energy. Then to calculate the emissions of CO_2 in PgC/year from this country, K_I will be the emission factor, which is the amount of CO_2 emitted per unit of energy produced in this

Global Climate Change and Human Life, First Edition. M. A. K. Khalil.

Companion Website: www.wiley.com/go/khalil/Globalclimatechange

country, and may be expressed as PgC/BTU (1 British Thermal Unit = BTU = 1055 j). C_I is the per capita consumption rate of energy in BTU/person – y. P_I is the population of the country in number of people ("persons"). In the product then, we get the emissions from the country in PgC/year. The same expression can be used to describe emissions of CH_4 and N_2O, or any other gas, instead of CO_2, with the appropriate measures and units for K and C. If for instance, the emissions are from food production, such as rice agriculture for methane or fertilizer use for nitrous oxide, then K is Tg emitted/ton food produced, C is tons of food consumed per person per year, and P is the population (persons) giving E in Tg/year.

Equation 12.1b expresses the change of global emissions at a time t in the future, due to changes in the underlying factors starting from an initial state with K, C, and $P = K_0$, C_0, and P_0, respectively. Its utility comes from two applications. First, it tells us that we have just three factors (K, C, and P) that can be used to reduce future emissions, if that is what we want. The means of controlling emissions can be to select different blends of energy sources, to reduce the effective K. Inessential uses of material goods can be regulated to reduce C, as was done for the chlorofluorocarbons. Management of population is the final factor; it has not been included in any plan to control global warming but has been most intractable when attempted for other reasons. In the end, Eq. 12.1 informs us that any plan to reduce emissions as a means of managing climate change, has to deal with all three components. A lack of commitment to any one of these would offset the gains from reducing the others. We will return to this subject in Chapter 14. Second, it can be used to evaluate future emissions of greenhouse gases, based on our expectations about how the world may change in terms of population, affluence, and consumption patterns, including efforts to manage the climate.

Although the variables in Eq. 12.1a are definitive, the transition to the global Eq. 12.1b requires further discussion, staying with the idea that "I" represents a country and we are looking at CO_2 emissions. The main use of Eq. 12.1b is to look at the effect on global emissions of greenhouse gases as these three factors change in the future, either by design or under uncontrolled circumstances. An intermediate variable can be introduced representing the total energy consumption in a country $U_I = C_I\ P_I$. The global emissions can be calculated from Eq. 12.1a by using data from each country and adding up the country emissions as $E = K_1U_1 + K_2U_2 + .. + K_N\ U_N$ for N countries in the world. A global effective K (PgC/BTU) is defined as $= E/U$; it is the global emissions (PgC/yr) divided by total energy used U (in BTU/yr $= U_1 + .. + U_N$). The global per capita consumption rate is C (BTU/yr-person) $= U$ (BTU/yr)/ P (persons) where P is the global population. Then, $C \times K = U/P \times E/U = E/P$; from which we get $E = C\ K\ P$ and a change in emissions due to these factors is the difference between an initial condition with $E_0 = C_0\ K_0\ P_0$ and the future emissions, that is, Eq. 12.1b (Endnote 12.1). It is convenient to incorporate the effects of non-fossil energy consumption in the definition of K since U can then represent the total energy consumption in the world, and U_I for each country instead of using a U_I that is just from fossil fuels. Therefore, $K = f \times K(CO_2)$ where f is the fraction of fossil fuel energy, and $K(CO_2)$ is the emission factor for it. To use Eq. 12.1b, a benchmark E_0 is calculated first, usually representing a base period or the present time. For the data from period around the year 2000, $K_0 \approx 1.5 \times 10^{-17}$ PgC/BTU, $C_0 \approx 6.9 \times 10^7$ BTU/y-person, and $P_0 \approx 6.2 \times 10^9$ people and $E_0 = C_0\ K_0\ P_0 \approx 6.4$ PgC/year. It may be noteworthy that this literally amounts to a "ton of carbon emissions per year per person."

Example Suppose by 2050 we want to reduce the emissions of CO_2 by half, relative to the base in 2000, by shifting to non-fossil energy sources. The population rises to 13 billion, the per capita consumption rate rises to double of present to offset poverty. (a) By how much do you have to reduce the emission factor? (b) What fraction of your energy can still come from fossil sources assuming that all of it fossil based to start with?

Solution: (a) Initially $E_0 = 6.4$ PgC/y. In 2050, according to Eq. 12.1b, $K = E/(C\,P)$ and we are told that the E must $= 3.2$ PgC/y, $C = 13.8 \times 10^7$ BTU/y-person, and $P = 13 \times 10^9$ persons. Therefore, $K = 1.8 \times 10^{-18}$ PgC/BTU, and before it was $K_0 = 1.5 \times 10^{-17}$ PgC/BTU (based on the benchmarks established above). We must reduce the emission factor by $\Delta K = K - K_0 = (1.8 - 15) \times 10^{-18} = -13 \times 10^{-18}$ PgC/BTU. (b) The allowed fraction of fossil fuel energy is: $K = f \times K_0$ or $f = 1.8/15 = 0.12$, or 12%. The shift of emission factor comes from switching to non-fossil sources of energy.

12.2 Societal Factors in Climate Change

Societal factors are the causes behind the causes of climate change. Any means to control the climate will have to accommodate them. Affluence is not precisely defined, but it is useful nonetheless because it incorporates many, or perhaps all aspects of a prosperous society. We will consider it to mean the abundance of material goods and necessities of life accessible to most people in the society. In any society, affluence can be related to the monetary wealth of the majority of the people, but the availability of material goods is also required. It can be represented by the per capita gross domestic product (GDP/c). Basic human desires lead to more children, lower infant mortality, and longer lives. In Figure 12.1a,b, we see that

Figure 12.1 Markers of affluence. (a) The benefits of affluence are illustrated in the increase life expectancy and (b) reduced infant mortality. (c) Energy consumption is closely related to affluence as measured by the per capita gross domestic product (from Our World in Data).

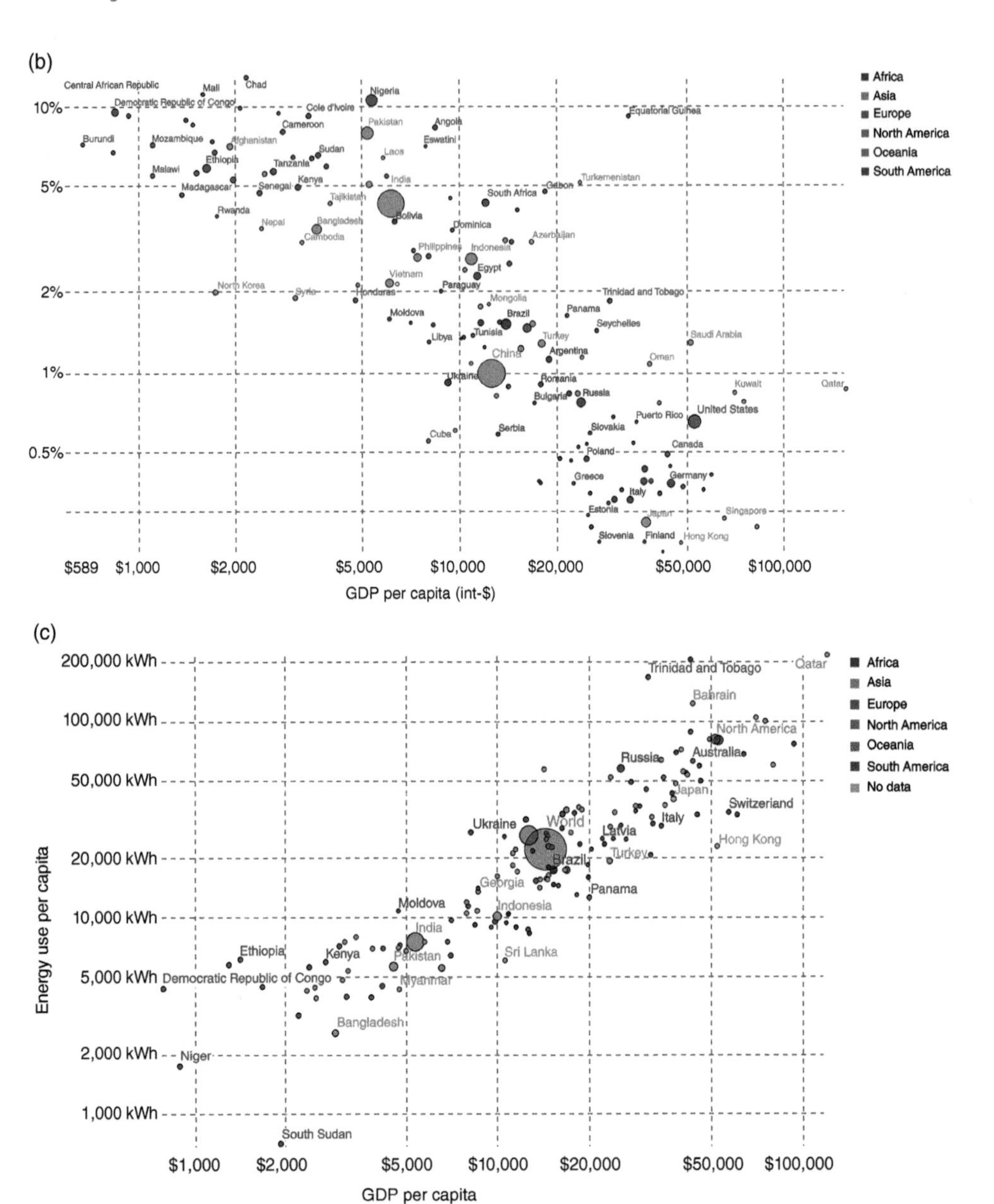

Figure 12.1 (Continued)

the GDP/c (representing affluence) is correlated with these desires and aspirations. These results are associated with greater consumption rates of food and material goods which leads to a per capita energy consumption that is proportional to GDP/c (Figure 12.1c). Societies strive to, and move toward, more affluence and higher populations, leading to increasing consumption rates, and global greenhouse gases in the process. Increases of affluence and population have brought us to the present state of the climate. If these cannot be controlled, because they penetrate deep into the human psyche, then managing climate seems to be a daunting, if not an impossible task. We will return to this matter later in Chapter 15. The end result of affluence on emissions of CO_2 are illustrated in Figure 12.2.

The salient point from this discussion and Figures 12.1 and 12.2 is that affluence plays a major role in the emissions of greenhouse gases and has done so for a long time, and will

Who has contributed most to global CO_2 emissions?

Cumulative carbon dioxide (CO_2) emissions over the period from 1751 to 2017. Figures are based on production-based emissions which measure CO_2 produced domestically from fossil fuel combustion and cement, and do not correct for emissions embedded in trade (i.e. consumption-based). Emissions from international travel are not included.

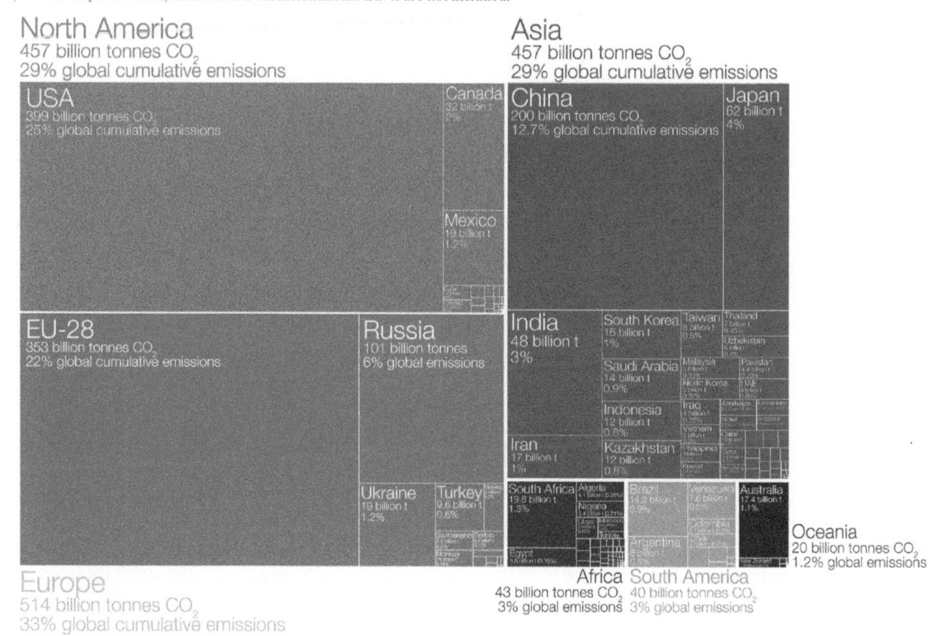

Figures for the 28 countries in the European Union have been grouped as the 'EU-28' since international targets and negotiations are typically set as a collaborative target between EU countries. Values may not sum to 100% due to rounding.
Data source: Calculated by Our World in Data based on data from the Global Carbon Project (GCP) and Carbon Dioxide Analysis Center (CDIAC).
This is a visualization from OurWorldinData.org, where you find data and research on how the world is changing.

Figure 12.2 Cumulative CO_2 emissions from various countries and regions. It shows that emissions from affluent countries are much greater than poorer countries even though the poorer countries are more populated.

likely do so into the distant future. A recent case study is the change in the socio-economic conditions of China. Starting around 1980, the relatively inexpensive Chinese labor supplied an enormous amount of consumer goods to the world. It built wealth rapidly as seen in their GDP/c, which went from $2,000 to $12,000 between 1980 and 2015; a 500% increase. The population changed by only 40% during this time, but the CO_2 per capita emissions went from 0.4 tons C/yr to 1.9 tons C /yr comparable to the GDP/c increases. China became the largest contributor to present-day annual emissions for the first time in history. For decades, this championship was held by the United States.

The example of China emphasizes that major increases of CO_2 and other greenhouse gases can, and do occur without a commensurate increase of population because they are also driven by affluence. Indeed, the same causes have played out with North American and European emissions during the last 100–200 years. The populations of these regions increased much less than the world average, but they account for most of the current CO_2 in the atmosphere. Affluence, however, is limited, as are consumption rates. After a certain point, they do not increase very much. From this we can conclude that the main increases of greenhouse gases occur as the poorest sectors of the world rise toward greater affluence and this change is essentially disconnected from population increases. At the other end, emissions from the richest countries change at rates closer to their population increases, which are generally slower than the poor countries. For the rich countries, the emission factors can be lower since they can afford more expensive sources of non-fossil energy and higher efficiency power generation; but these are easily offset by higher per capita consumption.

Fig 12.2 shows that over the twentieth century and now, the increase of greenhouse gases is mostly a consequence of rising affluence within a small segment of the population represented by the developed countries. Increasing affluence has been supported by the growth of the remaining segments of the global population that have provided inexpensive "goods and services" which also contribute to global change. While it is apparent that a very large but poor population can cause as much global warming as a small affluent one, questions arise whether it is possible to have a large population without the technological advances that drive the small affluent populations at the core, with high per capita greenhouse emissions. With its various connections, population change has a complex role in the increase of the greenhouse gases and the consequent increase of temperature. However, the simultaneous increase of temperatures and population will interact to produce synergistic impacts on human life which we will examine next.

12.3 Population Growth and Resources

12.3.1 Population Dynamics

The growth of population is one of the variables that affects the modern climate and its impacts on human life (Eq. 12.1). Let's examine the aspects that can illuminate the potential for the future population. The rate at which the population will increase can be written as: dP/dt = Birth Rate – Death Rate. It seems intuitive that both depend on the existing population. The Birth Rate is proportional to the fraction of women within the child-bearing ages (α_0); keeping our attention on human population for now, α_1 is the number of children born to each mother during her lifetime (fertility) divided by the average number of child-bearing years. Then, the Birth Rate = $\alpha_0 \times \alpha_1 \times P$. α_0 is dimensionless and α_1 has units of 1/time because it is persons (children born) per person (mothers)/(child-bearing years). The Death Rate = $(1/\tau)\,P$ where τ is the average life span or life expectancy (readers may note the similarity to the concept of lifetime for atmospheric constituents introduced earlier). We get:

$$\frac{dP}{dt} = \left(\alpha_0\alpha_1 - \frac{1}{\tau}\right)P \qquad (12.3)$$

If the α's and τ are constant, we get the equation dP/dt = Constant × P. This equation has the solution we have seen before, that is: $P = P_0\,e^{\lambda t}$ where $\lambda = \alpha_0\,\alpha_1 - 1/\tau$. The population increases exponentially if $\lambda > 0$ and falls toward extinction if $\lambda < 0$. For increasing populations as with us, the doubling time can be found to be: $T_{\text{Double}} = \ln 2/\lambda$ (Endnote 12.2). We see that the shorter the average life span is the longer it will take to double the population, and the lower the fertility is, or the fewer women there are in the population, the longer it will take. These results make sense and may be applied to specific populations such as countries or the whole world.

Example World data in 2020 are that the fertility rate is 2.4 children/woman, child-bearing age range = 30 years, fraction of women of child bearing ages = 0.3, and the average life expectancy = 75 years. (a) How fast should the population be increasing? Compare it with the measured rate of 1.1% increase per year. (b) If this rate persists, when will the population double to 15.6 billion people?

Solution: (a) $\alpha_1 = 2.4/30$ years $= 0.08/\text{yr}$, $\lambda = (\alpha_0 \alpha_1 - 1/\tau) = 0.3 \times (0.08/\text{yr}) - (1/75\,\text{yrs}) = 0.011/\text{year}$. That's 1.1% per year, which is the growth observed in 2020. (b) $T_{\text{Double}} = \ln 2/\lambda = 0.69/(0.011/\text{year}) \approx 60$ years. That's within the time frame of climate management.

Under the assumption of constant but unequal birth and death rates this theory does not predict any stable point for the population. For a population to exist at all, the birth rate has to be higher than the death rate to start with. With constant rates the population will rise indefinitely. If at some point the rates change and the death rate exceeds the birth rate, and this condition persists, then the population will fall to extinction. For a population to be stable, the birth and death rates have to come into balance and stay that way for the time scales of our interest. Primary mechanisms that cause stability can be thought of as negative feedbacks. That is, the increase of population itself leads to a reduction in the birth rates or an increase in the death rates. It is observed that the fertility rate declines with increasing population. The causes, although of considerable scientific interest, will have to be set aside here except to note that these may have biological as well as cognitive origins. Regardless, the outcome can be expressed mathematically as: $\alpha_1 = \beta_0 - \beta_1 P$. The other aspect is that as the population increases, the life span may decrease due to causes such as limitation of food and health resources, and increase in infant mortality, some of which may be exacerbated by climate change. These factors would also be reflected collectively in a loss of affluence. We can write this as: $1/\tau = 1/\tau_0 + \gamma P$. In these equations, τ_0 and β_0 are the portions of the life expectancy and fertility that may still change in time, but from causes that are independent of the population. γ and β_1 are taken to be constants that describe the strengths of the population-related feedbacks. Substituting these population-dependent τ and α_1 into our original Eq. 12.3 gives:

$$dP/dt = AP\left[1-(D/A)P\right] \tag{12.4a}$$

$$A = \alpha_0 \beta_0 - 1/\tau_0 \tag{12.4b}$$

$$D = \gamma + \alpha_0 \beta_1 \tag{12.4c}$$

Equation 12.4 is the logistic equation in which the A expresses the birth and death processes and is essentially the same as the right-hand-side coefficient in the exponential growth equation (Eq. 12.3). The D represents the processes that limit population from increasing indefinitely. Equation 12.4 has a more complex solution than our simpler exponential equation. Assuming that the α's β's, γ, and τ_0 are constant leads to a sigmoid-shaped solution as shown in Figure 12.3 and a maximum predicted population (Endnote 12.3). When the population is low, the second term $(-D \times P^2)$, because it goes as the square of the population, is much smaller than the first term $(A \times P)$ so the population increases exponentially as in Eq. 12.3 ($dP/dt \approx A\,P$). Eventually, however, as the population grows, the second term starts to exert an increasing influence and we see that the population will stop increasing when $P = P_{\text{Max}} = A/D$ because then $dP/dt = 0$. The maximum population is:

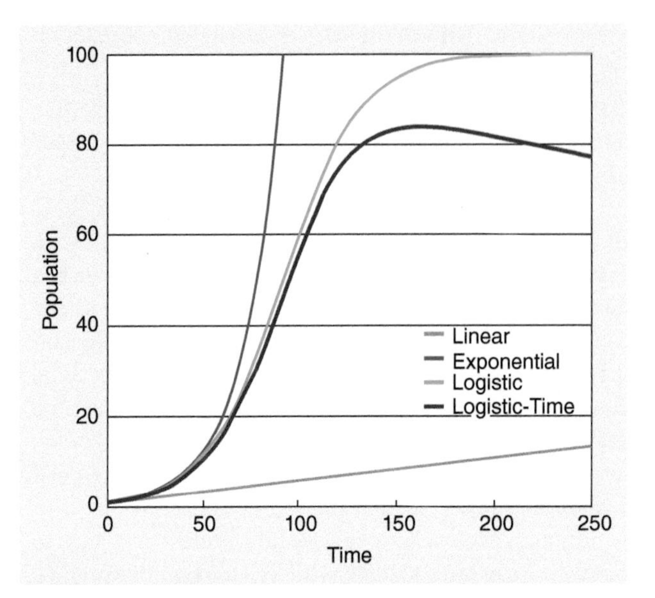

Figure 12.3 Population Growth Functions. Population may follow the exponential growth for some periods of time, but it is not sustainable. The logistic growth can occur when population-dependent phenomena limit the maximum possible population. These can arise from limitations of resources or from population management. In "logistic-time" growth curve we see the effect of reduction in time of the base in fertility or life span that starts around time = 120 units. This is an effect of changes in population characteristics that come from external influences not related to the size of the population itself. A linear growth is shown for comparison. It starts at the same point and grows at a constant rate. Although it can be slow, it also has no upper limit.

$$P_{Max} = \frac{\alpha_0 \beta_0 - 1/\tau_0}{\gamma + \alpha_0 \beta_1} \tag{12.5}$$

This result is valid when $(\alpha_0\,\beta_0 - 1/\tau_0) > 0$; otherwise, the population goes to extinction in Eq. 12.4 because the birth rate is lower than the death rate. Equation 12.5 is an important relationship because it shows the interaction of the only two variables that can lead to a stable population whether naturally or by management. These variables are fertility and life expectancy. That leaves fertility as the only marginally realistic variable we have at our disposal since there is a human desire for a longer life that takes precedence.

Example Assume that $\alpha_0 = 0.3$, $\beta_0 = 0.08$ /yr, and $\tau_0 = 75$ years, as in the previous example. Suppose $\gamma = 0.0002$ per billion people per year and $\beta_1 = 0.0025$ per billion people per year. What will be the stable maximum population? What will be the new average life span and fertility at that time?

Solutions: Using Eq. 12.5, $\alpha_0\,\beta_0 - 1/\tau_0 = 0.0107$/yr. $\gamma + \alpha_0\,\beta_1 = 0.001$/billion people – year. Therefore, $P_{Max} = 10.7$ billion. At that time the life span will be $1/\tau = 1/$ (75 yrs) + 0.0002 per billion per year × 10.7 billion makes $\tau \approx 65$ years and the fertility will be 1.6 children per mother, while it was 2.4 children per mother previously.

12.3.2 Logistic Theory and Carrying Capacity

In the logistic theory, when the growth of population reduces the growth rate due to limitation of resources, the maximum attainable population in Eq. 12.5 is called the *carrying capacity* of the environment. The carrying capacity is not fixed a priori, but rather by the prevailing circumstances of a time or a period, perhaps a century long for human populations. As the circumstances change, the carrying capacity may become larger or smaller. The concept of a maximum sustainable population can be extended in what I call the "principle of maximum utility." It can explain aspects of the natural environment and its changes over time. The idea of the principle is that as long as there are no constraints, there is no reason to expect a population to stop growing. Since all population growth must have limits, populations tend toward the maximum consistent with prevailing environmental conditions, P_{Max} in Eq. 12.5.

The population theory we have discussed can be modified to apply to almost all living things, including bacteria and plants. While the details of birth and death processes may be different, the broad conclusions will be the same. In this concept then, the total plant biomass, which controls atmospheric CO_2 concentrations, would adjust to the prevailing conditions even as we alter them by burning fossil fuels or by deforestation. These processes will cause the biomass to take up more carbon from the atmosphere, because more is available, and will re-grow wherever the forest is allowed to return. If the availability of carbon in the atmosphere is the limiting factor for the maximum growth of plants, adding more extends the carrying capacity. In the new adjustment a maximum biomass will result consistent with the altered carbon balance. It will, in turn, reduce the climate impact of the added carbon constituting a negative biological feedback. Similarly, bacteria that produce and consume greenhouse gases, such as methane and nitrous oxide, will adjust to the changing climate conditions and will therefore provide a feedback that either reduces or increases the concentrations of these gases relative to the pre-industrial, undisturbed conditions depending on how the climate change affects their carrying capacities.

Further, it is instructive to connect the carrying capacity with the temperature. Earlier we discussed a means of relating the population with the temperature of the environment in which a species lives (Eq. 10.6). In the context of the population theory we have discussed here, the connection arises when the temperature influences either the birth or death rates as reflected in the variable D of Eq. 12.4. Recall that the variable D represents impedances to an unfettered growth of population. Therefore to represent the temperature effect, we can modify it as follows:

$$D = \frac{D_0}{\left[1 - \left(\delta T / \Delta\right)^2\right]} \tag{12.6}$$

Here $\delta T = T - T_0$ and Δ are as in Eq. 10.6. The new D is at a minimum when the temperature is at its optimum value, representing the lowest death rate. As the temperature changes in either direction from the optimum, the death rate increases. Since P(steady state) $= A/D$, it is $= P_{Max}(1 - \delta T^2/\Delta^2)$ (Eq. 12.5), $P_{Max} = A/D_0$. We see that in steady state the resulting population equation in the logistic theory will be the same as in Eq. 10.6. Further, we see that the carrying capacity is now expressed with a dependence on the temperature. It is maximum at the optimum temperature and can go to zero at temperatures that are lower by Δ (°C) or higher by the same amount, relative to the optimum. It should be noted that the effect of temperature may be direct or indirect. For instance, if a food crop depends on temperature, and people depend on that food crop, the effect of climate change on human populations can still be represented by Eq. 10.6. It expresses the general idea that for any species, or perhaps for life on earth as a whole, there is a range of temperatures when it can exist and an optimum at which it will thrive. It gives the earth's temperature an important place in the story of living things. Climate change, driven by global warming, may itself be a limiting factor for human population as one of the γ's of Eqs. 12.4 and 12.5.

Example Suppose a variety of bamboo follows the growth curve according to the parabolic function of temperature. Suppose further that giant pandas live only on this bamboo. Does the population of pandas follow a parabolic function as in Eq. 10.6, even if the environmental temperature does not directly affect them, that is $P(\text{pandas}) = P_{Max}\,[1 - (\delta T/\Delta)^2]$?

Solution: The answer is "yes." The amount of bamboo available is F (kg) $= F_0$ (kg) $[1 - (\delta T/\Delta)^2]$. The simplest connection we can make is that the death rate of pandas from population-related effects is inversely related to the food supply. $D = \alpha/F$, α is a proportionality constant (kg/y). The less food there is the higher the death rate. The population of pandas in equilibrium will be (Eq. 12.5): $P = A/D = AF/\alpha = (AF_0/\alpha)\,[1 - (\delta T/\Delta)^2]$ and $P_{Max} = AF_0/\alpha$, as in Eq. 10.6. The maximum panda population occurs when the food supply is at its maximum. The panda population goes to extinction if the bamboo abundance F goes below α/A.

12.3.3 Neo-Malthusian Ideas and Sustainability

In 1798 the Reverend Thomas R. Malthus first articulated the idea that the growth of human population follows an exponential increase and recognized that it was not sustainable given the available resources, including food production, which he described as increasing linearly with time. Under this assumption, it is straightforward to calculate that a time will come, rather too soon, when the population will outstrip the food needed, leading to a catastrophe. The Malthusian theory was the beginning of an acceptance that human life and natural resources were interconnected in such a way that unfettered population growth would lead to significant social problems. Although the assumption of a linear increase of food production is overly pessimistic and not founded in basic science, the idea that increasing population will inevitably lead to environmental degradation, shortage of resources, and social issues has emerged from the original theory to become established in the environmental sciences (Endnote 12.4). The concept of sustainability has become increasingly popular; however, it means quite different things to different people. In general terms, it can be seen as a desire to avoid the Malthusian catastrophe by exercising controls over the environmental system and adopting social practices that would ensure a stable world.

12.4 Vulnerability Theory

A theory of vulnerability delineates how a growing population interacts with climate change. The interaction comes from two causes – expansion of warm climate zones and the impact of extreme events. We will look at the latter and examine how natural extreme events and how changes in the frequency or intensity of those events contributes to potential damages and fatalities in a more populated world. In this idea, the increasing population puts more people in harm's way of all extreme events whether occurring naturally or altered by global warming. It applies to flooding from rain events, heat waves, storms, or another climate event. The vulnerability to climate change is illustrated with an allegorical model involving storms.

Consider a place where cyclones hit every year during the monsoon season. These storms bring high winds which push waves of sea water inland. Together with heavy rains the cyclones cause damage by flooding and high winds in the coastal zone. There are N storms per year with varying degrees of impacts that can be labeled as $i = 1,2,..,N$. When one of these storms strikes, let's label it $i = 1$, there is a probability of death for each person affected by it: p_1 = probability/person. The storm affects an area A_1 (km^2/storm). The population density in this area is P_D (people/km^2), so the number of vulnerable people or people affected is $A_1 \times P_D$ and therefore the number of fatalities per year for this storm will be:

$$F_1 = p_1 A_1 P_D \tag{12.7}$$

The simplest assumption we can make is that p and A are proportional to the intensity of the storm. This would say that as the intensity (I) increases, the probability of death to exposed people increases proportionally, that is, $p = \alpha I$ and the affected area increases as $A = \beta I$. The α and β are constants specific to the location. The damage from a storm is approximately proportional to the square of the wind speed. This is because the pressure of the wind on structures such as buildings or houses is generally proportional to the square of the wind from basic physics. Moreover, as we know from experience, the height of water waves also increases with wind. The height of the bigger waves under storm conditions tends to increase close to the square of the wind speed. The intensity of a storm as it affects human vulnerability can therefore be represented as:

$$I = \gamma U^2 \tag{12.8}$$

γ is another constant. For notational convenience we set $\alpha_0 = \alpha\beta\gamma$ and Eq. 12.7 becomes:

$$F_1 = \alpha_0 P_D U_1^4 \tag{12.9}$$

Since there are many storms each year, we can write the total annual fatalities as $F = F_1 + F_2 + .. + F_N$ where the subscripts differentiate the fatalities from one storm to another as would be the case because Eq. 12.9 was written for one storm ($i = 1$). Therefore:

$$F = \alpha_0 P_D (U^4)_{Avg} N \tag{12.10}$$

Here $(U^4)_{Avg} = (1/N) \times (U_1^4 + U_2^4 + ... + U_N^4)$. Let's look at how these ideas can be used to understand the confluence of population and extreme events.

In our region of interest, in the base year 1920, P_D = 100,000/km^2, N = 10 storms/year, U = 170 km/hr for all events, and $\alpha_0 = 1.2 \times 10^{-14}$ (fatalities – hr^4 /person -km^2 -storm) which gives F = 10 fatalities/year from Eq. 12.10. Now, 100 years later, the population has increased by a factor of 4.3, following the same rate as the increase of global population during this time (1.8 billion to 7.8 billion). Assume that the density P_D has increased by the same amount as the population and that the number of storms or the wind speeds has

Table 12.1 Outcomes of the Interaction between Population Increase and Climate Impacts.

Action	ΔN	ΔI	ΔP	F	ΔF
	# Storms	Intensity	Population	Fatalities	Change
Base	0	0	0	10	0
More Storms	10%	0	0	11	1
More Intense	0	10%	0	15	5
Both More	10%	10%	0	16	6
Population Increase	0	0	430%	43	33
All Causes	10%	10%	430%	63	53

increased by 10% due to the 1°C change of temperature. We can find the effect of climate change by calculating the increase in fatalities from an increase of either the number of storms N, or the increase of intensity I, or both while holding the population the same. And we can calculate the effect of population increase by holding the number of storms and their intensities the same. The results of the several calculations are summarized in Table 12.1.

We can draw several lessons from these results. The most important one is that the population has increased a lot during the last century (430%) while the temperature change is relatively small (< 1%). Even with no climate change, the natural extreme events will have a more damaging effect on people than a century ago; an increase of 33 fatalities or 330%. In our example we took the increase of the population density as that of the global population increase. In the real world, the shifts of population can be much more dramatic. Areas where extreme events occur may be desirable in other ways drawing more people than the average population increase. Second, note that climate change has a relatively small effect by itself and that more intense storms have a greater effect than just the increase in number. In other words, the number of storms doesn't have to increase at all to cause significant additional damage if some storms just become more intense due to climate change. This is because of the non-linear connections between intensity and damage. Finally, it is noteworthy that the confluence of climate change and population growth creates a synergistic effect by which the damage is amplified. In the example the synergistic effect is 14 additional fatalities (53 – 33 – 6), which is about 25% of the total impact. These concepts can be extended to many types of extreme climate events (Endnote 12.5).

A real-world example from which this model is drawn is the Sundarbans of Bengal, which is the delta at the confluence of three huge rivers of the Indian sub-continent. It is one of the most remarkable and unique ecological regions of the world. It is famous for its mangrove forests and as the habitat of the Royal Bengal Tiger. Despite the tigers, according to recent estimates, between 1950 and 2010 the population increased from 1.15 to 4.44 million or about a factor of 4, while the world population increased by less than a factor of 3 during the same time. The area is attractive because of its extreme agricultural fertility, forest products, and supply of fishes. It is also famous for facing powerful cyclones every year during the monsoon season which have caused tens of thousands of deaths in recent years which were unheard of in the earlier years of the twentieth century. The situation demonstrates the effects we have discussed here.

Review of the Main Points

1. We make the transition to the last piece of the climate change story – the direct connection with human life. A basic relationship is introduced to describe the emission rates of environmentally important constituents as the product of affluence, population, and the emission factors. The purpose is to separate these influences so that their impacts can be examined further and to recognize that several factors together contribute to man-made climate change. It will require managing all of them to achieve a stable climate; otherwise, the uncontrolled causes will overwhelm our attempts.
2. It is argued that climate change is a consequence of continuously rising affluence and, so far, to a lower extent by population increases. These aspects are propelled by forces deep in the human psyche that may be extremely difficult to manage or control.
3. A basic theory of population growth was presented. It is one of the three ingredients in the emissions formula for greenhouse gases and climate change. Population is a key factor in many environmental issues facing people all over the world, and will remain so.
4. Building on the growth of population, a vulnerability theory was formulated. It shows that the growth of population, which has been some 400% over the last century, will be the main cause of fatalities from extreme climate events even without climate change at this time. However, climate change would not only add more fatalities from extreme events, but there is a synergistic increase caused by the intersection of climate change and population growth.

Exercises

1. Justify the statement in Endnote 12.1 about the connection between changes in the emissions of a gas and the variables that affect it: $\% \Delta E/\Delta t = \% \Delta K/\Delta t + \% \Delta C/\Delta t + \% \Delta P/\Delta t$ for instantaneous changes in the variables. Suppose that at this time there is a move to reduce the emission factors by energy efficiency and switching to non-fossil sources that amounts to 5% per year. At the same time the consumption rate, or affluence, is increasing at a rate of 1.5% per year and the population is rising at 1% per year. What is the percentage rate of change of the emissions of CO_2?
2. Take the population variables to be (as in the example illustrating Eq. 12.5): $\alpha_0 = 0.3$, $\beta_0 = 0.08$ /yr, and $\tau_0 = 75$ years, as in the previous example. Suppose $\gamma = 0.0002$ per billion people per year and $\beta_1 = 0.0025$ per billion people per year. Suppose we want to hold the maximum population at 10.7 billion but at the same time we want the life span to remain at 75 years.
   (a) By how much do you need to reduce the fertility?
   (b) Suppose the life span rises to 100 years. What will be the maximum population based on the fertility rate in part (a)?
3. In a region storms cause about 10 fatalities/year. Suppose the average wind speeds during storms increase by 30% and all else is the same. How many fatalities will there be per year? How much will be the increase in percentage?
4. The emissions of CO_2 from India are 0.5 tons C/y – person in 2020 (1.3 billion population) and from the United States emissions are 4.2 tons C/y-person (331 million population).

(a) If the Indian consumption rose to the US level and the rest of the world remained the same, by how much would the global CO_2 (as carbon) emissions increase relative to today (~10 PgC/y)? Give your answer in PgC/y increase and percentage more than present worldwide emissions.

(b) Calculate the increases in PgC/y and percentage of present, if the whole world's per capita consumption rose to US levels (take the present world population as 7.8 billion).

Endnotes

Endnote 12.1 Equation 12.1b has a useful implication – the percentage rate of increase of emissions at any time is sum of the percentage increases of the population, consumption rates, and emission factors that are taking place at that time, or $\% dE/dt = \% dC/dt + \% dP/dt + \% dK/dt$. This result follows readily from Eq. 12.1b by using small incremental changes in the variables over small increments of time around the present.

Endnote 12.2 Doubling time for an exponential growth is a commonly used concept because it remains the same as long as the process continues. It is fundamentally related to the concept that the rate of change of the variable is proportional to the variable itself ($dX/dt = \lambda X$). Because of this, the rate of increase constantly goes up resulting in explosive growth. The solution of the rate equation is: $X = X_0 e^{\lambda t}$ as discussed several times. Readers can prove to themselves that the value of X will double over every period $T_{\text{Double}} = \ln 2/\lambda$. Some bacteria have a λ around 0.7/hr, which means their populations can double in one hour. If you start with just one bacterium, you will have two in one hour, four in two hours, and ... yada yada yada ... 240 trillion in 2 days! Most natural phenomena that follow such a pattern do so over a limited time. For bacteria populations the limitations of food or space may curtail growth making it more logistic.

Endnote 12.3 Real populations may be described by Eq. 12.3 or 12.4, however the coefficients that define birth and death processes (β_0, β_1,τ_0, and γ) may be functions of time determined by external influences unrelated to the population. For instance, a deadly disease may reduce τ_0, the average life span over a period of time. In such cases the equations may not be solvable in simple algebraic forms such as the exponential $P = P_0 e^{\lambda t}$ for Eq. 12.3. For the logistic Eq. 12.4, if the A and B are constant, which in applications would come from the constancy of β_0, β_1, τ_0, and γ, then the solution is: $P = P_0 e^{At}/\{(B/A)P_0 e^{At} + [1 - (B/A) P_0]\}$. For human populations the equations can represent the growth starting with population P_0 and calculating the future numbers for up to several decades later. In this way they can fit the time scales we have adopted.

Endnote 12.4 Further reading: Neo-Malthusian ideas about the nature of population growth and its environmental consequences can be found in: Paul R. Ehrlich, *The Population Bomb*, Ballantine Books, 1968; Paul R. Ehrlich et al., *Ecoscience: Population, Resources and the Environment*, W.H. Freeman & Co., 1977; D.H. Meadows et al. *The Limits to Growth*, Universe Books, 1972; *Limits To Growth: The 30-Year Update*, Chelsea Green Publishing, 2004.

Endnote 12.5 The concept of vulnerability discussed in the text applies to many extreme climate events such as storms, flooding, and heat waves, starting with Eq. 12.7, which is stated for the i-th event. There are N in the year ($I = 1,..,N$): $F_i = p_i A_i$ P_D = (probability of death per person – i-th event of the year) × (area affected by this event) × (persons per area). Then, the total fatalities during the year are: $F = P_D N (p A)_{Avg}$ where $(p A)_{Avg} = (A_1 p_1 + .. + A_N p_N)/N$ for the N events of the year. Simplifications can be made for special cases. If p's are the same for all events, $F = P_D N p A_{Avg}$; if the area affected is the same for all events, $F = P_D N p_{Avg}$ and if both A's and p's are the same for all events, then $F = P_D N p$. In these formulas it is assumed that the average values such as $(pA)_{Avg}$ are estimates of underlying environmental characteristics. The total fatalities are seen to be proportional to the number of events and the average probability or the probability modified by areal expansion. Both the number of events and the probability of death are characteristics that can change with global warming and the nature of the human or animal populations in harm's way. For instance, the probability of death from an extreme event, such as a cyclone, may be significantly reduced by modern technology and buildings compared with a century ago. The idea can be further extended to any natural disaster including secondary effects of climate change such as wild fires. An increase in wild fires may be expected due to significantly more hot days in a warmer world that are discussed in Chapter 13.

13

Impacts of Climate Change on Human Life

13.1 Impacts Classified

Since climate is an integral part of how plants and animals live and die on earth, any change will affect global habitability in myriad ways. Climate changes cascade through multitudes of phenomena to eventually touch human life. Many climate-change impacts, such as the spread of a disease, like malaria, have counteracting influences making them increase in one place and decrease in another. This makes a global average calculation unreliable. It may be a net zero but could have profound effects on human life nonetheless.

As stated before in Figure 2.5, global warming not only includes changes in the averages of the defining parameters but may also alter the variabilities. The well-documented slow increase of the earth's average temperature does not directly affect human life; it does so through two broad mechanisms: increases in the natural extreme events and through large-scale environmental change that includes shifts of climate zones. From the latter, impacts on human life arise when climate change alters the physical environment, or disturbs the ecology on which we have a dependency. The compound effects represent a wide range of regional-scale phenomena that are easy to articulate, but extremely difficult to quantify. These effects are hard for people to internalize because they are outside normal human experiences, unlike hot weather or high winds. A widely known example of a physical environmental change from global warming is sea-level rise. An expected and possibly observed consequence of ecological change is a northward shift in the habitats of pests and disease-causing agents. As you would expect, the direct effects are the most unequivocal and the ecological effects and their consequences are the most uncertain. These concepts are summarized in Figure 13.1. In the end, the complex interactions between climate change and human life can be channeled into a few areas shown in Figure 13.1 – habitability, health, and socioeconomic (including the availability of food and energy). The direct effects, as defined in Figure 13.1, influence the area of "health" because they occur with extreme events and result in a certain number of fatalities. But after the events, there can be residual effects on economic conditions from the losses.

We can alternatively classify the impacts as beneficial or harmful to segments of society as shown in Figure 13.2. The main point of this classification is to recognize that no climate perturbation will result in a uniform effect on everyone. *The impacts of climate change, as they influence human life, are fundamentally regional in extent in which some people will benefit and others will not.*

Global Climate Change and Human Life, First Edition. M. A. K. Khalil.

Companion Website: www.wiley.com/go/khalil/Globalclimatechange

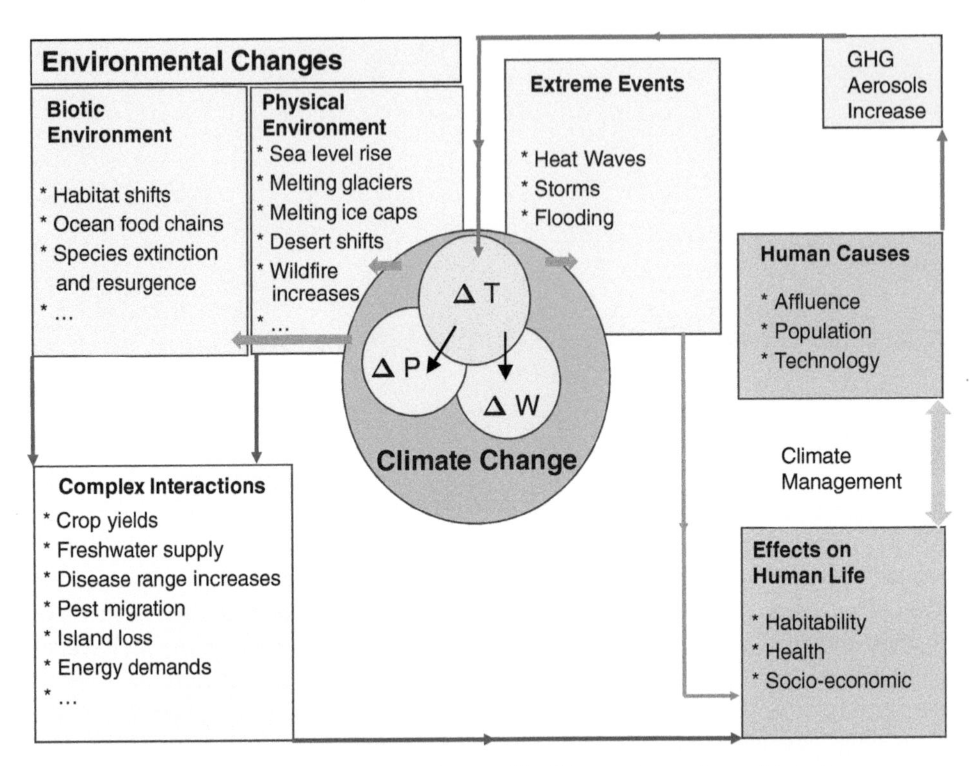

Figure 13.1 Impacts of climate change on human life. Climate change is manifested in temperature, precipitation, and wind changes shown as Δs *T*, *P*, and *W*, which we chose at the start to represent the climate. Temperature changes are at the core since they drive changes of precipitation and winds. The direct effects of climate change are manifested in extreme events that affect human life, such as heat waves, storms, and floods. Simultaneously, climate change causes environmental and ecological changes over large parts of the world which cascade through complex interactions before they touch human life.

13.2 Health

Extreme events occur naturally for all climate variables and are a major fraction of natural disasters that have plagued human societies for millennia. These events are defined by the variance from the average values. As such, the idea doesn't apply to the global average of any variable and of necessity describes events over some smaller region of the earth.

13.2.1 Heat Waves

The frequency distributions of temperature for a season or the year at a location can be determined as discussed in Endnote 13.1. If we represent our data in this way by frequencies of occurrence, over the entire range of temperatures that are encountered, a graph such as the one in Figure 13.3 is obtained. It is from data taken at Yanbu – a hot desert environment. Similar graphs can be constructed for "heat waves" which are defined as sustained hot temperatures over a number of days. Occurrences of hotter and cooler temperatures spread out around the average. The further they spread, the more variable the temperature is said to be. An extreme event may be defined, for instance, as the hottest 1% of the summer daily

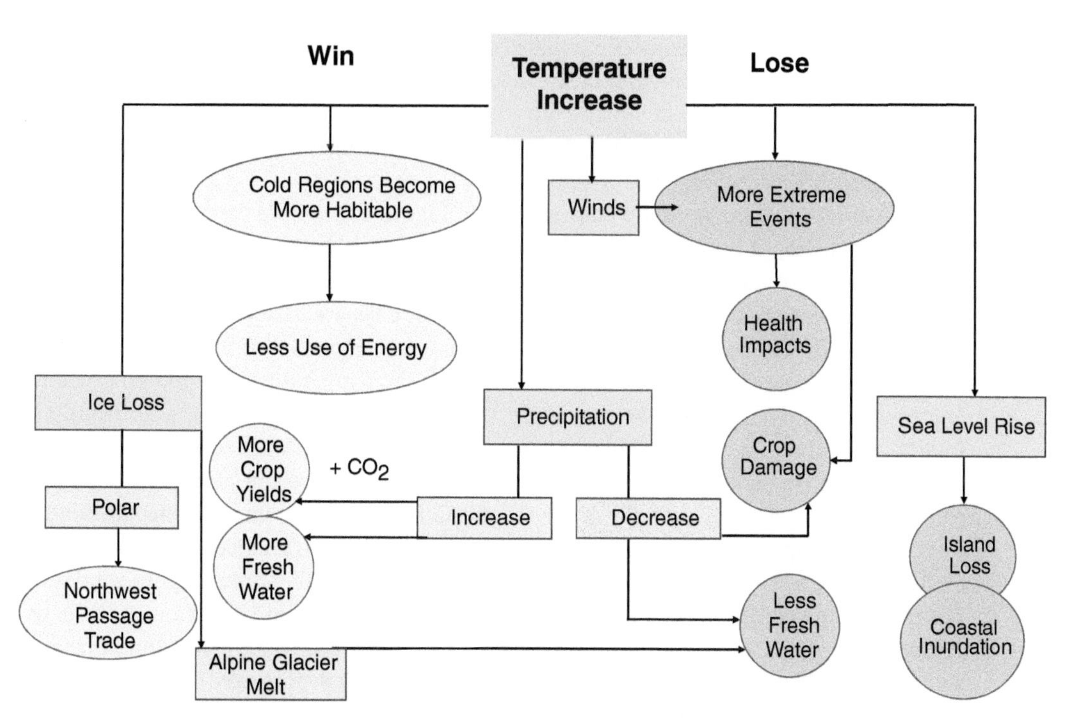

Figure 13.2 Win-lose diagram of climate impacts. The diagram illustrates that climate change impacts will not be uniform or global, but will affect regions, countries, and peoples differently.

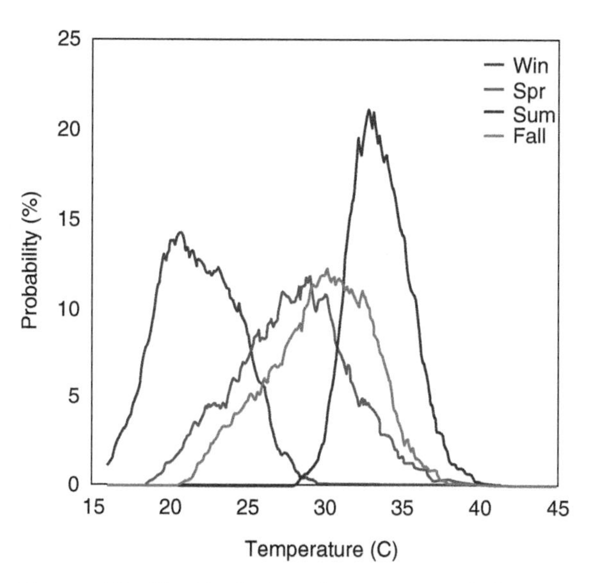

Figure 13.3 The distribution of temperatures during seasons in a desert environment. The pattern of more sharply peaked summer distributions is seen at many diverse locations. It determines how global warming may affect human life.

temperatures in the environment undisturbed by global warming, or perhaps more incisively, by a threshold over which the temperature may be harmful to crops, or human health. How the number of extreme events changes is evaluated from this natural base. Although global warming may be characterized by the increase of the average temperature, how it is manifested adds a layer of complexity. It can come about by an upward shift of the entire distribution so that only the mean changes but the variability stays the same, or by the more complex situation in which both the mean and variability change (Figure 13.4). Regardless, there will be an increase in many types of extreme climate events, which is an inevitable manifestation of global warming.

It is instructive to examine the observations in Figure 13.3 a little further. Let's look at how the frequency of the extreme hot days changes as we shift the average temperature by 1, 2, or more degrees simulating global warming. This is shown in Figure 13.5 for measurements taken at Yanbu (Figure 13.3). The extreme hot days are defined as when the day's average temperature is over 39°C (102°F). This happens about 1% of the time before climate change (Figure 13.3). Now we add 0.5°C to all the temperatures to represent global warming; that is, we shift the average temperature by 0.5°C. It will shift the distribution forward toward the high end and now there will be more days that will exceed the threshold of 39°C. Then we can shift by 1°C and so on. For each shift we can plot the number of extreme hot days according to our original definition. We will see that the effect depends on the underlying distribution and can be highly non-linear. Even relatively small increases of average temperature cause a substantial increase in the number of extremely hot days. In this case, the percentage of hot days would rise from the original 1% to around 7% for a 2°C warming. This can be interpreted to mean that what was seen once in a year may occur six or seven times every year in a warmer world. Even rarer events may be seen, but this requires more complex means of estimation.

This amplification of the extreme events with greater warming is caused by the steep drop in the probability of hot days in the case we have considered. In Figure 13.5 it is also noteworthy that some extreme hot days will penetrate into the transition seasons, when there were none before. The characteristics of the temperature distribution in which the summer temperatures are more sharply focused around the mean compared with other seasons is seen frequently and all over the world.

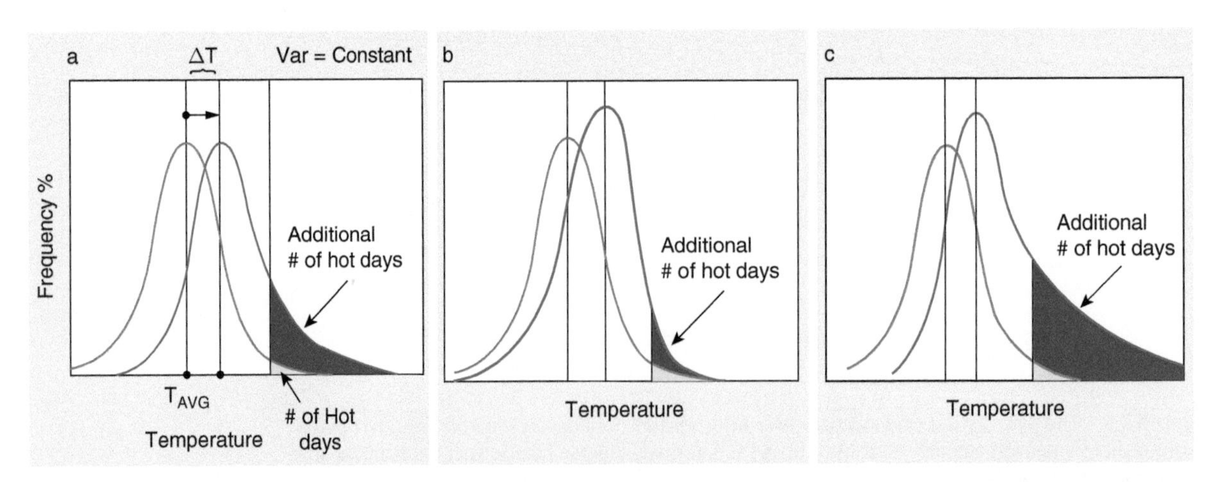

Figure 13.4 How extreme events increase with global warming. Extreme events are affected by the changes of the mean and the variance of the temperature distributions. In most expected situations the number of extreme events increases (designated by #), and some are very intense as they become less rare. The change in the variability can either amplify the impact on extreme hot days or reduce it. (a) The mean shifts but variability stays the same. (b) The mean and variability both shift, but the shift is skewed to the cooler than average temperatures. In this case, extreme hot events are reduced. (c) As in "b," but the variability shifts toward the hotter side of the distribution, amplifying the number of extreme events.

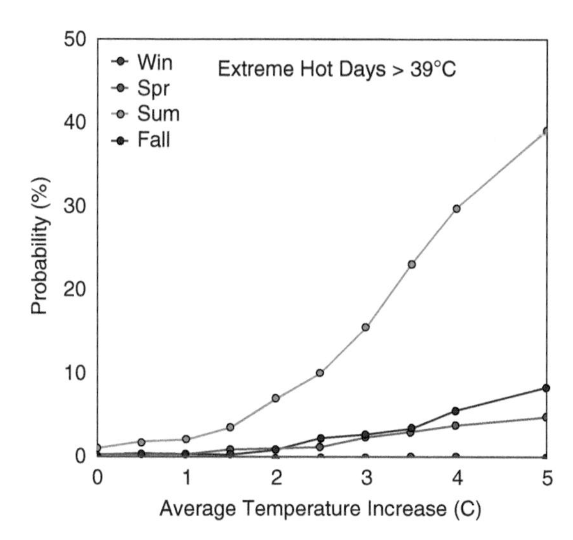

Figure 13.5 How extreme hot days increase with global warming. The extreme hot days are represented by temperatures over 39°C (102°F). As the distribution mean shifts, more days exceed this temperature. How much the increase of hot days is, depends on the distribution and how sharply it falls at the high end.

The idea of a uniform shift of temperature is supported by the fact that global warming caused by increased greenhouse gases adds a nearly constant daily radiative forcing over a season. But other factors and feedbacks can not only shift the middle values of the distribution further but also change the variability. If for temperature, as in Figure 13.3, the summer frequency distribution remains bunched up toward the high end, then an increase in variability would arise from the increase of more events on the cooler side of the middle. While this doesn't mean there won't be any more extreme hot events with global warming, the additional such days would be fewer than if the variability stayed the same (Figure 13.4b). In the hot environments, the sensible and latent heat fluxes may increase with increasing temperature as the surface gets very hot. This may bring in cooler air from elsewhere, possibly from above by a local air circulation cell and thus limit the increase of surface temperature. Such a situation would cause a negative lapse rate feedback. Wherever this happens, the extreme events will be reduced relative to the case of uniform temperature increase at all frequencies. A similar effect occurs if the annual increase of temperature is not distributed uniformly over the seasons. If winter temperatures rise more than in summer, the increase of extreme hot events is reduced. If the change of variability stretches the distribution at the higher end then the increase in the extreme hot days will be amplified (Figure 13.4c).

The most serious direct effect of warming is deaths from exposure to prolonged and intense heat about which there is much human experience from naturally occurring extreme events. There is even more experience with heat-related illnesses and heat stress. It can be represented by the vulnerability theory discussed earlier (Section 12.4 and Endnote 12.5). Observations show that in many environments, the death rate has a relationship with temperature, although the causes remain nebulous. In it there is an optimal temperature at which the least number of deaths occur during the year. During days when the temperatures are above this optimum, death rates climb. These observations can be interpreted to define the probability of death from heat at each temperature. This probability increases non-linearly for increasing temperature. For days that are cooler, death rates increase but much more slowly than on the upside. With global warming, the number of hotter days will increase during the summers and this will convolve with the probability of death from heat waves to increase the number of fatalities. The effect of the larger populations in harm's way would interact with this, exacerbating the effect.

13.2.2 Storms and Flooding

We saw earlier that based on the Clausius-Clapeyron equation, higher temperatures will certainly increase the absolute humidity and it will increase the rainfall as well if the lifetime of water vapor in the atmosphere remains the same or decreases. Normal rainfall, without the high winds of a hurricane or cyclone, can cause flooding during extreme events, and these would increase due to the increase of the average rainfall by a similar reasoning as for temperature. Often, flooding is associated with cyclones or hurricanes. These high winds affect coastal areas and are accompanied by torrential rains that are not seen without them. Along the coasts, the rain, wind-driven waves and sea-level rise can synergistically add to an increased flooding potential in a warmer world.

For winds, the extreme events may be defined when there is significant potential for damage to trees or homes. Wherever this cutoff is, gale force winds, hurricanes, and cyclones would be included. The impact of winds on buildings and trees, and on the high waves they generate near coastal areas is proportional to the square of the wind speed as mentioned earlier. So even if the environment is fairly windy with 15–20 mph winds, the weakest hurricane at about 75 mph will have 15 times the potential for causing damage and may well stimulate flooding by added rainfall when the normal high winds would not.

Tropical storms and cyclones can be considered a natural part of climate extremes since they have been occurring for millennia. The formation and intensity of these cyclones depend on thermodynamic and environmental conditions, among them being the sea-surface temperature (SST). A threshold SST is needed to start the storm and its magnitude affects the strength. Since increase of SST is a manifestation of global warming, it has been theorized that hurricanes will increase in strength if not in number and will carry more moisture and therefore a greater potential for flooding. Although this effect remains a major fear for the public and especially the sizable coastal populations, it will take more time before observational records can determine whether such storms are being affected by global warming and by how much.

13.3 Habitability

It is certain that global warming will lead to many compound influences but whether they will be of sufficient magnitude to cause harm, or be of benefit, at least for some people, remains to be seen. There are some that have been clearly observed and their impacts have been evaluated. The best known is sea-level rise which we consider next.

13.3.1 Sea-Level Rise and Melting Ice

Sea-level rise is an inevitable consequence of global warming. It occurs because of a combination of mechanisms, primarily thermal expansion and melting of land ice either from the polar regions or alpine glaciers. Of these, thermal expansion has been the largest contributor and will persist for as long as there is a global warming. It is a fundamental physical phenomenon by which liquids and solids expand as they are heated. The added heat increases the agitation of the molecules causing a further separation between them, expanding the material. It is described by a coefficient of thermal expansion β which has a value of $2.1 \times 10^{-4}/°C$ for a volume of water. Thermal expansion for water can be written as:

$$\Delta V = \beta V \Delta T \tag{13.1}$$

Here V is the volume of the oceans being heated and ΔT is the average temperature change of the volume. This equation can be converted to sea-level rise by using $V = DA$ where D is the average depth of the ocean water that has been heated and A is the area of the oceans. The change of the volume is $\Delta V = h\,A$, where h is the sea-level rise. Sea-level rise by thermal expansion can therefore be written as:

$$h = \beta D \Delta T \tag{13.2}$$

Equation 13.2 is a simplified representation that assumes that the oceans are equivalent to a box, ignoring depth variations. The volume being heated includes the mixed layer, 50–200 m at the top of the oceans. The heat moves rapidly in this region which has a uniform temperature, but global warming has been going on for a long time, so the heat has penetrated to deeper layers by now, making it difficult to assign a depth of warming.

It should be noted that sea volume expands by the addition of heat into the oceans. Whether the heat is retained only in the mixed layer or transferred deeper or moved around does not affect the amount of sea-level rise. Since there has been a steady rise of the greenhouse gases over the last century, it must have supplied a commensurate steady increase of heating to the surface, including the oceans. We would expect, therefore, that the sea-levels should have risen steadily to reflect the increase of greenhouse gases. We see in Figure 1.1 that such is the case. The surface temperature does not increase steadily (Figure 1.1). As we discussed earlier, there are periods when the temperature does not increase, even though the greenhouse gases were increasing unabated. The oceans, like the atmosphere, are warm at the top because of the sunlight and atmospheric radiation and cooler as you go deeper beyond the mixed layer. Waters are exchanged by flows, currents, and over-turning events that can move the heat in the oceans around and temporarily decrease or increase surface temperatures if surface waters are exchanged with deeper layers. This would be consistent with the surface temperature changing unsteadily, but the thermal expansion following the trends of the greenhouse gases.

We can further advance this idea by re-casting Eq. 13.2 in an alternate form that does not involve the depth of warming (D), recognizing that the addition of heat into the oceans is related to the temperature change by Eq. 8.7, we get:

$$h = \left[\frac{\beta}{\rho_W A C_W}\right] \Delta Q \tag{13.3}$$

Here ρ_W is the density of water, C_W is the heat capacity, and A is the area of the oceans. Determining how much heat is being stored in the oceans is more difficult than the surface temperature measurements but it is more reliable since it takes into account the temperature changes of the deeper ocean.

The actual sea-level rise has additional contributions from melting of glaciers, both alpine and polar and possibly from a reduction of the water stored on or below the land.

The significance of thermal expansion is that it has been the largest contributor to sea-level rise, but the other factors have caught up. With global warming, it will persist unabated, regardless of how much the glaciers or polar ice-caps melt, which will eventually diminish or even disappear, reducing their contributions. During the recent decades between 1990 and 2010, thermal expansion is estimated to have contributed 2.2 cm, glaciers 1.6 cm, and other factors, mostly ice sheets, amount to 2.2 cm making a total of around 6 cm sea-level rise. During the same time the energy storage in the oceans is estimated to have increased by about 150×10^{21} j, with about two-thirds in the upper ocean from 0 to 700 m and a third in the ocean 700 m down to 2000 m (IPCC-AR5 in Endnote 1.1). It is time to consider an example to get our bearings.

Example A time has come when the surface temperature of the ocean has gone up by 5°C compared with the present. Heat has penetrated to 3000 m making the average temperature of the oceans 1.5°C warmer down to this depth. What is the expected sea-level rise due to thermal expansion? How much heat has been added to the oceans? If the mixed layer has an average depth of 150 m and has the uniform temperature measured at the surface, how much would it contribute to the sea-level rise?

Solution: From Eq. 13.2, $h = (2.1 \times 10^{-4}/°C) \times 3000\ m \times 1.5°C \approx \mathbf{1\ m}$. From Eq. 3.3, $\Delta Q = \rho_W D A C_W \Delta T = (10^6\ g/m^3) \times (3000\ m) \times (3.6 \times 10^{14}\ m^2) \times (4.2\ j/g\text{-}°C) \times (1.5°C) =$ **6800 zillion j** (1 zillion $= 10^{21}$). The role of the mixed layer is: h (mixed layer) $= (2.1 \times 10^{-4}/°C) \times 150\ m \times 5°C \approx$ **0.16 m**, which is only 16% of the total thermal expansion in this hypothetical example, even though it was warmed more.

While thermal expansion is directly related to changes of radiative forcing and it is immediate, the effects of melting ice on sea levels are more complex and the timing is difficult to estimate. Of the ice, alpine glaciers are the fastest to melt and have a maximum potential of about 40 cm of sea-level rise. The polar ice caps have an enormous amount of water that, if it is all melted, would raise sea levels by 65 m not accounting for thermal expansion. For this to happen, the climate would have to change much more than is expected from man-made global warming, and it would bring greater issues for human habitability than sea-level rise.

There are several reasons why the melting of polar ice is a slow process and may take longer than the time-scale of man-made climate change. As with all earth system processes, the ice sheets are in a balance between mechanisms that melt them and those that continuously replenish them. On the sink side, ice can break off at the edges and float toward warmer waters where it melts. At these edges, there is plentiful sea ice which is formed mostly in the very cold waters of the oceans and so is not part of the permanent land-based polar ice caps. The waters at the edge are warmed not only by the increased greenhouse effect but by the transport of heat in the oceans. Warming of these waters does two things. Not as much sea ice is produced and the heat can melt the more permanent land-based ice at the edges of the poles. Dramatic reductions in the amount of sea ice have been reported but these have no effect on sea levels if it doesn't come from the ice caps that are fixed to the land. You can see it yourself by floating an ice cube in a glass of water and marking its level; the level will remain the same when the ice is melted. Similarly, when the ice is formed, it displaces only the amount of water it contains causing no decrease in sea level. Although sea ice doesn't cause sea levels to rise, its decline is a cause of higher-than-average warming at the polar regions because of the ice-albedo feedback, which can accelerate the loss of polar ice caps that will increase sea levels (see Figure 1.1a). Melting can occur on the top layers of the ice-caps, but with the subfreezing temperatures, the ice can absorb heat, warm up, but not melt. At the bottom of the polar ice caps, melting occurs all the time because of the high pressure that is part of the balance of the ice sheets and is not affected by climate change. On the replenishment side is the continual snowfall. Global warming may increase the snowfall and thus further slow down the loss of the ice caps and under some conditions, may even increase the ice volume for a while. The time scales are lengthened further because the ice runs deep making the ratio of the surface area to its depth rather small, which reduces the amount of energy that is being absorbed and extends the time of meltdown.

The primary effect of sea-level rise is that low-lying places are particularly vulnerable to inundation. Most coastal cities are by their nature, close to the sea level and would be greatly affected. Additionally, islands are particularly vulnerable to inundation. A well-known example is the Maldives, which is a collection of a thousand or so islands in the Indian Ocean that have no ground higher than about 3 m and some 80% of their area is within 1 m of the sea level. There is concern that many of the lower islands would disappear entirely under the rising seas while others would be reduced in area. The reduced area is expected not only in the Maldives but also in other places. Many volcanic islands have a graded elevation with gentle slopes of a few percent at the coasts. The less steeply the land rises from the sea, the more area will be inundated by sea-level rise. Since sea-level rise is a slow process; it allows people to move in time, but the economic costs of such movement can be enormous.

13.3.2 Global Habitability

In a broader sense, the matter of climate change is related to global habitability for human life. We know that there are many fundamental variables and their complex interactions determine habitability; two we can agree on are temperature and availability of freshwater including precipitation. These are also the key climate variables that will change with global warming. A region is more habitable if it has a moderate climate and an abundance of natural resources to sustain large human populations including fertile soils and freshwater. Additional factors certainly exist that may also contribute to where the present human populations are thriving; these may be a lack of natural diseases, agricultural pests, and reduced competition for resources with other species. With this context, let's look at the global connection between present climate and population density in Figure 13.6 (Endnote 13.2).

In Figure 13.6, the dry climates are generally in the desert zones of the latitudes beyond the tropics; temperate climates reflect middle latitudes and continental climates represent the cooler and dryer higher latitudes. The most inhabitable parts of the world lie within about 20 and 50 degrees North latitude where most of the people live. The favored regions are temperate climates followed by the tropical. The population distribution is also heavily skewed toward the northern hemisphere partly because of the high fraction of land area and abundance of temperate climates (Endnote 13.1).

The vast areas of current continental climate are in Russia and Canada which border temperate climate zones to the south. There are no comparable land masses in the southern hemisphere. These conditions favor the greatest movement of human habitability into the regions of the northern hemisphere as global warming shifts both tropical and temperate climates higher up. This will replace the less populated continental zones in Russia and Canada with the more habitable temperate climates.

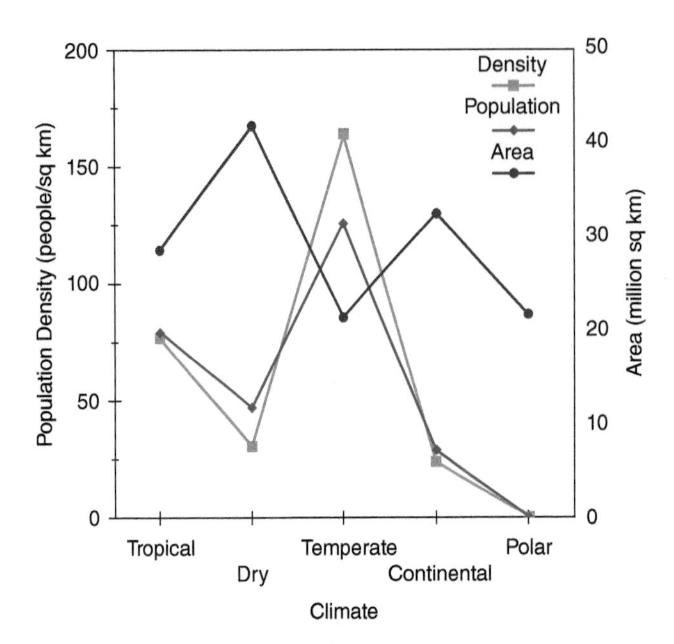

Figure 13.6 Population density and climate. Favorable climates create a higher density of human populations. Climate on the horizontal axis approximately represents the transect from equatorial toward polar regions. Dry climates have the largest area and temperate climates have the largest density. The population in billions is 1/30 times y-axis values.

13.3.3 Many Effects

There are many compound effects mentioned in Figure 13.1; most of them have large uncertainties under the present state of climate change. These uncertainties are due to the build-up as we look through the cascade of events that takes global warming to its effects on human life as shown in Chapter 2 (Figure 2.7). The compound effects lead to compounding of uncertainties. Moreover, the non-global nature of impacts makes it difficult to evaluate the consequences at each inhabited location and in some cases, even where there will be a detriment and where it may be a benefit. It is worthwhile going back to Figure 2.7. We have reached the top, which is the uncomfortable area of maximum uncertainty. An example about a fictitious crop illustrates the point.

Example There is a crop in our environment that has a yield represented by Eq. 10.6, so that $Y = Y_{Max} - \beta\,(T - T_0)^2$, where Y is the yield, let's say in bushels/hectare-yr, β represents the sensitivity of the crop to temperature change (bushels/ha-yr-°C^2), T is the ambient temperature, and T_0 is the optimal temperature for the crop. At present: $T = 289$°K, $CO_2 = 400$ ppm, $Y_{Max} = 1000$ b/h-y, $\beta = 35$ b/h-y-°C^2, and $T_0 = 290$°K. The uncertainty in T_0 is within a fairly narrow range of 287.5–290.5°K. At a future time, some 70 years from now, we estimate that CO_2 will most likely be 600 ppm with a potential range due to uncertainties of 500–900 ppm. (a) What is the expected effect of global warming on the yield? (b) What is the range of uncertainty?

Solution: We will use equilibrium conditions and round numbers. The current yield is: $Y = 1000$ b/h-y – 35 b/h-y-°C^2 $(289 - 290)^2 = 965$ b/h-y. (a) Best estimate for future yield: $CO_2 = 600$ ppm or $\Delta RF = 2.2$ w/m^2 (Eq. 9.10a). $\Delta T = 0.7$°C/(w/m^2) × 2.2 w/m^2 = 1.5°C (Eq. 9.9). Therefore, the yield will be: 1000b/h-y – 35 b/h-y-°C^2 $(290.5 - 290)^2 \approx 990$ b/h-y, or a net gain of about 25 b/h-y. That is encouraging, but how reliable is our result? (b) The uncertainties can be calculated by looking at the effect of each uncertain variable, and then the combined impact. For just the effect of uncertainty in the prediction of CO_2, and keeping all the other values the same as in the previous calculation, you will get: Y(for $CO_2 = 500$ ppm) $\approx$ 1000b/h-y – 35 b/h-y-°C^2 (289.8 – 290)^2 $\approx$ 1000 b/h-y or $\Delta Y \approx$ 34 b/h-y, and Y(for $CO_2 = 900$ ppm) $\approx$ 860 b/h-y or a net loss of $\Delta Y \approx -110$ b/h-y (in round numbers)! Even with just the uncertainty in the future levels of CO_2, we are not able to say whether climate change will be a benefit or detriment. Similarly the uncertainty in the climate sensitivity $\lambda = 0.4$ - 1.1°C/(w/m^2) results in a range of $\Delta Y = -30$ to + 30 b/h-y and the uncertainty in T_0 gives a range of + 40 to –300 b/h-y, which readers can easily verify by calculating the yield as done above. Finally, the range of yield that we get by looking at the extreme cases based on the uncertainties in all three of these variables, namely, λ, ΔRF and T_0, the expected range of yield is + 30 to –340 b/h-y.

We can conclude from the example that uncertainties in our understanding of global warming make it difficult to say whether we should expect an increase or decrease in the yield of this crop, which may also happen for predicting yields of real crops. In a warmer world, we may be able to shift farming to other locations where the temperature may be more suitable for any particular crop. For some cases, perhaps a larger area will become more suitable for raising the crop, but there is a limit, however, beyond which there may be no suitable location.

Two further points should be noted about uncertainties supported by the example. First, that when several variables are involved, the effect of uncertainties in each cannot be used to easily estimate the combined effect of all. Second, that we can calculate a range of yield in our example, which is consistent with the uncertainties in the individual variables involved, but that the extremes in this range are unlikely to occur. This is because we are taking the nexus of extreme limits of the variables (three in our example). Each limit is unlikely, so all three to be at one end or another, is even more unlikely. It is as if we are flipping three coins, the extreme outcomes are three tails or three heads, and we are likening our range to these extreme cases. Most probably something more in the middle will occur. Even the uncertainties are uncertain.

Review of the Main Points

1 We considered the impacts of climate change and related them to effects on human life. Two categories emerge: localized natural disasters that can be affected by climate change, such as extreme events, and the larger-scale seasonal and latitudinal changes. The former can be addressed by the vulnerability theory discussed in the previous chapter. In the latter category there are many impacts, including on crop yields, health, and water resources.
2 All impacts of climate change are regional to varying degrees. Although global warming can be described well by an average increase of the earth's temperature, the impacts are not as easily represented. The global average change in the impacts may be negligible but the effect on human life can still be significant wherever it is harmful.
3 There are enormous uncertainties in establishing how the compound influences will play out and, in some cases, even whether they will be detrimental or beneficial. This chapter has taken us to the top of Figure 2.7 where our confidence is the least. It is an area of scientific limbo and we have no recourse but to leave it that way and consider our next subject – the possible human response.

Exercises

1 Calculate the sea-level rise for the period between 1990 and 2010 based on the estimated increase of heat content of 150 zillion joules (1 zillion = 10^{21}). Estimate the fraction of the observed sea-level rise in Figure 1.1 explained by thermal expansion.

2 It is estimated that the world's glaciers in 2010 had the following volumes: Alpine glaciers = 1.7×10^5 km^3, Greenland ice sheet = 2.9×10^6 km^3, Antarctic ice sheet = 2.5×10^7 km^3. For each of these, calculate the sea-level rise in meters if all the ice in each of them melted. The following information may be useful: density of ice = 0.9168 g/m^3, area of the oceans = 3.6×10^8 km^2.

3 Assume that you have a low-lying island that is in the shape of a conical frustum. Its base at sea level has a radius R_B and the top has a radius R_T. The vertical height is H and sea-level rise is h.
   (a) Show that the area of low-lying islands of this type is: $A \approx \pi R_B^2$ (when $H \ll \Delta R$, where $\Delta R = R_B - R_T$). Show next that with sea-level rise, the radius of the base will be: $R_{B\text{-New}} = R_B - (h/H)\,\Delta R$, and therefore the area of inundation will be: $A(\text{Lost}) \approx \pi (R_{B\text{-New}}^2 - R_B^2)$.
   (b) Suppose $R_B = 20$ km, $R_T = 5$ km, and $H = 3$ m. Calculate the area for a 0.3 m and 1 m sea-level rise. What fraction do these represent of the original sloping area of the island?

4 At the location of our interest, the summer temperatures are very hot ranging from 30°C to 50°C (86°F–122°F). The natural distribution is described by "extreme hot days" if the temperature is between 40°C and 50°C, and "normal days" if it is 30°C to 40°C. The frequency of hot days is $f_H = 0.3$ (30%) and so the frequency of normal days is $(1 - f_H) = f_N = 0.7$ (70%). Global warming raises the average summer temperature by 5°C.
   (a) What is the average summer temperature before global warming?
   (b) Suppose with global warming the frequency of hot and normal days does not change, and the warming comes entirely from increased temperature of hot days. What will be the average temperature of the extreme hot days?
   (c) If the entire warming comes from increased frequency of hot days, but the range of temperatures of both the normal and extreme hot days remains the same, how many more extreme hot days will there be?

Endnotes

Endnote 13.1 Although a frequency distribution of almost any environmental variable can be defined abstractly, it may be clearer if, for the sake of discussion, we chose temperature at a given location. Let's say it is measured every hour for each season for many years. For one of the seasons, we can find the maximum and minimum temperatures. Then we can collect all the times that the temperature was found to be between the minimum and the minimum plus 0.5°C. On a graph of frequency (number of times in a season) vs the temperature, we can put the number found in the middle of this range (minimum + 0.25°C). Then we count how many times the temperature was between the minimum plus 0.5°C and the minimum + 1°C and plot that as the second point on the aforementioned graph. We can keep doing this until we have covered all the temperatures to the maximum. The result is the frequency distribution of the temperature for that season and location. The distribution may be a bell shaped curve or something else, but it will always have an average value and the measurement will be spread around this average. The more they spread, and how they spread, determine the environmental effects. The increments of 0.5°C are arbitrary and can be selected for convenience or determined by how much data you have. The upper end will represent the extreme hottest days of the season and the lower end will be the coldest. It is common practice to report the normalized frequency distribution, which is calculated by dividing the frequencies by the total number of observations taken during the season. Then it represents the fraction of days during the season when the temperature is in one of the increments, rather than the number of days it is within that band.

Endnote 13.2 Climates can be characterized by temperature and precipitation. As noted in Chapter 2, a transect from the tropics to the poles represents the major changes of climate. The Koppen climate classification is commonly used to distinguish climates of various regions. It has five main types: A: Tropical: in the tropics wet and warm. B: Dry: Mostly the desert zones, including semi-arid regions, most of which are hot. C: Temperate: Subtropical and middle latitudes with well-defined seasons and including monsoon-driven sub-tropical climates. It includes most of the United States, China, and Europe. D: Continental climates mostly in Russia, Canada, and parts of northern China. E: Polar and Alpine: Tundra and polar regions including mountainous areas. The data used in Figure 13.6 is from M. Kummu and O. Varis, "The World by Latitudes," *Applied Geography*, Vol. 31, 495–507, 2011. This paper also contains Koppen climate zone maps and various aspects of human habitability across latitudes that are of considerable importance for global change science.

14

Climate Management

14.1 Tragedy of the Commons

"Tragedy of the Commons" is a wide class of environmental problems in which several groups, or individuals, share a common resource without forces of management. In many such situations, individuals do not benefit from controlling their usage, but there is always some benefit in taking more from the commons. As long as the users are few and the common resource plentiful, everything works well and everyone gets what they want. Problems arise when the number of users increases and there continues to be no incentive to curtail individual consumption. Then the tragedy happens as the commons get overused and resources are either depleted or otherwise degrade to a point where they can no longer provide the desired benefits to anyone. Tragedy of the commons adds another dimension to the clash between population, consumption, and the environment. The idea, though recognized for a long time, was articulated in this form by Garett Hardin in 1968 (Endnote 14.1). He illustrated it with a situation of a common pasture where shepherds from the surrounding area were free to graze their sheep. For a few sheep the commons works well, but as the number of shepherds increases, or the number of sheep increases the supply of grass and space available to each sheep decreases and so do the profits from grazing. Eventually, there are so many sheep and they each produce so little profit, the shepherds can no longer survive from their income. What drives the tragedy of the commons is that if a shepherd adds one more sheep the benefits come entirely to him, but the environmental price is shared by everyone grazing sheep on the pasture. Since the pasture is the commons, not only can one shepherd increase his herd, but new shepherds can constantly come in. Hardin points out that there is no solution to the tragedy, except to abandon the concept of the commons when people can no longer afford it.

Let's take this thought a little further and see what we can learn from it about climate change, which seems a bit distant from grazing sheep. As long as the shepherds realize that there is a limitation to the total number of sheep that can be grazed on the pasture and if they can talk with each other, they can, and probably will, reach some understanding of what each of them should do. If a wealthy shepherd believes that the other shepherds will continue to increase their flocks, at his expense, because it will still benefit them, he may agree to reduce his flock somewhat to avoid conflict and maintain the pasture in return for limitations from the others. Then again, he may resort to conflict as a means to maintain or increase his wealth. Hardin points out that a social approach would be for the shepherds to

Global Climate Change and Human Life, First Edition. M. A. K. Khalil.

Companion Website: www.wiley.com/go/khalil/Globalclimatechange

come together and hire a manager with the understanding that everyone can only graze as many sheep as the manager approves. The manager, on the other hand, brings in environmental knowledge and studies the state of the pasture at any given time, determines how many sheep can be profitably grazed and then allocates them among the shepherds under an agreed upon and presumably fair system.

Climate change belongs to this class of problems. The commons is the global environment (pasture), or to be specific – the atmosphere, which flows across all countries. It is "used" to dilute and remove by-products of our technology, energy use, or food production (CO_2, CH_4, N_2O, HCFCs, aerosols, and others). The benefit comes from building power-generating stations, raising animals, and increasing the food supply (affluence). As with the pasture, the reduction of profits arises only when the resulting global warming reduces our affluence (by reducing crop yields, for instance). If left unchecked, there is an implication of dire consequences. International treaties to manage climate change, such as the Kyoto Protocol, are designed to get a manager to limit the amount of greenhouse gases each country can emit, or how much developed and developing countries can emit to maintain a climate target under a presumed fair allocation.

We can explore the concepts further using a simple mathematical model. It can be cast with the variables involved in climate change, but we will stay with the pasture story because it gives simpler and clearer insights into the nature of such problems, including the role of global warming in human life. The model is written as:

$$P(t) = P_0 \qquad N \leq N_0 \tag{14.1a}$$

$$P(t) = P_0\left[1 - N(t)/A\right] \qquad N > N_0 \tag{14.1b}$$

$$P_J(t) = N_J(t)P_J(t) \tag{14.1c}$$

$J = 1,..., M$ is the identification number for each of the M shepherds.
$P(t)$ = Pasture profit per sheep (dollars /sheep) at time t.
P_0 = Maximum profit per sheep when number of sheep is below the critical value N_0.
N = Total number of sheep on the pasture.
N_0 = Number of sheep above which pasture profits per sheep decline.
A = Pasture capacity; profits go to zero when $N = A$.
P_J = Profit for shepherd J.
N_J = Number of sheep grazed by shepherd J.
$P_{TOT} = P_1 + P_2 + ... + P_N$ is the total profit from the pasture which can be written as:

$$P_{TOT} = P_0\left(1 - N/A\right)N \tag{14.2}$$

A is presumed to be near the carrying capacity of the pasture. The maximum total profit occurs when: $N = ½\,A$, that is, when the total number of sheep is half as many as the pasture can profitability support. This relationship is a reflection of the fact that we have constructed the simplest possible model of this situation and has no profound implications.

Let's take $P_0 = 1$, $N_0 = 8$, $A = 28$ and look at what happens to two shepherds, one rich and one poor as they use the commons – a disturbingly small pasture. The poor shepherd has only one sheep and not enough resources to add more. The rich shepherd has one sheep to start with, but has enough resources to add one more every year. From our model (Eq. 14.1), the profits for the two shepherds are shown in Figure 14.1a.

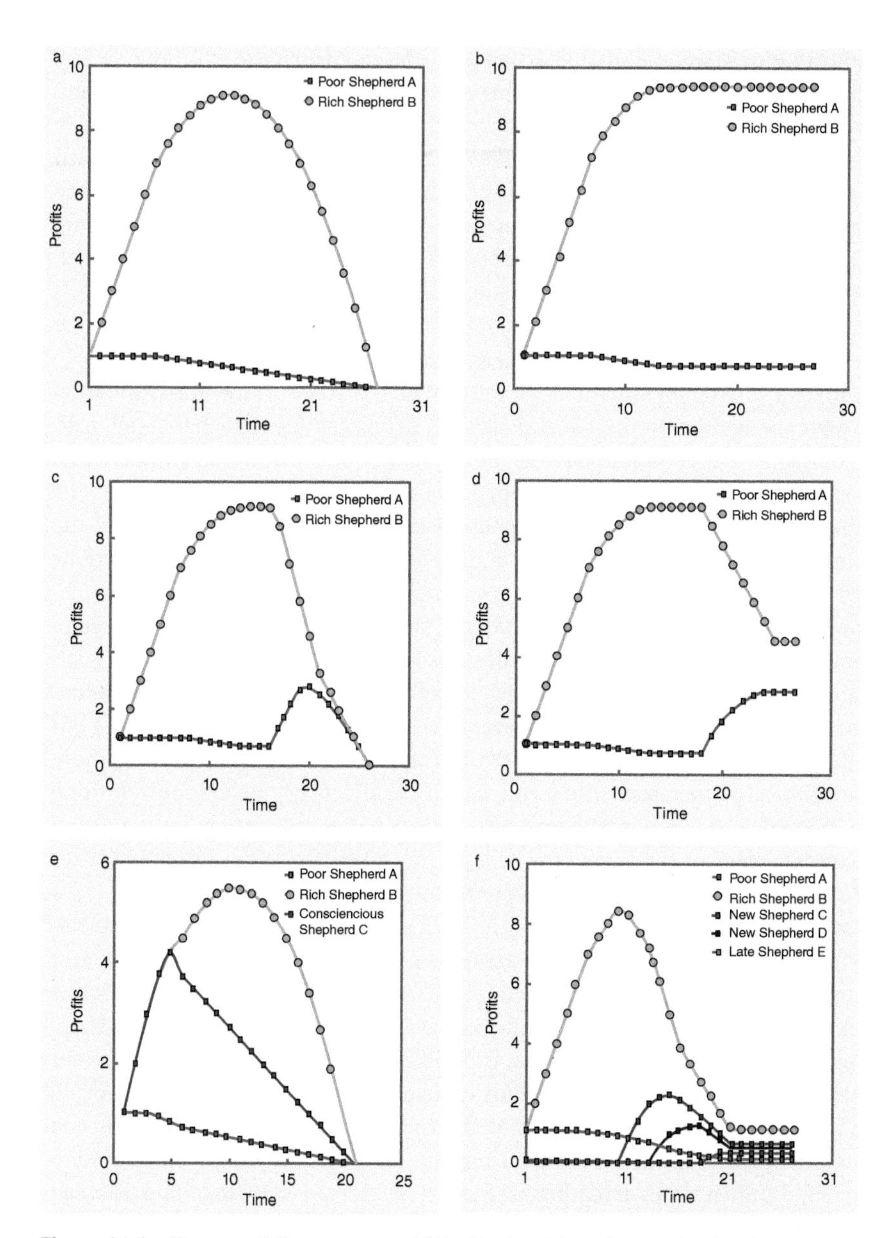

Figure 14.1 Tragedy of the commons. (a) Profits for rich and poor shepherds unmanaged. (b) Rich shepherd maximizes profits. (c) Poor shepherd adds sheep following situation in part (b). (d) Conscientious shepherd comes in; rich shepherd has uncontrolled flock. (e) Poor shepherd optimizes flock for profit. (f) Many shepherds come in and grow flocks. Some profits for all, but all are poor.

We see that in the non-interacting and unmanaged situation, even though the poor shepherd does nothing, his profits go to zero as the rich shepherd uses up the resources. In the real world, this outcome won't happen even without a manager. The reason is that when it becomes obvious to the rich shepherd that adding more sheep is not making more profits, he would stop adding them, and maybe even pull back to the number when he had the most profit and the outcome will be as shown in Figure 14.1b.

Let's complicate the situation a little. Starting with the case in Figure 14b, suppose the poor shepherd realizing that his profit from his one sheep has declined, gets a loan and adds more sheep at one per year. In the beginning his profits will rise proportionately although the rich shepherd will notice a decline in his profits. But if the poor shepherd keeps adding sheep, the pasture will decline just the same as before (Figure 14.1c). So, the poor shepherd too will pull back to a level when he got the most for his investment. In that steady state he will have more than before he started adding a sheep a year, and the rich shepherd will have less than his maximum profits, but still quite a lot (Figure 14.1d). In this situation both shepherds share, somewhat disproportionately, the optimum benefit of the pasture as might happen from interactions and agreements between shepherds.

Next, we will look at two new situations. In the first one, we assume the initial situation of a rich and poor shepherd with the addition of a third conscientious shepherd – all in an unmanaged commons. The conscientious shepherd adds one sheep every year, but after he has four sheep, he stops adding more to prevent damage to the pasture (Figure 14.1e). The result illustrates that the conscientious shepherd has no effect on the outcome of the tragedy and in fact he still reduces the profits of the poor shepherd as much as the actions of the original rich shepherd.

The last case starts with the original rich and poor shepherds and adds additional shepherds over time. In this scenario, each shepherd stops adding sheep when the profits don't increase and could represent a weakly managed situation (Figure 14.1f). As additional shepherds come in, the profits deteriorate for everyone and in the end, everyone is poor but functioning despite the fact that every shepherd has acted rationally by not adding more sheep if the profits didn't increase. This result shows the effects of "population" on finite resources. In the real world, how limited resources are is not always easy to evaluate, and often technological advances can extend the benefits much further, as say, by adding faster growing or more nutritious hybrid varieties of grass to the pasture.

The translation of the results to climate science has a number of features and nuances. The most important result is that unilateral action is not effective. To solve the matter of potential harm from climate change, international action is necessary in which everyone participates and is motivated to exercise limits on the activities that generate greenhouse gases. That motivation is difficult to achieve, especially since the point at which greater consumption is no longer profitable is difficult to determine and is not the same for all countries and people. If we extend the idea to environmental management within each country, the principles remain the same. Consider the case of urban air pollution management in the United States; it required the Clean Air Act of 1970 to mandate enforced compliance with manufacturing, testing cars and other rules to manage emissions to have cleaner city air. Such regulations also exist in most developed and developing countries because it is not expected that voluntary reductions would be sufficient. For climate change, the voluntary actions of a few individuals within a country have virtually no effect on global warming; the entire country, as a block, has to reduce greenhouse gas emissions. Even then, some countries with low populations or small per capita emissions will have virtually no effect on climate even if they reduce emissions to zero.

There are two sides to the idea of the tragedy of the commons that "increased number of sheep benefits the owner and the cost is shared by all." Both are only approximately valid for climate change. As increasing industrialization and technological advances are the main reasons behind climate change, they benefit all societies, including the "non-industrialized" ones, by trade and a "trickle down" effect. The damage from climate change is also not shared equally by all. Instead, there is a wide range as we discussed in the last chapter. Countries can be lost, such as the example of the Maldives, or some can benefit, as may be the case for countries in the middle and higher northern latitudes. These disparities in both the benefits and costs are an impediment to strong and binding international agreements. There are other impediments too.

Some further aspects of management policies may be of interest to the readers. As discussed for the case of the overused pasture, management may take place if there is a recognition of a potential problem. This is the case with climate change but there is no present mechanism to implement international control on such important matters as the consumption rates, population, energy use, or food technologies. A loose understanding has been documented in the Kyoto Protocol (1997) and the more recent Paris Accords (2016) to reduce emissions voluntarily. The success of these agreements is not assured because they are missing two characteristics of successful environmental regulations, such as those for urban pollution in the Clean Air Act in the United States. One is verification that the targets have been met, and second is enforceability, that is, the remedies if the targets have not been met. Countries can readily commit to reductions, but are not capable of delivering them, especially since their political and social situations change more rapidly than the climate does. Moreover, since managing population or aspirations of affluence are not mentioned in any treaty, even the best intentions will go awry.

No one knows whether it is possible within the laws that govern human social behavior to have sufficiently stringent international agreements that can deliver on the goals of keeping the earth's environment sustainable. It seems though that for an agreement to be successful it must have the following broad characteristics. (1) Universality: A lesson from the tragedy of the commons discussion is that a viable and predictable management of the global climate will not happen without an agreement between all the countries and especially those that have the largest emissions of greenhouse gases. Unilateral actions are futile, and even more so if undertaken by small groups within any country. (2) Verifiability: Agreements must include a means to verify that countries are meeting goals. More than that, the effect of the agreements must be observed in the decrease or stabilization of greenhouse gas concentrations in the atmosphere. Policies that lead to no measurable change in the atmosphere may be neither verifiable nor capable of reducing global warming. (3) Enforceability: There must be a means to enforce violations of the agreement. All treaties are for mutual benefit but for climate change they will often conflict with national economic goals. There is no mechanism for enforceability in the current agreements. (4) Flexibility: Each country or region must have flexibility to contribute in ways that are suited to their specific consumption patterns and economic aspirations. One of the ideas in current agreements is the allowance of trading the control of one gas for another. A country may make its contribution to the management of the global climate by reducing methane emissions instead of carbon dioxide. How much is that worth relative to another country that is reducing CO_2? This question has been difficult to answer; however, the idea of the global warming potential (GWP) has been used as a means to trade gases. Since the idea is also useful in understanding the warming characteristics of different gases, we will discuss it in the next section. (5) Holistic approach: The agreement or the understanding has to target the root causes of climate change. Although managing energy efficiency and using alternative fuels will reduce emissions, there are the other two elements that, if left unchecked, will overwhelm such efforts. These are the increasing per capita consumption rates, or affluence and an unmanaged population increase. Ignoring them is tantamount to the unchecked tragedy of the commons. As we saw, the population, affluence, and emissions are connected in Eq. 12.1 as Emissions = $K \times C \times P$. The emissions are directly tied to temperature change. To keep the temperature stable or below a target warming, the emissions have to be kept below set levels. This fixes the left-hand side as long as fossil energy is used and likewise for the non-CO_2 greenhouse gases. After optimizing the emission factor, we are compelled to keep the combination of affluence and population constant. If the population is allowed to increase unchecked, it will prevent the stabilization of the earth's temperature as reflected in the emissions of the greenhouse gases, or lead to increasing poverty by forcing reduced consumption, or both. (6) Comprehensive: A plan to manage global warming will require an approach that uses all the possible "leverage points" in the climate system. Any one method is unlikely to produce the desired result, but using many methods together may.

14.2 Compounding Forces of Resistance

Several forces compound the effects from the tragedy of the commons on climate management. One of the strongest is the competition with other societal issues that require more immediate attention. In every country, at any time, there are a number of issues that engage people's attention. Many of them are related to the dynamic nature of economic conditions, such as international competition in trade or social services such as health care. Organizations, including countries, focus their attention on issues that can be resolved from right now to about five years out. Rarely do governments articulate plans beyond five years; and even those plans rarely work out as projected. Most human institutions and programs are confined to planning for periods shorter than five years. College degrees, for example, are designed to be completed in four years. Institutions make annual budgets and businesses look to solutions over a quarter of a year and rarely beyond. Climate change is an issue, at least in the beginning, that is decades in the future. UN's IPCC puts evaluation benchmarks that are 30 years to 100 years forward in time. These time scales are well beyond what we and our societies have evolved to consider and plan for. It is difficult to justify investing financial and other resources to address a potential problem that is so far away, when other problems require immediate action. In this context a matrix may be useful that looks at the potential risks of addressing societal problems of the distant future (Table 14.1).

If we consider our chances of success in planning, it looks like major investments made now would only be considered beneficial in one of the three possible outcomes we have put here. It is the last category of "no-regrets options" describing actions that may be taken that will reduce potential climate change but also provide another immediate benefit. Energy and resource conservation belongs in this group and has been promoted by governments and environmental organizations alike. The 2007 Energy Independence and Security Act in the United States is a case in point. It led to the replacement of incandescent light bulbs with more efficient CFL and now LED bulbs that reduce energy consumption and thereby reduce greenhouse gas emissions. Although the cost is higher, it can be recovered by the lower power consumption and consumers still get the same lighting benefits, so it does not compromise their affluence. In addition to providing a tangible benefit at the present, many people feel satisfied that they have done something to benefit the environment. The no-regrets options don't require international agreements because their main benefits are assured to

Table 14.1 A Dilemma of Climate Change.

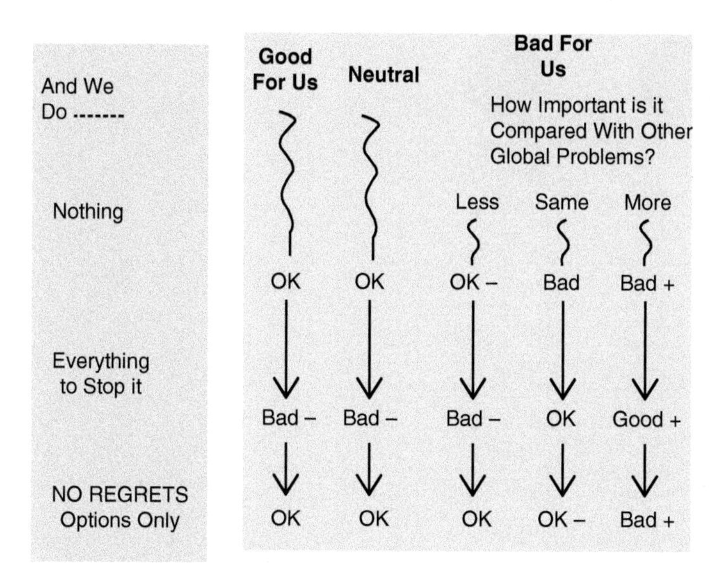

the country that exercises them and the climate effect is a bonus. It is an example of a technological fix that can reduce the impacts of consumption and population, like genetically engineering sheep that can grow fatter on less grass to connect with the pasture commons.

When the long-time scale connected with climate change is taken together with the uncertainty about how much change there will be, when it will occur, and what it will do to human life, the forces preventing enforceable agreements are further compounded. We considered the issue of uncertainty earlier in Chapter 2 and said that a significant amount is irreducible. That is to say, regardless of advances in science it is likely that prediction of climate change into the decades to centuries in the future will remain highly uncertain. This leads to the thought that maybe it won't be a problem at all. No one knows for sure. Moreover, there is an expectation that new energy technologies would lead to control of global warming since there is time available and scientists and industry are motivated to work on it. Finally, there is the impediment of cost. Although no-regrets options may be attractive, and there are other means of managing the climate as we will soon see, the only method that assures control over global warming is reduction of greenhouse gas emissions from energy, agricultural, and industrial sources. Reducing emissions, however, is likely to be a costly enterprise ranging up to several percentage of the global GDP, although some of it may be recovered in benefits as many proponents will argue.

14.3 Mechanisms for Managing the Climate

From the science we have learned in this book, a synopsis of the ways we have available to us to manage the climate can be constructed as shown in Figure 14.2.

Figure 14.2 is almost self-explanatory; however, a few highlights may encourage a closer look at it and serve to put the previously studied components together. The figure has

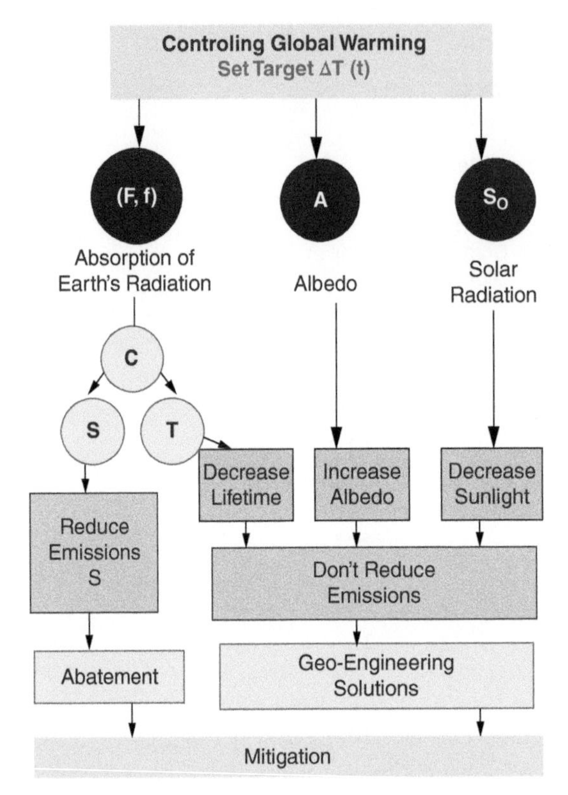

Figure 14.2 Mechanisms of climate management.

several levels. At the top is a goal we set to manage the climate. It is stated here in terms of a temperature change we do not want to exceed. Keep in mind that this remains more uncertain than a radiative forcing, but in the end, we don't care about how much radiative forcing is allowed to increase, only in how much global warming it will translate into as degrees C. At the next horizontal level, to meet the target ΔT there are only three main parts of the climate machine to manage internal to the earth – the absorption of solar (f) and earth radiation (F) streams as they pass through the atmosphere, the albedo (A) and from outside, it is the solar constant (S_0). In the next level, the absorption is controlled by the concentrations of the greenhouse gases written as C, which in turn are determined by emissions (S) and the lifetime (τ). The increasing emissions from human activities are the main cause of climate change we want to manage. Reducing emissions is *abatement* which is a part of the larger process of climate *mitigation*. Since there is reluctance to reduce the use of fossil energy, we can also achieve our goals without abatement, by controlling the remaining parts of the climate machinery. From the figure, we see that this amounts to decreasing the lifetimes of greenhouse gases, increasing the albedo or reducing the solar constant. We will consider actions to manage the climate by these mechanisms instead of reducing direct emissions to be *geo-engineering*. It will usually require engineering the global environment and include processes that leverage the natural system of the earth. That too would lead to mitigating climate change. It is usual to consider abatement as the primary means of controlling global warming, but here, our interest is wider in looking at all the options for planetary-scale climate management. Moreover, in recent years, the idea that we may actually use geo-engineering has gained favor and so we will look at some of the more practical possibilities next.

14.4 Geo-engineering

Geo-engineering allows control of global warming while we still burn fossil fuels, retain agricultural practices that emit methane and nitrous oxide, and continue processes that release the technological gases we discussed before. It usually departs from the known environmentally safe method of reducing man-made greenhouse gas emissions and allows for tinkering with the earth's climate system, including its natural components. This aspect makes people uneasy. However, if the climate changes sufficiently, it can be a means to bring it under control by direct and fast-acting processes. A large number of ways have been proposed, but we will consider the ones most likely to succeed or are the most entertaining. The criteria for such methods are that they have a low potential for side effects and may offer other benefits such as no-regrets options. Although we will not analyze the costs, they will always be a deterrent if too high.

Geo-engineering solutions may be divided into two broad categories: carbon capture (CC) and solar radiation management (SRM).

14.4.1 Carbon Capture

Carbon capture involves methods to remove CO_2 from sources and store it in deep underground geological reservoirs where it will stay for at least a few hundred years, and some of it may be incorporated into rocks and soils. It a way it attempts to return to "natural circumstances" by tying up carbon in geological reservoirs from which it came. CO_2 can be removed from three places: before fossil fuels are burned, from the streams of CO_2 emissions as from power plant chimneys, and finally from the atmosphere. The technologies for taking out carbon before fossil fuels are burned depend on shifting the main energy content into hydrogen and storing the carbon. When the hydrogen is burned, it makes water vapor and this will not affect the climate system because of the negligible effect on the water cycle. The second method is to capture carbon in the emission streams of power plants. It is efficient because there is a high concentration that is easier to capture than after it gets diluted in the atmosphere by many orders of magnitude. The geo-engineering part is storage in geological reservoirs.

The last method of sequestering man-made carbon is to take it out of the atmosphere. To do so by a direct means is virtually impossible because it is so diluted. The main method that has occupied the attention of the scientists and public alike is the idea of planting more trees to sequester the CO_2 from the atmosphere. Although it sounds very attractive, the practical aspects are daunting. The main impediments are that there is no vacant land available where this can be done. Either the land is being used for living or agriculture, or it is unsuitable for growing things, or there is already a forest on it, possibly from re-forestation, which is practiced if the land is not converted to other uses. The idea can work, however, if trees are planted, and then harvested and burned to produce energy. It cycles the carbon in the atmosphere instead of increasing it. Trees can be likened to either solar cells and rechargeable batteries. They capture energy from the sun, store it in the biomass (charged), we recover it by burning them, and replant more to repeat the cycle (re-charge). This practice generates energy that is "carbon neutral." Since there is no shortage of carbon dioxide in the atmosphere, this idea can be extended further by recovering the carbon from burning the trees and storing it as in the previous method, somewhere in geological formations. Then, the method will produce energy, offsetting fossil fuels and at the same time reduce the CO_2 that is already in the atmosphere. It is referred to as BECCS for bio-energy with carbon capture and storage.

14.4.2 Solar Radiation Management

The idea of solar radiation management (SRM) is to reduce the amount of sunlight from reaching earth's surface. From what we have discussed, there are three places where this can be done: near the surface, in the atmosphere or in space. Although there may be ideas that can reduce global warming by surface geo-engineering, we will set them aside because their impact is questionable. In the atmosphere there are two methods – make the clouds more reflective and put reflective or light scattering material in the stratosphere (Endnote 14.2). Of these, putting sulfur compounds that can generate sulfate particles or other forms of "dust" in the stratosphere is the most viable. As we discussed in Section 4.4 once we put fine particles in the stratosphere, it takes a long time for them to return to the surface partly because of the thermodynamic barrier at the tropopause. An important reason for its attractiveness is that it happens naturally every so often after volcanic eruptions (Section 11.2). During the aftermath, large parts of the world see temperatures reduced, whether we want it or not. As a geo-engineering method, it is cost effective and certain to work for some period of time, after which the sulfate aerosol can be replenished. There are additional possibilities of what we put in the stratosphere that includes virtually any type of reflective objects including shiny helium-filled balloons.

The last possibility is to block solar radiation from reaching the earth by putting reflectors in space (Endnote 14.2). The method of choice is to put reflectors at the Lagrange Point L1. This is a region in space, about a million miles from the earth between it and the sun in the ecliptic plane. The important characteristic of this point is that an object will stay there without requiring propulsion or other sources of energy. At L1 the force of attraction of the sun on this object is counteracted exactly by the attraction of the earth and the centrifugal force, both of which act in the opposite direction. If a giant mirror is placed at the Lagrange point L1 it would prevent some of the sunlight from getting to the earth, thereby modifying the solar constant, even though we have not altered the sun. A giant mirror is not needed for this idea to work; instead a cloud of small reflectors can be placed there that will have the same effect, be much less expensive, and can be replenished when needed. There are already several satellites at this location; especially noteworthy is NASA's Deep Space Climate Observatory (DSCOVR), so the technology to get reflectors there is not fantasy.

Example: Suppose we want to put a giant mirror disc at the Lagrange Point L1 so that it can stop global warming up to a doubling of CO_2. How big will it have to be? What if we used tiny mirrors only two cm in diameter (about an inch). How many will we need? How many grams of the tiny mirrors will we need? Is it practical?

Solutions: How big? The effect of doubling CO_2 is from Eq. 9.10, $\Delta RF = 5.35 \times \ln (C/Co) = 3.7\ \text{w/m}^2$. Energy to be removed from the solar stream = $E(\text{L1})\pi R_M^2$ where R_M is the radius of the mirror. The energy arriving at L1 is nearly the same as at the top of the earth's atmosphere, or $S_0 = 1360\ \text{w/m}^2$. So, $E(\text{TOA})\ \pi R_M^2 = \Delta RF \times (4\pi R_e^2)$ where R_e is the radius of the earth and E(TOA) is the energy absorbed at the top of the atmosphere $(1 - A) \times S_0 = 0.7 \times 1360\ \text{w/m}^2 \approx 950\ \text{w/m}^2$. Therefore, $R_M = 2\ R_e\ [\Delta RF/E(\text{TOA})]^{1/2} = 2 \times 6380\ \text{km} \times (3.7\ \text{w/m}^2/950\ \text{w/m}^2)^{1/2} \approx 800\ \text{km}$. A bit on the large side.

How many tiny mirrors of radius R_D? $N\ (\pi\ R_D^2) = \pi\ R_M^2$. $N = (R_M/R_D)^2 = (8 \times 10^5\ \text{m}/0.01\ \text{m})^2 = 6.4 \times 10^{15}$ tiny disks.

How much weight? Polycarbonate density is 1.2 g/cm^3. A tiny mirror with 1 cm radius and 0.1 cm thickness would have a mass of 0.38 g. Total mass = $6.4 \times 10^{15} \times 0.38\ \text{g} \approx 2.4$ Pg. The DSCOVR satellite has a mass of 570 kg.

Is it practical? No, but it is interesting. Offsetting doubling of CO_2 is a huge demand that would be difficult to implement with any one measure such as the mirrors at the Lagrange Point.

14.5 Trading Gases: The Global Warming Potential

The global warming potential is one of several possible methods to compare the effectiveness of one gas to cause global warming over another. It has been widely used by climate scientists and policymakers and has been adopted as a standard for trading gases under international agreements for reducing emissions. In the commonly used definition, it is the total energy a kilogram pulse of a greenhouse gas will trap per square meter over a selected length of time, relative to the energy trapped per square meter by a 1 kg pulse of CO_2 over the same time. It requires knowledge of the radiative forcing per kilogram and the gas concentration at all times up to the time horizon, T_H. The absorptivity or radiative forcing per kg (A_{gas}) is assumed to be constant over the times of interest and given in Endnote 14.3 for gases of our interest. For gases that follow Eq. 5.1, a pulse decays according to Eq. 5.3: $C = C_0 e^{-t/\tau}$. It can be shown that the total energy trapped over the time horizon of interest (T_H) is (Endnote 14.4):

$$AGWP_{gas} = A_{gas}\tau_{gas}\left[1 - \exp\left(-T_H/\tau_{gas}\right)\right] \tag{14.3}$$

This is the *absolute global warming potential* for the gas in j/m^2. The decay of a kg of CO_2 does not follow the simple exponential as was discussed earlier. From Section 7.2, it can be approximately written as: C (kg) = 1.188 $t^{-0.234}$ for t (in years) > 0 and then the absolute global warming potential for CO_2 is (Endnote 14.4):

$$AGWP_{CO_2} = 1.55 \times A_{CO_2} T_H^{0.766} \tag{14.4}$$

The global warming potential (GWP) is defined as the ratio of the two:

$$GWP_{gas} = \left(\frac{A_{gas}}{A_{CO_2}}\right)\frac{\tau_{gas}}{1.55T_H^{0.766}}\left(1 - e^{-T_H/\tau_{gas}}\right) \tag{14.5}$$

Equation 14.5 is written to emphasize that the GWP = effect of differences in absorptivities × effect of differences in the lifetimes; both expressed as ratios with respect to CO_2.

As an alternative, suppose instead of following a pulse emission, we consider the effect of putting 1 kg of a gas into the atmosphere every year, and ask: how much radiative forcing will it produce after a time T_H of sustained emissions? And compare it with the same process for CO_2. The ratio so calculated will be the same as Eq. 14.5. This interpretation may be more intuitive since it expresses the GWP as the radiative impact of the gases at the endpoint (Endnote 14.5).

Example: What is the global warming potential of methane at a time horizon of 100 years?

Solution: The definition of the AGWPs and GWP is in terms of radiative forcings per kg but the tables are for radiative forcings per ppb. The conversion is: A_{CH4} (w/m^2-kg)/A_{CO2} (w/m^2 - kg) = A_{CH4} (w/m^2 - ppb)/A_{CO2} (w/m^2-ppb) × (MW_{CO2}/MW_{CH4}). Therefore, in Eq. 14.5, A_{CH4} (w/m^2-kg)/A_{CO2} (w/m^2 - kg) = 3.7×10^{-4} (w/m^2 - ppb)/1.4×10^{-5} (w/m^2-ppb) × (44 g/mole/16 g/mole) ≈ 73. The term representing the lifetimes is: {12 y × $(1 - e^{-100/12})$/1.55 y × 100 $^{0.766}$} = 0.23. Then: GWP (100 y) = (73) × (0.23) ≈ 17. We see that the greenhouse potency factor is 73 for methane – it is that much more effective at absorbing the earth's radiation than CO_2, but the lifetime factor is 0.23 since it is not as long lived as our comparison molecule, carbon dioxide, reducing its final impact to 17.

The number calculated in the example is sometimes boosted higher for methane to 25 by including purported indirect effects. The result is often interpreted to mean that a kilogram of methane is about 25 times as effective at causing global warming as a kilogram of CO_2. Regardless of such effects, the GWP can vary significantly for different time horizons. For the example of methane, it ranges from 8 to 70 for 500-year to 20-year horizons (Endnote 14.3). The 100-year time horizon is a convention, but otherwise arbitrary. This horizon is also used to convert the emissions of greenhouse gases into CO_2 equivalents so that emissions from different countries can be compared with a single index (CO_2 Eq emissions per year). For instance, a 10 Tg/y emissions of methane from landfills is said to be 250 Tg/year of CO_2 equivalent emissions.

The global warming potentials of gases discussed here are given in Endnote 14.3. There you will see that the chlorofluorocarbons have very high values ranging from 1,000 to 14,000 and for sulfur hexafluoride (SF_6) it is 23,000 because of its long lifetime of 3,200 years. Such high GWP gases are included in the Kyoto Protocol as targets for controlling global warming and countries can trade them with this equivalence. Although there are many gases, such as SF_6, with extremely high GWPs, their usages and release to the atmosphere are limited and high concentrations are not likely whether they are targeted for control or not. The actual contribution of a gas to global warming is proportional to the product of its absorptivity per ppb times the concentration change (ppb) in the atmosphere to get radiative forcing, which then translates into a temperature change. We can see this effect in Figure 14.3 – gases with the highest GWPs have made the least contribution to present global warming. These shortcomings should be kept in mind when considering the GWP and its uses.

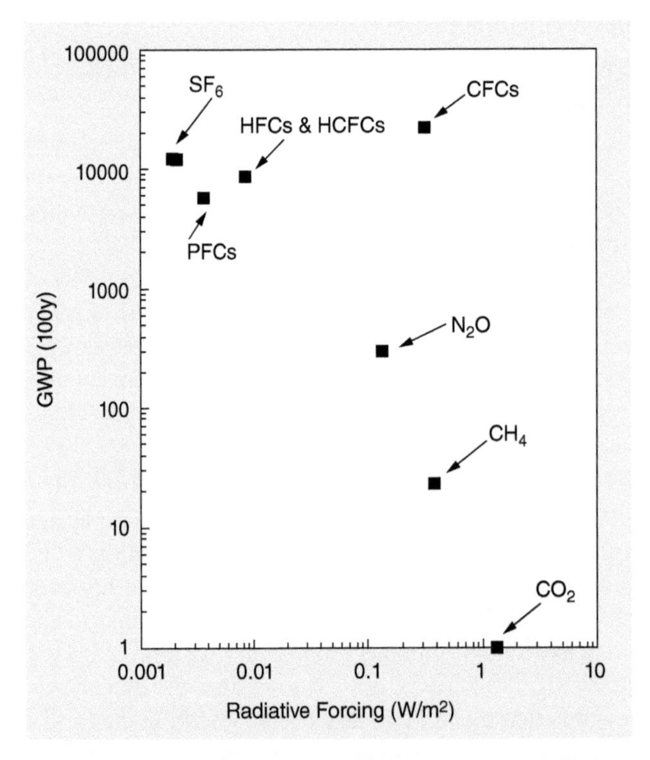

Figure 14.3 GWP and radiative forcing due to gases over the last 150 years.

Review of the Main Points

1 *Whether we can manage climate change or not will be determined by the rules governing human behavior and not climate science.* "Tragedy of the Commons" is a particularly intractable class of human social issues in which the common resources of the environment bring benefits to those who use them the most but the cost is shared by everyone, including those who use them less. Climate change falls into this wide class of human problems. Solutions require abandoning the idea of the commons.

2 Results that emerged from our study show that unilateral actions by a country or small groups would not result in a perceptible decrease in global warming or its feared consequences.

3 The ability to control global warming is affected by several compounding forces of resistance. One factor is that the impacts are in the distant future relative to human planning, and resources available at any time are channeled to resolve the most urgent issues.

4 Many tools are available to manage the climate and to live with climate change. But a sustainable solution requires a holistic approach that can balance population, affluence, resources, and environmental quality.

Exercises

1 Suppose we want to keep global methane concentrations at 1800 ppb. The lifetime is 8 years due to reactions with OH radicals and there are no other sinks.
(a) How many teragrams/year are emitted to the atmosphere?
(b) Suppose OH concentrations decrease by 20%. What would be the new lifetime of methane?
(c) How many teragrams can be emitted to keep the concentration at 1800 ppb with the new lifetime? Discuss what this case suggests about our ability to meet global warming targets based on fixing emissions.

2 Compare the lifetimes of a pulse of CO_2 between the formula in this chapter ($C = C_0 \times a\, t^{-bt}$) and Eq. 7.3. Use the % difference to state your results. At what time does the power law become unreliable?

3 Calculate the global warming potential for N_2O and CFC-11 for time horizons of 20, 50, 100, and infinite number of years.

4 Suppose you decide to reduce your "carbon footprint" from 10 tons C/y to 0. Your original consumption is about double the US average. You do this by reducing consumption and switching to alternative energies. Suppose also, that you convince 70 million others in the United States (~ 20% of the population), who were also consuming 10 tons C/y worth, to voluntarily do the same as you, that is, have zero carbon emissions. Calculate the annual effect in PgC reduction and % reduction of the global emissions
(a) if only you make the change and
(b) if the 70 million others make the change too.
(c) Calculate the effect on the earth's temperature in equilibrium if this process continues for the 70 million people for some time. For simplicity, assume that at the time of our interest the amount of CO_2 in the atmosphere is reduced from this activity by $\Delta C = -\,\Delta S \times 50$ yrs, where ΔS is the amount in part (b). Provide a brief conclusion about the results.

5 A geo-engineering solution for managing global warming is proposed in which reflective balloons are put into the stratosphere. Balloons currently used by NASA have a radius of up to 75 m. Assume they can reflect all the light that shines on them, and they are placed in the tropical zone because the most sunlight can be intercepted.
   (a) Estimate the number of balloons needed to compensate for the effect of doubling CO_2? How many balloons are needed to achieve a 10% compensation for doubled CO_2?
   (b) Estimate the percentage of the tropical sky area the balloons will occupy. Take the solar radiation to be 1360 w/m^2 at the top of the equatorial atmosphere (and take the day-night effect into account).
   (c) Discuss potential side effects for the global environment and equatorial regions.

Endnotes

Endnote 14.1 Garrett Hardin, The Tragedy of the Commons, *Science*, 162, pp. 1243–1248, 1968.

Endnote 14.2 Sources and further reading: National Academy of Sciences (NAS), Policy Implications of Greenhouse Warming, National Academy Press (NAP), Washington D.C., 1991. NAS, Climate Intervention: Reflecting sunlight to cool the earth, carbon dioxide removal and reliable sequestration, NAS, 2015.

Endnote 14.3 The efficiencies of gases for trapping the earth's infra-red radiation, expresses as A_{gas} in Eq. 14.5, lifetimes and the calculated global warming potentials of the gases we have discussed are recorded here as: **Gas**, Lifetime (yrs), Radiative Efficiency in w/m^2-ppb, GWP-20 years horizon, GWP - 100 years horizon. Main greenhouse gases: (**CO_2**, -, 1.4×10^{-5}, 1, 1), (**CH_4**, 10, 3.7×10^{-4}, 72, 25), (**N_2O**, 120, 3.03×10^{-3}, 289, 298); Super-greenhouse gases: (**SF_6**, 3200, 0.52, 16300, 22800), (**CF_4**, 50000, 0.1, 5210,7390), (**C_2F_6**, 10000, 0.26, 8630, 12200); Chlorofluorocarbons and replacement compounds: (**CFC-11**, 45, 0.25, 6730, 4750), (**CFC-12**, 100, 0.32, 11,000, 10,900), (**HCFC-22**, 12, 0.2, 5160, 1810), (**HCFC141b**, 9, 0.14, 2250, 725), (**HFC 134a**,14, 0.16, 3830,1430). These are taken from IPCC AR4 (2007), Table 2.14, where values for many other gases are also tabulated. See Endnote 1.1. The radiative efficiency has to be adjusted for molecular weights, as in the example of Section 14.5 to reflect the decay of 1 kg of gas versus 1 kg of CO_2.

Endnote 14.4 The total energy sent back is the integral of $A_{gas}\, C = A_{gas}\, C_0\, e^{-t/\tau}$ taken from the time of release t=0 to the time = T_H. This gives Eq. 14.3. The decay of a pulse of CO_2 is given by Eq. 7.3 and explained in Section 7.2. Equations 14.4 and 7.3 represent the same physical phenomenon, namely the decay of CO_2. Either can be used to calculate the global warming potential. Equation 7.3 is a sum of four exponential functions, and Eq. 14.4 is a much simpler equation with only two constants. It is derived by fitting a straight line to the natural logarithms of both sides of Eq. 7.3 creating a power law version for the GWP in Eq. 14.5.

Endnote 14.5 An alternative end point index is the global temperature potential (GTP) that is the ratio of the temperature change caused by a pulse of the gas to that from CO_2 at the time horizon T_H. Although somewhat more difficult to calculate, it is a more relevant index. It is proposed by Shine et al. "Alternatives to the global warming potential," *Climate Change*, Vol. 68, 281–302.

15

Possible Futures

15.1 Projections

No reliable predictions about future climate change can be made with present knowledge, as extensive as it is. Nonetheless, it is the expectation of future climate change that demands our attention at this early stage. A practice, adopted as a compromise, is to create possible story lines about how societies and countries may evolve over the next century and then associate emissions of greenhouse gases with these narratives. Let's look briefly at a range of possibilities under simple, transparent assumptions to complete the picture of climate and its ongoing change that has been our focus.

Going back to the factors that make up emissions of greenhouse gases (Eqs. 12.1, 12.2): Emissions $= K$ (emission factor) $\times$ C (per capita consumption rate) $\times$ P (population), we can consider the possibilities for each factor. The United Nations has evaluated the probable population to the end of the twenty-first century. We will rely on their projections (Endnote 15.1). Current data show that human fertility rates are declining due to various reasons that have come into play as the population has grown and gotten more urbanized, as well as societal changes, particularly in more equal rights for women. Although the global population still continues to rise, it does so at a slower rate than in the recent past. The UN median estimates for population are about 11 billion at 2100 AD with an upper limit of the estimate at about 13 billion. If the fertility rate is reduced by just 0.5 children per woman, it would bring the population into a stable balance at about the same as our present population ~ 8 billion (in 2021). The reduction in fertility rates may come about by global management planning and may include considerations of climate and poverty by the various nations, or it may occur naturally as part of the current trend.

Present benchmarks are: population $P = 7.7 \times 10^9$ people; per capita consumption of energy $C = 8 \times 10^{10}$ joules/person – year and an emission factor of carbon emissions $K = 1.7 \times 10^{-5}$ gC/joule. Currently about 85% of the energy generated worldwide is from fossil fuels and this emission factor includes that attribute and would be higher if only the fossil energy were considered. With these numbers the present emissions from fossil fuels are about 10 PgC/year worldwide (E = KCP = 1.7×10^{-5} gC/joule $\times$ 8×10^{10} joules/person – year $\times$ 7.5×10^9 people = 10 PgC/yr).

If we consider only the change of population and hold the other factors constant, then the expected emissions at the end of the twenty-first century would be 15 PgC/year for the median population estimate and a range of 10–17 PgC/year for the lower and upper estimates. The per capita global emissions were stable for several decades; however, in the

Global Climate Change and Human Life, First Edition. M. A. K. Khalil.

Companion Website: www.wiley.com/go/khalil/Globalclimatechange

recent times, the rate has risen from 1.1 to 1.3 metric tons of carbon per person per year driven mostly by the revolution in the Chinese economy. The European per capita emissions are about twice the world average at 2.6 tons C/y-person and the US emissions are about three times the world average at 4.2 tons C/y-person. As we discussed earlier, the consumption of energy is a strong indicator of affluence. If the world moved to the European average, which would be a significant reduction of poverty compared with present times, the emissions will rise to some 20 PgC/year, doubling the present rate (Eq. 12.2). This also assumes that the high-ranking countries such as the United States will come down to near the European average. Such a trajectory can be aided by a reduction in the emission factor as non-fossil sources are increased. Nuclear energy, for instance, is existing technology that is utilized most notably in France. It reduces their overall carbon emission factor close to the global average but with a high level of affluence. Nuclear energy however, brings a different set of risks that have made it difficult as a replacement for fossil energy.

We can estimate the worst-case scenario, in which the population will go to the upper estimates from the United Nations, and affluence will rise to the European standard and the emission factors will remain the same. Then, our model tells us that the global emissions will be about 30–40 PgC/year by the end of the twenty-first century, which is some three to four times the present rate. Such emissions would lead to significant global warming compared to present times. Our simplified formulas justify an increase of temperature of about 7°C for the worst-case calculations. The lower end is more difficult to calculate since many opposing forces may be effective. Still, with population managed and brought to near present levels, with mixes of energy to reduce dependence on fossil fuels and modest increases in consumption rates, this lower limit of emissions can be justified to be close to the present, or about 10 PgC/year.

From this analysis, we expect CO_2 emissions in 2100 to range from near present rates of about 10 PgC/y to about 40 PgC/y. The non-CO_2 greenhouse gases will most likely increase and exacerbate global warming. All of them – methane, nitrous oxide, and the technological gases – are closely tied to primary human needs for food, energy, and affluence. It is expected that nitrous oxide will increase substantially due to the use of nitrogen fertilizers. Methane too may rise to higher levels from the use of natural gas and fossil fuel production. But if the focus shifts to more non-fossil sources of energy, it will also reduce methane emissions. The agricultural sources of methane are already at their limits of growth, so future increases are expected to be driven by energy sources.

The “scenarios” and “representative pathways” that have been used to fill the gap in our ability to make predictions about the future climate are speculations at best. Some outcomes are inconsistent with human societal behavior, but this has been difficult to take into account and any association of credible probabilities with the various possible futures has not been done except in broad terms. Based on what we have discussed in this book, we can speculate on the likely paths, but estimates of crucial time scales remain elusive. What’s likely to happen is that the recognition of undesirable consequences from global warming will drive a period of transition to shift from the dependence on fossil energy to alternative sources. This requires building new infrastructure at considerable initial cost that will pay dividends only in some unspecified future. The speed of this transition will vary therefore, depending on two factors: the perceived damage being caused by on-going global warming at any given time, and the competition for societal resources for more immediate issues. How long it will take to reach a new approximate balance of climate is unknown, and perhaps incalculable. The changes will most likely occur in the richest countries of the world since they can better afford the cost of shifting the required infrastructure. During this time, the population and demands for greater affluence will continue to rise offsetting gains from shifting to non-fossil energy. Whether there will be successful efforts to manage the population along the way, remains to be seen. Regardless, in the near future of a few decades, there will be global warming and continued impacts on human life as these intersect with a growing population, but ameliorated by the actions underway, slow as they are. A shift away from fossil energy, as is currently occurring, in response to threats from global warming, has the additional benefit of extending its availability further into the future allowing more time to build both infrastructure and new ways of supplying the energy needs of the world. The less industrialized and poorer countries of the world will get more time to benefit from the available and inexpensive fossil energy. As this transition continues there is always the possibility of finding new inexpensive sources of readily available energy to rival the benefits of our present fossil sources.

15.2 The Metaphysics of Climate Change

The human response to the changing climate is often cast as *adaptation* and *mitigation*. Adaptation is a natural process that humans and all species follow to deal with changing environmental conditions and has often taken the form of migration to more suitable lands. This will still happen for many species in a warming world, but societal constraints will impede or possibly prevent the movement of people. Instead, adaptation will become a process of building more resilient infra-structures that will reduce the impacts of climate change on human life and to take advantage of beneficial possibilities if they arise. Examples of the former include building dikes to prevent flooding from sea-level rise for vulnerable cities, or reservoirs to store water if the supply cycle is expected to be disturbed. An example of the latter is a shift of agriculture to higher latitudes. Adaptation may be seen as a prudent and necessary action, but it does not alter the course of climate change. Because of that, we have focused instead on mitigation which has the goal of managing the earth's global climate and making it sustainable. Adaptation is not sustainable if global change continues unabated – it just buys more time for management to be implemented and made to work.

In the distant past the earth has sometimes been cold enough to be an ice ball and sometimes warm enough that there is no ice at all. It is clear that in the distant future it will not be the same as it is now, whether we affect it or not; or whether we are still here. Although the extremes in geological times lie far beyond the time scales of our interest here, they underscore the idea that climate will change as the various factors we have considered play out, including our own populations, which have never been this high, although human beings have been here for more than 200,000 years. It is inevitable that the boom of population that we are currently experiencing will have effects on the environment at all scales. Both the population and global warming are driven by the same cause – availability and utilization of plentiful energy from fossil fuels. It will not last. If we continue to use fossil fuels, we will run out soon in terms of human history, but in the meantime, we will experience substantial global warming putting up with whatever pains it causes and enjoying whatever benefits of affluence it brings. After a certain population is reached, the more people there are in the world, the more poverty there will be without inexpensive sources of energy and perhaps even with them (see Eqn 12.1).

It may seem overly simplified to suggest that the population dynamics of daisy world may apply to actual species in our world, including us! But it is certain that the earth's average temperature is one of the principal variables that determine the largest human population sustainable within the limits of prevailing technology and social state. It is undeniable that the early-twenty-first-century human population of almost 8 billion people would not survive in an ice age world that has occurred over time scales of tens of centuries. Likewise, it seems unlikely that such a population can exist in a world much warmer than the present. This limitation is due not only to the direct effects of the cold and warm climates on us, but also the production and management of food and perhaps energy. Food is controlled completely by plants which sequester the sun's energy and utilize the available water supplies. The welfare of plants in a warmer world may be the most stringent controlling factor on the human population unless viable technologies can create synthetic plants. The climate to sustain the maximum human populations is therefore closely tied to the temperatures that can sustain plants. It is not one number but a range because there are other variables that can move it up or down. Regardless, it is not known what that optimum temperature range is and how far it can be pushed with advancing technologies. This creates an understandable fear of rising global temperatures as myriad adverse effects of global warming on human life are articulated and promoted in the public view every day. That there are also benefits is not as important because it is feared that they are temporary and will be overcome by damages in the long run. All of this argues for a management of the climate if we are to maintain an optimum global population. Managing the global environment requires accepting that we now have more people than at any time in the human or even the geological history of the world. This aspect of the population and global change is often ignored, but it is no less than the recognition that the current levels of CO_2 are higher than during those same millennial times.

In the recent past of a few centuries there has been a growth of urbanization. This has brought global benefits, but in the big cities it led to a progressive increase of air pollution, especially with the growing use of automobiles, mechanization, and industrialization. This led to an increasing awareness of air pollution and that it was unhealthy for the residents. The impacts had to become clearly experienced before the governing organizations started to implement rules to regulate the sources and set targets for how clean the air of the cities had to be, balancing human health and economics. We now take those management controls on our cities for granted and have cleaner air but more expensive cars and other products that are required to meet the standards of emissions. In controlling city pollution, national regulations were required because each city could not be trusted to do what was necessary or to even have the means to do so. Some cities didn't need the regulations on how much pollution cars could emit, but they still had to comply and pay the financial cost to satisfy national interests. Global climate change is the same phenomenon taken to a different scale of human existence and governance. There is alarm that global change will be damaging to human life, but its effects are modest at this time. This prevents what people regard as a pre-emptive measure that has the undesirable consequence of limiting their aspirations for more affluence right now. The outcome at this time, the early twenty-first century, is that only modest, and mostly no-regrets options are palatable. In time, if the impacts of climate change have greater actual consequences for a large segment of the world's population, the motivation and means to take control of the climate will become more practical. It is argued, that when that happens, it will be too late because the lags between climate action and its result are several decades. This is indeed the case as we have seen from the science laid out in this book. But ultimately what this means is that even after we are sure that we must manage the climate or most of us will not fare well, there will be a period of many decades, various forms of environmental hardship, while the world transitions into a more sustainable state.

In the final analysis it can be said that readily available and plentiful fossil energy has increased affluence over more than a century and has brought us to where we are today. It has a feedback leading to longer life expectancies and thus a higher population, with greater energy demands. Global warming is an undesirable side-effect. Although reducing the use of fossil fuels will directly ameliorate climate change, the known alternatives will also reduce affluence, and perhaps indirectly, the global carrying capacity for human populations. There is a widely held expectation that this tightly woven connection between affluence, population, and climate change can be circumvented by technological advances that can generate our energy needs while maintaining affluence but without global environmental side effects. Whether this is so, remains to be seen.

Endnote

Endnote 15.1 See the United Nations Population Division, World Population Prospects 2019. The estimates are revised periodically and are available to the public in numerical tables and figures.

List of Symbols Used

α: In this text, it is usually used to represent a constant for some specific context. If one variable is related to another by proportionality, α can represent the proportionality constant. Its units and meaning will depend on which variables it connects.

β: Used similarly to α when more than one constant is present. In Ch. 13 it is the coefficient of volume expansion due to an increase of temperature.

Γ: The adiabatic lapse rate of temperature in the atmosphere. This has a specific value for wet and dry air. The actual lapse rate in the atmosphere is not always adiabatic, but can be so when taken as an average over a long time and over large spatial scales.

Δ and δ: An finite increment of a variable. Or a difference of the values of a variable between two stages or points as in $\Delta x = x_2 - x_1$. In Ch. 8, δ is a parameter that measures the difference of how much energy the earth's atmosphere sends to outer space relative to the amount it sends back to the surface.

θ: Various uses as an angle, including latitude.

Λ: Actual temperature lapse rate in a given environmental circumstance.

λ: Various usages: Wavelength of light, climate sensitivity parameter, variable in population dynamics. The usages are in different contexts.

ν: Frequency of oscillation in cycles per second. Applied to light, solar radiation and the earth's radiation. In Ch. 6 it is the kinematic viscosity. Context makes the usage clear.

ρ: Density of air. Usually in grams per cubic centimeter.

σ: Stefan-Boltzmann constant (most common usage). Also used for standard deviation which is a statistical measure of how dispersed the values of some variable are.

τ: Lifetime. It is a measure of the persistence in the atmosphere of a gas, or another atmospheric constituent, such as a particle. In some cases a subscript is used to designate the lifetime in some location of the atmosphere; for example: "n" and "s" for northern and southern hemispheres.

τ_T: Transport time. For instance, it can be the time it takes to exchange the number of molecules of air in the northern hemisphere with the same number of molecules of air from the southern hemisphere.

Global Climate Change and Human Life, First Edition. M. A. K. Khalil.

Companion Website: www.wiley.com/go/khalil/Globalclimatechange

τ_ν: Optical depth. Used to determine how much light is absorbed while traveling through some part of the atmosphere (See Ch. 8).

φ: Latitude on earth's surface.

ω: Angular frequency of oscillation in radians per second = $2\pi\,\nu$.

Ω: The rotational speed of the earth. That is, it goes around $2\,\pi$ radians in 24 hours, or one complete day-night cycle.

A: Various uses: Area of the earth's surface or some part. Area in general, as for example that of an aerosol particle. It is used as a constant in some cases such as the Arrhenius Equation. It can represent a molecule in chemical reactions. In the climate segment, it is the albedo.

a: Acceleration

a, b, c: Used as constants, similar to α and β described earlier.

c: Speed of light.

B: Generic molecule in a chemical reaction. Example A+B —> C + D.

C: Concentration or mixing ratio of a gas in the atmosphere. Subscripts may be used to designate concentrations in specific regions. Example: C_T for concentration in the troposphere. Concentration can be expressed as molecules per cubic centimeter, of grams per cubic centimeter, whereas a mixing ratio is the concentration of a gas divided by the "concentration" of air, or molecules of a gas per cubic centimeter divided by the number of molecules of air in the same cubic centimeter. A mass mixing ratio is also used, which is grams of a gas per cubic centimeter divided by the grams of air in the same cubic centimeter.

C_V: Heat capacity. Subscripts other than v are also used for heat capacity in specific conditions.

D: Diffusion constant. Depth. Distance. Death rate. Depending on context.

E: Various usages: Energy (joules). In the Arrhenius Equation it is an activation temperature in some writings (°K). In the discussion of the population and climate change, E is the emissions of greenhouse gases in Tg/y or similar units.

e: Mathematical constant. Base of natural logarithms (ln(x)). It is used in exponential solutions as e to some power x (e^x), which is the same as exp (x).

F, F_ν: Absorptivity of the atmosphere to the earth's radiation. The latter is for absorption at frequency ν. In Ch. 12 it represents fatalities from extreme environmental events.

f(x): f as a function of x where f and x can stand for virtually any two variables of our interest. Usually, a different symbol is used to be more specific, such as T(t), which is temperature as a function of time.

f, g: May represent constants similar to a, b, c. In the discussion of climate science, f is the fraction of the solar radiation that is absorbed by the atmosphere before it can reach the surface.

GWP: Global warming potential.

g: Acceleration due to gravity at or near the earth's surface (9.8 m/s^2). Also valid throughout the atmosphere including the troposphere and stratosphere.

H: Scale height. It represents the rate at which the pressure or density of air fall with altitude.

H_O: Solubility coefficient for gases in water as an inverse of Henry's Law constant.

H^+: Hydrogen ions in water. Used to define pH and ocean acidity.

h: Several usages: Generic symbol for height. Planck's Constant.

I_ν: Intensity or amount of radiation of frequency ν that is passing through some part of the environment. See Beer's Law.

J, j: Several usages. An index is shorthand for writing a set of variables. Example: There are 4 temperatures T1, T2, T3 and T4. We can write this as T_J (J = 1,...,4). J is used in Ch. 6 to describe the interaction of light with molecules (photochemistry).

K: Transport efficiency related to transport times. In this context it has the units of cm^2/s. In the discussion of the role of population in climate change, K is an emission factor. K or °K are used for units of Kelvin temperature.

K_V: Transfer velocity (cm/s) in ocean-air exchange of gases.

K_H, K's with other subcripts: Rate constants for chemical processes in water (see Ch. 7).

k: Rate constant, as for example in a chemical reaction of two gases.

k_ν: Absorption coefficient when light is being absorbed by molecules in the environment.

L: Several usages: Losses of a gas due to any process. Gases can be lost from the atmosphere if they are destroyed in a chemical reaction, or attach themselves to objects such as particles floating in the atmosphere, or the surface. L is also used for a generic length (see Ch. 5).

L_V: Latent heat of vaporization.

l: Lapse rate. Rate at which temperature decreases with altitude. Ch 2.

M: Molecular weight. M_A, M_G molecular weight of air (A), or a gas (G). In some chemical reactions M is used to designate any molecule of air and its concentration [M] is then the same as the density of air at that altitude.

N: Density of air in molecules per cm^3, or a similar unit. In some contexts N is used as a generic for "number" (see Ch. 12).

N_0: Density of air at the earth's surface in molecules per cm^3.

N_A: Number of molecules of air in the atmosphere (about 10^{44}).

N_T, N_S: Number of molecules of air in the troposphere (T) and stratosphere (S).

N_a: Avogadro's number (6.02×10^{23} molecules per mole)

n: Number of moles of a gas.

P: Several usages. Pressure - usually air pressure, but also the pressure of specific gases. P is production in Ch. 6 and population in Chapter. 10 and later.

p: Probability.

ppt: parts per trillion. Commonly used units of mixing ratio for a gas. Similarly ppb and ppm for parts per billion and million respectively.

Q, q: Heat energy.

q_S: Water vapor mass mixing ratio.

R: Various usages: Universal gas constant in Ch 2. The "ratio" of concentrations in Ch. 4. Radius of the earth or of a generic sphere. The context determines the usage.

RF: Radiative forcing.

S: Various usages: Sources - it is a number that represents emissions into the atmosphere, oceans or soils in mass per unit time. In Ch. 6 it is also used as a saturation ratio. In the discussion of climate it is the solar constant.

T: Temperature. Units used are °C, °K. These can be converted to °F to get a more intuitive feel for the temperatures under consideration. °K = 273.15 + °C and °F = 9/5°C + 32.

T_e: Radiating temperature of a planet. It is the temperature of an equivalent black body that is radiating the amount of energy that is arriving at the top of the atmosphere of the planet. If there is no atmosphere, this would be the average surface temperature.

T_{IN}, T_{OUT}, T_{NET} and similar symbols: Here the T's represent "transport". That is, the amount of a gas that is flowing per unit time from one region to another. Here the IN and OUT represent flows into a region and out of the region and the NET is the difference between the two.

u, U, v: Wind speeds, usually in the horizontal directions.

V: Volume.

W: Work defined as force applied to some object times the distance moved.

x, y, z: Variables. Often arbitrary, but also specify coordinates of an object. In the later case, x and y are horizontal coordinates and z is the vertical distance from the origin.

z: Altitude, usually in km. In some uses, a subscript, as in z_1 or z_a, is used to distinguish one altitude from another.

Index

Global Climate Change and Human Life, First Edition. M. A. K. Khalil.

Companion Website: www.wiley.com/go/khalil/Globalclimatechange

p

r

s

Λ

ρ

Σ

τ

Φ

Ω